MONOGRAPHS AND TEXTBOOKS IN PURE AND APPLIED MATHEMATICS

1. *K. Yano,* Integral Formulas in Riemannian Geometry (1970)
2. *S. Kobayashi,* Hyperbolic Manifolds and Holomorphic Mappings (1970)
3. *V. S. Vladimirov,* Equations of Mathematical Physics (A. Jeffrey, editor; A. Littlewood, translator) (1970)
4. *B. N. Pshenichnyi,* Necessary Conditions for an Extremum (L. Neustadt, translation editor; K. Makowski, translator) (1971)
5. *L. Narici, E. Beckenstein, and G. Bachman,* Functional Analysis and Valuation Theory (1971)
6. *S. S. Passman,* Infinite Group Rings (1971)
7. *L. Dornhoff,* Group Representation Theory (in two parts). Part A: Ordinary Representation Theory. Part B: Modular Representation Theory (1971, 1972)
8. *W. Boothby and G. L. Weiss (eds.),* Symmetric Spaces: Short Courses Presented at Washington University (1972)
9. *Y. Matsushima,* Differentiable Manifolds (E. T. Kobayashi, translator) (1972)
10. *L. E. Ward, Jr.,* Topology: An Outline for a First Course (1972)
11. *A. Babakhanian,* Cohomological Methods in Group Theory (1972)
12. *R. Gilmer,* Multiplicative Ideal Theory (1972)
13. *J. Yeh,* Stochastic Processes and the Wiener Integral (1973)
14. *J. Barros-Neto,* Introduction to the Theory of Distributions (1973)
15. *R. Larsen,* Functional Analysis: An Introduction (1973)
16. *K. Yano and S. Ishihara,* Tangent and Cotangent Bundles: Differential Geometry (1973)
17. *C. Procesi,* Rings with Polynomial Identities (1973)
18. *R. Hermann,* Geometry, Physics, and Systems (1973)
19. *N. R. Wallach,* Harmonic Analysis on Homogeneous Spaces (1973)
20. *J. Dieudonné,* Introduction to the Theory of Formal Groups (1973)
21. *I. Vaisman,* Cohomology and Differential Forms (1973)
22. *B.-Y. Chen,* Geometry of Submanifolds (1973)
23. *M. Marcus,* Finite Dimensional Multilinear Algebra (in two parts) (1973, 1975)
24. *R. Larsen,* Banach Algebras: An Introduction (1973)
25. *R. O. Kujala and A. L. Vitter (eds.),* Value Distribution Theory: Part A; Part B: Deficit and Bezout Estimates by Wilhelm Stoll (1973)
26. *K. B. Stolarsky,* Algebraic Numbers and Diophantine Approximation (1974)
27. *A. R. Magid,* The Separable Galois Theory of Commutative Rings (1974)
28. *B. R. McDonald,* Finite Rings with Identity (1974)
29. *J. Satake,* Linear Algebra (S. Koh, T. A. Akiba, and S. Ihara, translators) (1975)
30. *J. S. Golan,* Localization of Noncommutative Rings (1975)
31. *G. Klambauer,* Mathematical Analysis (1975)
32. *M. K. Agoston,* Algebraic Topology: A First Course (1976)
33. *K. R. Goodearl,* Ring Theory: Nonsingular Rings and Modules (1976)
34. *L. E. Mansfield,* Linear Algebra with Geometric Applications: Selected Topics (1976)
35. *N. J. Pullman,* Matrix Theory and Its Applications (1976)
36. *B. R. McDonald,* Geometric Algebra Over Local Rings (1976)
37. *C. W. Groetsch,* Generalized Inverses of Linear Operators: Representation and Approximation (1977)
38. *J. E. Kuczkowski and J. L. Gersting,* Abstract Algebra: A First Look (1977)
39. *C. O. Christenson and W. L. Voxman,* Aspects of Topology (1977)
40. *M. Nagata,* Field Theory (1977)
41. *R. L. Long,* Algebraic Number Theory (1977)
42. *W. F. Pfeffer,* Integrals and Measures (1977)
43. *R. L. Wheeden and A. Zygmund,* Measure and Integral: An Introduction to Real Analysis (1977)
44. *J. H. Curtiss,* Introduction to Functions of a Complex Variable (1978)
45. *K. Hrbacek and T. Jech,* Introduction to Set Theory (1978)

46. *W. S. Massey,* Homology and Cohomology Theory (1978)
47. *M. Marcus,* Introduction to Modern Algebra (1978)
48. *E. C. Young,* Vector and Tensor Analysis (1978)
49. *S. B. Nadler, Jr.,* Hyperspaces of Sets (1978)
50. *S. K. Segal,* Topics in Group Kings (1978)
51. *A. C. M. van Rooij,* Non-Archimedean Functional Analysis (1978)
52. *L. Corwin and R. Szczarba,* Calculus in Vector Spaces (1979)
53. *C. Sadosky,* Interpolation of Operators and Singular Integrals: An Introduction to Harmonic Analysis (1979)
54. *J. Cronin,* Differential Equations: Introduction and Quantitative Theory (1980)
55. *C. W. Groetsch,* Elements of Applicable Functional Analysis (1980)
56. *I. Vaisman,* Foundations of Three-Dimensional Euclidean Geometry (1980)
57. *H. I. Freedan,* Deterministic Mathematical Models in Population Ecology (1980)
58. *S. B. Chae,* Lebesgue Integration (1980)
59. *C. S. Rees, S. M. Shah, and C. V. Stanojević,* Theory and Applications of Fourier Analysis (1981)
60. *L. Nachbin,* Introduction to Functional Analysis: Banach Spaces and Differential Calculus (R. M. Aron, translator) (1981)
61. *G. Orzech and M. Orzech,* Plane Algebraic Curves: An Introduction Via Valuations (1981)
62. *R. Johnsonbaugh and W. E. Pfaffenberger,* Foundations of Mathematical Analysis (1981)
63. *W. L. Voxman and R. H. Goetschel,* Advanced Calculus: An Introduction to Modern Analysis (1981)
64. *L. J. Corwin and R. H. Szcarba,* Multivariable Calculus (1982)
65. *V. I. Istrăţescu,* Introduction to Linear Operator Theory (1981)
66. *R. D. Järvinen,* Finite and Infinite Dimensional Linear Spaces: A Comparative Study in Algebraic and Analytic Settings (1981)
67. *J. K. Beem and P. E. Ehrlich,* Global Lorentzian Geometry (1981)
68. *D. L. Armacost,* The Structure of Locally Compact Abelian Groups (1981)
69. *J. W. Brewer and M. K. Smith, eds.,* Emily Noether: A Tribute to Her Life and Work (1981)
70. *K. H. Kim,* Boolean Matrix Theory and Applications (1982)
71. *T. W. Wieting,* The Mathematical Theory of Chromatic Plane Ornaments (1982)
72. *D. B. Gauld,* Differential Topology: An Introduction (1982)
73. *R. L. Faber,* Foundations of Euclidean and Non-Euclidean Geometry (1983)
74. *M. Carmeli,* Statistical Theory and Random Matrices (1983)
75. *J. H. Carruth, J. A. Hildebrant, and R. J. Koch,* The Theory of Topological Semigroups (1983)
76. *R. L. Faber,* Differential Geometry and Relativity Theory: An Introduction (1983)
77. *S. Barnett,* Polynomials and Linear Control Systems (1983)
78. *G. Karpilovsky,* Commutative Group Algebras (1983)
79. *F. Van Oystaeyen and A. Verschoren,* Relative Invariants of Rings: The Commutative Theory (1983)
80. *I. Vaisman,* A First Course in Differential Geometry (1984)
81. *G. W. Swan,* Applications of Optimal Control Theory in Biomedicine (1984)
82. *T. Petrie and J. D. Randall,* Transformation Groups on Manifolds (1984)
83. *K. Goebel and S. Reich,* Uniform Convexity, Hyperbolic Geometry, and Nonexpansive Mappings (1984)
84. *T. Albu and C. Năstăsescu,* Relative Finiteness in Module Theory (1984)
85. *K. Hrbacek and T. Jech,* Introduction to Set Theory: Second Edition, Revised and Expanded (1984)
86. *F. Van Oystaeyen and A. Verschoren,* Relative Invariants of Rings: The Noncommutative Theory (1984)
87. *B. R. McDonald,* Linear Algebra Over Commutative Rings (1984)
88. *M. Namba,* Geometry of Projective Algebraic Curves (1984)
89. *G. F. Webb,* Theory of Nonlinear Age-Dependent Population Dynamics (1985)
90. *M. R. Bremner, R. V. Moody, and J. Patera,* Tables of Dominant Weight Multiplicities for Representations of Simple Lie Algebras (1985)

91. *A. E. Fekete,* Real Linear Algebra (1985)
92. *S. B. Chae,* Holomorphy and Calculus in Normed Spaces (1985)
93. *A. J. Jerri,* Introduction to Integral Equations with Applications (1985)
94. *G. Karpilovsky,* Projective Representations of Finite Groups (1985)
95. *L. Narici and E. Beckenstein,* Topological Vector Spaces (1985)
96. *J. Weeks,* The Shape of Space: How to Visualize Surfaces and Three-Dimensional Manifolds (1985)
97. *P. R. Gribik and K. O. Kortanek,* Extremal Methods of Operations Research (1985)
98. *J.-A. Chao and W. A. Woyczynski, eds.,* Probability Theory and Harmonic Analysis (1986)
99. *G. D. Crown, M. H. Fenrick, and R. J. Valenza,* Abstract Algebra (1986)
100. *J. H. Carruth, J. A. Hildebrant, and R. J. Koch,* The Theory of Topological Semigroups, Volume 2 (1986)
101. *R. S. Doran and V. A. Belfi,* Characterizations of C*-Algebras: The Gelfand-Naimark Theorems (1986)
102. *M. W. Jeter,* Mathematical Programming: An Introduction to Optimization (1986)
103. *M. Altman,* A Unified Theory of Nonlinear Operator and Evolution Equations with Applications: A New Approach to Nonlinear Partial Differential Equations (1986)
104. *A. Verschoren,* Relative Invariants of Sheaves (1987)
105. *R. A. Usmani,* Applied Linear Algebra (1987)
106. *P. Blass and J. Lang,* Zariski Surfaces and Differential Equations in Characteristic p > 0 (1987)
107. *J. A. Reneke, R. E. Fennell, and R. B. Minton,* Structured Hereditary Systems (1987)
108. *H. Busemann and B. B. Phadke,* Spaces with Distinguished Geodesics (1987)
109. *R. Harte,* Invertibility and Singularity for Bounded Linear Operators (1988)
110. *G. S. Ladde, V. Lakshmikantham, and B. G. Zhang,* Oscillation Theory of Differential Equations with Deviating Arguments (1987)
111. *L. Dudkin, I. Rabinovich, and I. Vakhutinsky,* Iterative Aggregation Theory: Mathematical Methods of Coordinating Detailed and Aggregate Problems in Large Control Systems (1987)
112. T. Okubo, *Differential Geometry* (1987)
113. *D. L. Stancl and M. L. Stancl,* Real Analysis with Point-Set Topology (1987)
114. *T. C. Gard,* Introduction to Stochastic Differential Equations (1988)
115. *S. S. Abhyankar,* Enumerative Combinatorics of Young Tableaux (1988)
116. *H. Strade and R. Farnsteiner,* Modular Lie Algebras and Their Representations (1988)
117. *J. A. Huckaba,* Commutative Rings with Zero Divisors (1988)
118. *W. D. Wallis,* Combinatorial Designs (1988)
119. *W. Więsław,* Topological Fields (1988)
120. *G. Karpilovsky,* Field Theory: Classical Foundations and Multiplicative Groups (1988)
121. *S. Caenepeel and F. Van Oystaeyen,* Brauer Groups and the Cohomology of Graded Rings (1989)
122. *W. Kozlowski,* Modular Function Spaces (1988)
123. *E. Lowen-Colebunders,* Function Classes of Cauchy Continuous Maps (1989)
124. *M. Pavel,* Fundamentals of Pattern Recognition (1989)
125. *V. Lakshmikantham, S. Leela, and A. A. Martynyuk,* Stability Analysis of Nonlinear Systems (1989)
126. *R. Sivaramakrishnan,* The Classical Theory of Arithmetic Functions (1989)
127. *N. A. Watson,* Parabolic Equations on an Infinite Strip (1989)
128. *K. J. Hastings,* Introduction to the Mathematics of Operations Research (1989)
129. *B. Fine,* Algebraic Theory of the Bianchi Groups (1989)
130. *D. N. Dikranjan, I. R. Prodanov, and L. N. Stoyanov,* Topological Groups: Characters, Dualities, and Minimal Group Topologies (1989)
131. *J. C. Morgan II,* Point Set Theory (1990)
132. *P. Biler and A. Witkowski,* Problems in Mathematical Analysis (1990)
133. *H. J. Sussmann,* Nonlinear Controllability and Optimal Control (1990)
134. *J.-P. Florens, M. Mouchart, and J. M. Rolin,* Elements of Bayesian Statistics (1990)
135. *N. Shell,* Topological Fields and Near Valuations (1990)
136. *B. F. Doolin and C. F. Martin,* Introduction to Differential Geometry for Engineers (1990)

137. *S. S. Holland, Jr.,* Applied Analysis by the Hilbert Space Method (1990)
138. *J. Okniński,* Semigroup Algebras (1990)
139. *K. Zhu,* Operator Theory in Function Spaces (1990)
140. *G. B. Price,* An Introduction to Multicomplex Spaces and Functions (1991)
141. *R. B. Darst,* Introduction to Linear Programming: Applications and Extensions (1991)
142. *P. L. Sachdev,* Nonlinear Ordinary Differential Equations and Their Applications (1991)
143. *T. Husain,* Orthogonal Schauder Bases (1991)
144. *J. Foran,* Fundamentals of Real Analysis (1991)
145. *W. C. Brown,* Matrices and Vector Spaces (1991)
146. *M. M. Rao and Z. D. Ren,* Theory of Orlicz Spaces (1991)
147. *J. S. Golan and T. Head,* Modules and the Structures of Rings: A Primer (1991)
148. *C. Small,* Arithmetic of Finite Fields (1991)
149. *K. Yang,* Complex Algebraic Geometry: An Introduction to Curves and Surfaces (1991)
150. *D. G. Hoffman, D. A. Leonard, C. C. Lindner, K. T. Phelps, C. A. Rodger, and J. R. Wall,* Coding Theory: The Essentials (1991)
151. *M. O. González,* Classical Complex Analysis (1992)
152. *M. O. González,* Complex Analysis: Selected Topics (1992)
153. *L. W. Baggett,* Functional Analysis: A Primer (1992)
154. *M. Sniedovich,* Dynamic Programming (1992)
155. *R. P. Agarwal,* Difference Equations and Inequalities: Theory, Methods, and Applications (1992)
156. *C. Brezinski,* Biorthogonality and Its Applications to Numerical Analysis (1992)
157. *C. Swartz,* An Introduction to Functional Analysis (1992)
158. *S. B. Nadler, Jr.,* Continuum Theory: An Introduction (1992)
159. *M. A. Al-Gwaiz,* Theory of Distributions (1992)
160. *E. Perry,* Geometry: Axiomatic Developments with Problem Solving (1992)
161. *E. Castillo and M. R. Ruiz-Cobo,* Functional Equations and Modelling in Science and Engineering (1992)
162. *A. J. Jerri,* Integral and Discrete Transforms with Applications and Error Analysis (1992)
163. *A. Charlier, A. Bérard, M.-F. Charlier, and D. Fristot,* Tensors and the Clifford Algebra: Application to the Physics of Bosons and Fermions (1992)
164. *P. Biler and T. Nadzieja,* Problems and Examples in Differential Equations (1992)
165. *E. Hansen,* Global Optimization Using Interval Analysis (1992)
166. *S. Guerre-Delabriere,* Classical Sequences in Banach Spaces (1992)

Additional Volumes in Preparation

TENSORS AND THE CLIFFORD ALGEBRA

Application to the Physics of Bosons and Fermions

Alphonse Charlier
Alain Bérard
Marie-France Charlier
Danièle Fristot

University of Metz
Metz, France

Translated by Amanda Taylor

Marcel Dekker, Inc. New York • Basel • Hong Kong

Library of Congress Cataloging-in-Publication Data

Tensors and the Clifford algebra; application to the physics of bosons and fermions/Alphonse Charlier . . . [et al.]: translated by Amanda Taylor.
p. cm.—(Monographs and textbooks in pure and applied mathematics; 163)
Includes bibliographical references and index.
ISBN 0-8247-8666-1 (acid-free paper)
1. Bosons—Mathematics. 2. Fermions—Mathematics. 3. Calculus of tensors. 4. Clifford algebras. 5. Mathematical physics.
I. Charlier, Alphonse. II. Series.
QC793.5.B62T46 1992
539.7'21—dc20 92-12642
CIP

This book is printed on acid-free paper.

MARCEL DEKKER, INC.
270 Madison Avenue, New York, New York 10016

Current printing (last digit):
10 9 8 7 6 5 4 3 2 1

PRINTED IN THE UNITED STATES OF AMERICA

Preface

This book had its origin in a series of lectures presented by Alphonse Charlier in the autumn of 1989 at the University of Metz. The extensive task of transforming lecture notes into manuscript was accomplished with the help of Alain Bérard and for some of the appendixes, by Marie-France Charlier and Danièle Fristot.

The book deals principally with tensors and Clifford algebra with applications to the physics of bosons and fermions. Despite the widespread interest in and the importance of the subjects of tensor analysis, Minkowski space, and covariant formulation of electromagnetism, no comprehensive modern textbook on these subjects exists in the literature. The present book is meant to fill that void.

As a textbook it offers instruction in these fields presupposing only a minimum background in physics. Clifford algebra is a generalization of geometry permitting the construction of a vector algebra. The importance of Pauli and other derived matrices is exhibited in a quaternion formulation. The mathematical tools developed in earlier sections of the book are applied to the physics of scalar bosons (null spin) and the physics of fermions (spin $\frac{1}{2}$).

This book is neither a text on pure mathematics nor a book of theoretical physics on tensors and Clifford algebra; it is an account of the simpler applications of mathematics to physics.

We have become aware through our teaching that the majority of physics students at beginning graduate level are incapable, without prior preparation, of reading a book of theoretical physics and will therefore probably never do so. It is for this reason that we began to write up our notes, in the hope of arousing greater interest in the subject. We have therefore not hesitated to repeat ourselves, to deal with the same subjects in various ways, and above all to develop even the simplest calculations in detail. We have not assumed the mathematical knowledge usually required for the study of these subjects, but have tried to make them vivid by introducing the language of the physicist.

If readers hasten to widen their knowledge of the ideas we have expounded, through a study of the literature at an even higher level, we will have achieved our objective.

We are grateful to Mr. Graham Garratt, of Marcel Dekker, Inc., who initiated this volume; to Professors Earl Taft and Zuhair Nashed, of the Editorial Board of Dekker's Pure and Applied Mathematics program; and to Ms. Maria Allegra of Marcel Dekker, Inc. We are indebted to Miss Amanda Taylor for translation of the entire manuscript and to Dr. C. Crochard, head of UFR SciFA (University of Metz), and Rémy Vincent, head of UFR MIM (University of Metz), for financial support of the translation. We wish to thank Ms. A. Bérard for setting up and rendering the illustrations. Last, but certainly not least, we are indebted to Miss Pierina Leonora for her superb typing of both the French and English texts, and to Alphonse Charlier's wife and Alain Bérard's wife for their patience and assistance.

Alphonse Charlier
Alain Bérard
Marie-France Charlier
Danièle Fristot

I remember Clifford
Stan Getz (ts), Eddy Louiss (o),
René Thomas (g), Bernard Lubat (d)
London, March 15, 16, or 17, 1971

Contents

1

Tensor Analysis

1.1 UNITARY TRANSFORMATION

This section will serve as a prerequisite for what is to follow. We use the Dirac notation, familiar in quantum mechanics. We work in a space of dimension N. Note that $\{|\vec{e}_i\rangle\}$, the set of unitary orthonormal vectors, form a basis in this space. The relations

$$\langle\vec{e}_i|\vec{e}_j\rangle = \delta_{ij} \tag{1.1}$$

$$\sum_{m=1}^{N} |\vec{e}_m\rangle\langle\vec{e}_m| = \hat{1} \tag{1.2}$$

translate not only the orthogonality and normalization of the unitary vectors but also the fact that the basis chosen is *complete*; (1.2) is called the *closure relation*.

In this basis a vector is written

$$\vec{V} = \sum_{i=1}^{N} V_i\vec{e}_i \tag{1.3}$$

$\vec{V}$ can also be expressed on an orthonormal, complete basis $\{|\vec{e}'_i\rangle\}$. If we project one of the vectors $|\vec{e}'_j\rangle$ in this new basis onto the old basis $\{|\vec{e}_i\rangle\}$,

$$|\vec{e}_j'\rangle = \sum_{m=1}^{N} S_m^{(j)} |\vec{e}_m\rangle \tag{1.4}$$

the coefficients $S_m^{(j)}$ are associated with a change of basis

$$\{|\vec{e}_i\rangle\} \xrightarrow{\hat{S}} \{|\vec{e}_i'\rangle\} \tag{1.5}$$

The transformation $\hat{S}$ being linear [cf. (1.4)], we then check that (1.4) is equivalent to writing

$$|\vec{e}_j'\rangle = \hat{S}|\vec{e}_j\rangle \tag{1.6}$$

To be sure, we simply use the closure of (1.2),

$$|\vec{e}_j'\rangle = \sum_{m=1}^{N} |\vec{e}_m\rangle\langle\vec{e}_m|\hat{S}|\vec{e}_j\rangle = \sum_{m=1}^{N} S_{mj}|\vec{e}_m\rangle \tag{1.7}$$

by definition of the elements of the operator matrix $\hat{S}$ in the basis $\{|\vec{e}_i\rangle\}$. Clearly,

$$S_m^{(j)} = S_{mj} \tag{1.8}$$

The orthogonality of the unitary vectors of the new basis $\{|\vec{e}_i'\rangle\}$ leads us to the identity

$$\langle\vec{e}_i'|\vec{e}_j'\rangle = \delta_{ij} \tag{1.9}$$

which becomes, with the help of (1.6),

$$\langle\hat{S}\vec{e}_i|\hat{S}\vec{e}_j\rangle = \langle\vec{e}_i|\hat{S}^+\hat{S}\vec{e}_j\rangle = \delta_{ij} \tag{1.10}$$

If we compare (1.10) to (1.1), by identification we see that

$$\hat{S}^+\hat{S} = \hat{1} \tag{1.11}$$

that is,

$$\hat{S}^+ = \hat{S}^{-1} \tag{1.12}$$

by definition of the inverse operator. The relation (1.12) defines a *unitary* transformation. To clarify relation (1.12), we must go back to (1.10) by using (1.7):

$$\langle\hat{S}\vec{e}_i|\hat{S}\vec{e}_j\rangle = \left\langle \sum_n S_{ni}\vec{e}_n \middle| \sum_m S_{mj}\vec{e}_m \right\rangle = \delta_{ij}$$

That is,

$$\sum_{n,m} S_{ni}^* S_{mj} \delta_{nm} = \delta_{ij} \tag{1.13}$$

which is written

$$\sum_n S_{ni}^* S_{nj} = \delta_{ij} \tag{1.14}$$

If the matrix elements are real and $N = 3$, the relation (1.14) is clarified by

$$\begin{aligned} S_{11}^2 + S_{21}^2 + S_{31}^2 &= 1 \\ S_{12}^2 + S_{22}^2 + S_{32}^2 &= 1 \\ S_{13}^2 + S_{23}^2 + S_{33}^2 &= 1 \\ S_{11}S_{12} + S_{21}S_{22} + S_{31}S_{32} &= 0 \\ S_{11}S_{13} + S_{21}S_{23} + S_{31}S_{33} &= 0 \\ S_{12}S_{13} + S_{22}S_{23} + S_{32}S_{33} &= 0 \end{aligned} \tag{1.15}$$

We have six conditions for nine coefficients; only three of S_{ij} are independent (cf. Euler angles, Cayley–Klein parameters).

We can easily check that

$$(\hat{S})_{\{|\vec{e}_i\rangle\}} = \begin{bmatrix} \cos\theta & \sin\theta & 0 \\ -\sin\theta & \cos\theta & 0 \\ 0 & 0 & 1 \end{bmatrix} \tag{1.16}$$

The matrix (1.16) associated with a rotation of angle θ around axis $0z$ represents a unitary transformation. The same thing happens for the matrix associated with the operator representing the inversion of the unitary vectors of the basis $\{|\vec{e}_i\rangle\}$:

$$(\hat{S})_{\{|\vec{e}_i\rangle\}} = \begin{bmatrix} -1 & 0 & 0 \\ 0 & -1 & 0 \\ 0 & 0 & -1 \end{bmatrix}$$

Note. We show the invariance of the dot (*scalar*) product of two vectors under the action of a unitary transformation.

$$\text{Old basis } \{|\vec{e}_i\rangle\}: \quad \vec{A}\cdot\vec{B} = \left\langle \sum_i A_i \vec{e}_i \middle| \sum_j B_j \vec{e}_j \right\rangle = \sum_i A_i^* B_i \tag{1.17}$$

$$\text{New basis } \{|\vec{e}_i'\rangle\}: \quad \vec{A}\cdot\vec{B} = \left\langle \sum_i A_i' \vec{e}_i' \middle| \sum_j B_j' \vec{e}_j' \right\rangle = \sum_i A_i'^* B_i' \tag{1.18}$$

But also

$$\vec{A}\cdot\vec{B} = \left\langle \sum_i A_i'\hat{S}\vec{e}_i \middle| \sum_j B_j'\hat{S}\vec{e}_j \right\rangle = \left\langle \sum_i A_i'\vec{e}_i \middle| \sum_j B_j'\vec{e}_j \right\rangle \tag{1.19}$$

which we identify with (1.17) since the basis vectors used are those of $\{|\vec{e}_i\rangle\}$, from which

$$\sum_i A_i^* B_i = \sum_i A_i'^* B_i' \tag{1.20}$$

The dot product is invariant under the unitary transformation $\hat{S}$.

1.2 INTRODUCTION OF TENSOR NOTATION

1.2.1 Physical Properties and Tensors

Traditionally, we recognize two types of physical properties:

1. *Scalars*. These require only a single datum of definition (their intensity) and are invariable throughout any unitary transformation of the space. An example of scalar magnitude is the scalar product $\vec{A}\cdot\vec{B}$. Observe that pseudoscalars exist which are characterized by a change of sign during the unitary transformation that reverses the direction of the frame. This is actually the case for the mixed product $\vec{A}\cdot(\vec{B}\wedge\vec{C})$ (cf. Figure 1).

$$\begin{bmatrix} \vec{e}_1{}' \\ \vec{e}_2{}' \\ \vec{e}_3{}' \end{bmatrix} = \begin{bmatrix} -1 & 0 & 0 \\ 0 & -1 & 0 \\ 0 & 0 & -1 \end{bmatrix} \begin{bmatrix} \vec{e}_1 \\ \vec{e}_2 \\ \vec{e}_3 \end{bmatrix}$$

Take the volume v made from the three vectors $\vec{A}$, $\vec{B}$, $\vec{C}$. We know that

$$v = \vec{A}\cdot(\vec{B}\wedge\vec{C})$$

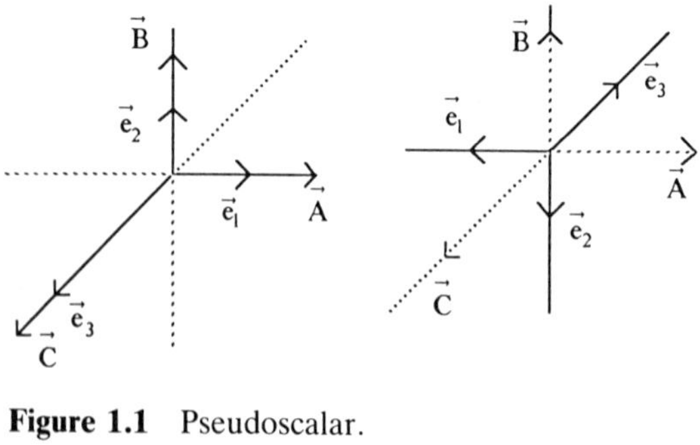

Figure 1.1 Pseudoscalar.

To simplify the calculations, we choose

$$\vec{A} = \vec{e}_1 A$$
$$\vec{B} = \vec{e}_2 B$$
$$\vec{C} = \vec{e}_3 C$$

in which case

$$v = ABC\vec{e}_1 \cdot (\vec{e}_2 \wedge \vec{e}_3) = ABC\vec{e}_1 \cdot \vec{e}_1 = ABC$$

If the direction of the axes is reversed,

$$v = ABC(-\vec{e}_1) \cdot [(-\vec{e}_2) \wedge (-\vec{e}_3)] = -ABC$$

2. *Vectors*. These are characterized by a set of three data: either

$$\begin{cases} \text{direction} \\ \text{sign} \\ \text{magnitude} \end{cases} \quad \text{or} \quad \begin{cases} \text{three components} \\ \text{in a basis} \end{cases}$$

whose behavior during a unitary transformation of the space follows very precise rules described in the next section.

Among the vectors we single out the following:

1. The real or polar vectors, denoted $\vec{A}$, of which the components change sign when an inversion transformation reverses the direction of the frame. This is the case, for example, for position and gradient vectors:
   a. The vector $\vec{r} = \Sigma_{i=1}^{3} x_i \vec{e}_i$; if $\vec{e}_i \to -\vec{e}_i$, then $\vec{r} \to -\vec{r}$.
   b. The gradient vector $\vec{\nabla} f = \Sigma_{i=1}^{3} (\partial f/\partial x_i)\vec{e}_i$, which transforms itself into $-\vec{\nabla} f$ during the inversion of the coordinate axes.
2. The axis vectors, denoted $\overleftrightarrow{C}$, whose components do not change sign when there is an inversion of space. The example given is that of the cross product

$$\vec{A} \wedge \vec{B} = \overleftrightarrow{C}$$

To simplify, we take

$$\vec{A} = \vec{e}_1 A$$
$$\vec{B} = \vec{e}_2 B$$
$$\vec{A} \wedge \vec{B} = AB\vec{e}_1 \wedge \vec{e}_2 = AB\vec{e}_3$$

and if the direction of the unitary vectors is reversed,

$$\vec{A} \wedge \vec{B} = AB(-\vec{e}_1) \wedge (-\vec{e}_2) = AB\vec{e}_3$$

Examples

1. The kinetic momentum at point 0 of a particle of mass m and of speed $\vec{v}$:

$$\overset{\hookrightarrow}{L_0} = \vec{r} \wedge \vec{p} = m\vec{r} \wedge \vec{v}$$

2. The momentum of a force calculated at point 0:

$$\overset{\hookrightarrow}{M_0} = \vec{r} \wedge \vec{F} = m\vec{r} \wedge \vec{\gamma}$$

3. According to the definition of kinetic momentum, the rotation vector will be an axial vector

$$\overset{\hookrightarrow}{L_0} = I\overset{\hookrightarrow}{\omega}$$

if I is the inertia matrix. Note that

$$\overset{\hookrightarrow}{\omega} \wedge \vec{r} = \vec{v}$$

The linear speed $\vec{v}$ is a true vector.

In mathematical space, which is the space of calculation, scalars and vectors correspond to the same category of variables: tensors.

1.2.2 Definition of a Tensor

A *tensor* of rank α is an ordered set of N terms (diverse functions or expressions) classified with the help of α indexes or exponents whose transformation during a change of frame follows very precise rules.

Among the ordered sets of N terms we can single out the following:

Tensor of rank 0. This is made up of a single term and thus does not require a classification index. It can represent the scalars in the condition that the laws of transformation must satisfy, that is, invariance under a unitary transformation.

Tensor of rank 1. Composed of N terms classified with the help of an index or exponent, $\alpha = 1, 2, 3, \ldots, N$. The matrix representation of this kind of tensor is a line or column matrix of N terms. These tensors can represent generalized vectors (N components) or real vectors (three components).

Tensor of rank 2. Its terms are classified with the help of two indexes or two exponents, or one index and one exponent. It is represented by a table, the term being identified by its line and column number. It can represent physical properties such as inertia or elasticity tensors:

$$(I)_{\{|\vec{e}_i\rangle\}} = \begin{bmatrix} I_{xx} & I_{xy} & I_{xz} \\ I_{yx} & I_{yy} & I_{yz} \\ I_{zx} & I_{zy} & I_{zz} \end{bmatrix}$$

There are $N \times N$ terms.

Tensors of rank higher than 2. These cannot be represented by a table—the tensor notation becomes indispensable.

Not every ordered set of terms is a tensor. Examine the behavior of the tensor components during a unitary transformation of space. A tensor of rank 0 must be invariant during every unitary transformation of the space, whether or not the transformation maintains the position of the referential. There are two types of rank 1 tensors:

1. *Contravariant tensors*, characterized by an exponent and represented by a column matrix:

$$|A\rangle = \begin{bmatrix} a^1 \\ a^2 \\ \vdots \\ a^N \end{bmatrix}$$

   The terms a^i transform themselves as the components of the position vector (contravariant law).

2. *Covariant tensors*, characterized by an index and represented by a line matrix:

$$\langle A| = (a_1, a_2, \ldots, a_N)$$

   The terms a_i transform themselves as the components of the gradient vector (covariant law).

Tensors of rank 2 can be:

1. *Contravariant*. The terms are identified by two exponents and transform themselves according to the contravariant law; they are represented by the matrix (a^{ij}).
2. *Covariant*. The terms are identified by two indices and transform themselves according to the covariant law; they are represented by the matrix (a_{ij}).
3. *Mixed*. The terms are identified by an index and an exponent and transform themselves according to the law specific to mixed terms; they are represented by the matrix (a_i^j).

1.3 CONTRAVARIANT COMPONENTS OF A VECTOR

Consider a vector space **E** of N dimensions. The N vectors of the basis $\vec{e}_\mu$ are linearly independent:

$$\sum_{\mu=0}^{N-1} \alpha_\mu \vec{e}_\mu = \vec{0}$$

$\mu = 0, 1, 2, \ldots, N - 1$ leading to $\alpha_\mu = 0$ regardless of μ.

Minkovski time-space, for example, used primarily in relativity, is four-dimensional ($\mu = 0,1,2,3$) space in which the index $\mu = 0$ corresponds to the time coordinate and the indexes $\mu = i = 1, 2, 3$, correspond to coordinates of space x, y, and z, respectively.

We then use the Greek letters $\mu, \nu, \ldots$ for the indexes or exponents of a space of dimension $N > 3$ and Latin letters $i, j, k, \ldots$ for the indexes or exponents of ordinary geometric space.

The dot product $\vec{e}_\mu \cdot \vec{e}_\nu$ of two basis vectors expresses the *metric* $g_{\mu\nu}$ of this space:

$$\vec{e}_\mu \cdot \vec{e}_\nu = g_{\mu\nu} \tag{1.21}$$

It is clear that if

$$g_{\mu\nu} = g_{\nu\mu} \tag{1.22}$$

the metric is said to be *symmetrical*.

Example 1. Consider the cartesian space of three dimensions in which

$$\vec{e}_i \cdot \vec{e}_j = g_{ij} = \delta_{ij} \tag{1.23}$$

The basis vectors are orthonormal.

Example 2. This well-known example concerns the metric of Minkovski space-time, which is usually denoted

$$g_{\mu\nu} = \eta_{\mu\nu} \tag{1.24}$$

and defined by authors (two possible choices) as

$$\eta_{00} = \mp 1 \tag{1.25}$$

$$\eta_{\mu\nu} = \pm 1 \tag{1.26}$$

for $\mu = \nu \neq 0$ while

$$\eta_{\mu\nu} = 0 \tag{1.27}$$

for $\mu \neq \nu$, that is, by

$$(\eta_{\mu\nu}) = \begin{bmatrix} \mp 1 & 0 & 0 & 0 \\ 0 & \pm 1 & 0 & 0 \\ 0 & 0 & \pm 1 & 0 \\ 0 & 0 & 0 & \pm 1 \end{bmatrix} \tag{1.28}$$

A mathematical object, which will be identified later and called vector V (without an arrow), can be described in the system of coordinates of N dimensions as a weighted linear combination of vectors of basis $\vec{e}_\mu$. The coefficients of weight are called *contravariant components* of vector V and we write

$$V = \sum_{\mu=0}^{N-1} V^{\mu}\vec{e}_{\mu} \tag{1.29a}$$

or, more simply,

$$V = V^{\mu}\vec{e}_{\mu} \tag{1.29b}$$

by using a convention created by Einstein which implies a summation on each repeated index in the superior and inferior positions (called *silent* indexes). We call V^{μ} the *contravariant component* of vector V projected onto $\vec{e}_{\mu}$.

Imagine simplifying the identification of vector V with position vector x of a point in a space of N dimensions recognized by its contravariant components x^{μ}, that is,

$$x = x^{\mu}\vec{e}_{\mu} \tag{1.30a}$$

and that a unitary transformation expresses the position of this point in space in another system of coordinates of N components by basis $\vec{e}'_{\mu'}$ vectors:

$$x = x^{\mu}\vec{e}_{\mu} = x'^{\mu'}\vec{e}'_{\mu'} \tag{1.30b}$$

Between x^{μ} and $x'^{\mu'}$ we can write the relations

$$x^{\mu} = x'^{\mu'}\vec{e}'_{\mu'}\cdot\vec{e}_{\mu} \tag{1.31}$$

and thus

$$\frac{\partial x^{\mu}}{\partial x'^{\mu'}} = \vec{e}'_{\mu'}\cdot\vec{e}_{\mu}$$

from which

$$x^{\mu} = \frac{\partial x^{\mu}}{\partial x'^{\mu'}}\, x'^{\mu'}$$

In a more general way,

$$x^{\mu} = f^{\mu}(x'^{0'},x'^{1'},x'^{2'},\ .\ .\ .\ ,x'^{N'-1}) \tag{1.32a}$$

$$x'^{\mu'} = g^{\mu'}(x^{0},x^{1},x^{2},\ .\ .\ .\ ,x^{N-1}) \tag{1.32b}$$

(equations of transformation), so

$$dx^{\mu} = \frac{\partial f^{\mu}}{\partial x'^{\mu'}}\, dx'^{\mu'} = \frac{\partial x^{\mu}}{\partial x'^{\mu'}}\, dx'^{\mu'} \tag{1.33}$$

We use $\Lambda^{\mu}_{\mu'} = \partial x^{\mu}/\partial x'^{\mu'}$, the element of the transformation matrix that enables us to go from the system of coordinates x^{μ} to the system $x'^{\mu'}$:

$$dx^{\mu} = \Lambda^{\mu}_{\mu'}\, dx'^{\mu'} \tag{1.34}$$

All the contravariant components of vector V, denoted V^{μ}, obey this law of transformation:

$$V^{\mu} = \Lambda^{\mu}_{\mu'} V'^{\mu'} \tag{1.35}$$

Note that the basis vector can be transformed directly by writing

$$V^{\mu}\vec{e}_{\mu} = V'^{\mu'}\vec{e}\,'_{\mu'}$$

with

$$V^{\mu} = \Lambda^{\mu}_{\mu'} V'^{\mu'}$$

[cf. (1.35)] or

$$\Lambda^{\mu}_{\mu'} V'^{\mu'}\vec{e}_{\mu} = V'^{\mu'}\vec{e}\,'_{\mu'}$$

and by identifying the coefficients of $V'^{\mu'}$ in the summation

$$\vec{e}\,'_{\mu'} = \Lambda^{\mu}_{\mu'}\vec{e}_{\mu}$$

or by changing the index μ to ν:

$$\vec{e}\,'_{\mu'} = \Lambda^{\nu}_{\mu'}\vec{e}_{\nu} \tag{1.36}$$

the relation of transformation of basis vectors.

If we go back to (1.32b),

$$dx'^{\mu'} = \frac{\partial g^{\mu'}}{\partial x^{\mu}}\, dx^{\mu} = \frac{\partial x'^{\mu'}}{\partial x^{\mu}}\, dx^{\mu} = \Lambda^{\mu'}_{\mu}\, dx^{\mu}$$

which enables us to write

$$V'^{\mu'} = \Lambda^{\mu'}_{\mu} V^{\mu}$$

and thus

$$V^{\mu}\vec{e}^{\,\mu} = V'^{\mu'}\vec{e}\,'_{\mu'} = \Lambda^{\mu'}_{\mu} V^{\mu}\vec{e}\,'_{\mu'}$$

or by identifying the coefficients of V^{μ},

$$\vec{e}_{\mu} = \Lambda^{\mu'}_{\mu}\vec{e}\,'_{\mu'} \tag{1.37}$$

an inverse relation of transformation of basis vectors. By substituting (1.36) into (1.37), we see that

$$\vec{e}_{\mu} = \Lambda^{\mu'}_{\mu}\Lambda^{\nu}_{\mu'}\vec{e}_{\nu}$$

that is,

$$\Lambda^{\mu'}_{\mu}\Lambda^{\nu}_{\mu'} = \delta^{\nu}_{\mu} \tag{1.38}$$

The matrixes transforming $\vec{e}_\mu$ into $\vec{e}'_{\mu'}$ and $\vec{e}'_{\mu'}$ into $\vec{e}_\mu$ are inverse.

1.4 CONTRAVARIANT COMPONENTS OF A TENSOR

We can generalize the vector notation by the introduction of mathematical entities, tensors, which transform themselves as do the products of vectors. The number of vectors concerned defines the tensor rank. A *rank* 1 *tensor* is thus a vector that clearly transforms itself like a single vector under a unitary transformation of the basis

$$V = V^\mu \vec{e}_\mu$$

Using two vectors V and W such that

$$V = V^\mu \vec{e}_\mu$$
$$W = W^\nu \vec{e}_\nu$$

we can define a new mathematical entity,

$$T = V \otimes W \tag{1.39}$$

called a *tensor product* of the two vectors V and W:

$$T = V^\mu \vec{e}_\mu \otimes W^\nu \vec{e}_\nu = V^\mu W^\nu \vec{e}_\mu \otimes \vec{e}_\nu = T^{\mu\nu} \vec{e}_\mu \otimes \vec{e}_\nu \tag{1.40}$$

or in the form of a table:

$$(T^{\mu\nu}) = ((V \otimes W)^{\mu\nu})$$
$$= \begin{bmatrix} V^0W^0 & V^0W^1 & \cdots & V^0W^{N-1} \\ V^1W^0 & V^1W^1 & & V^1W^{N-1} \\ V^{N-1}W^0 & V^{N-1}W^1 & & V^{N-1}W^{N-1} \end{bmatrix} \tag{1.41}$$

A rank 2 tensor thus transforms itself like a product of two vectors, and the contravariant components $T^{\mu\nu}$ of this tensor obey the law of transformation

$$T^{\mu\nu} = \Lambda^\mu_{\mu'} \Lambda^\nu_{\nu'} T'^{\mu'\nu'} \tag{1.42}$$

Formula (1.42) is found by a transformation of basis vectors:

$$T^{\mu\nu} = V^\mu W^\nu$$

and according to (1.35),

$$T^{\mu\nu} = \Lambda^\mu_{\mu'} V^{\mu'} \Lambda^\nu_{\nu'} W^{\nu'} = \Lambda^\mu_{\mu'} \Lambda^\nu_{\nu'} V^{\mu'} W^{\nu'} = \Lambda^\mu_{\mu'} \Lambda^\nu_{\nu'} T^{\mu'\nu'}$$

We note that $T^{\mu\nu}$ is twice contravariant.

We can also write

$$T = T^{\mu\nu} \vec{e}_\mu \otimes \vec{e}_\nu = T'^{\mu'\nu'} \vec{e}'_{\mu'} \otimes \vec{e}'_{\nu'}$$

Now

$$\vec{e}'_{\mu'} = \Lambda^{\mu}_{\mu'}\vec{e}_{\mu} \qquad \text{[cf. (1.36)]}$$
$$\vec{e}'_{\nu'} = \Lambda^{\nu}_{\nu'}\vec{e}_{\nu}$$

and

$$T^{\mu\nu}\vec{e}_{\mu} \otimes \vec{e}_{\nu} = T'^{\mu'\nu'}\Lambda^{\mu}_{\mu'}\Lambda^{\nu}_{\nu'}\vec{e}_{\mu} \otimes \vec{e}_{\nu}$$

or

$$T^{\mu\nu} = T'^{\mu'\nu'}\Lambda^{\mu}_{\mu'}\Lambda^{\nu}_{\nu'} \tag{1.43}$$

It must be noted that as we write the basis vectors one after another, we actually define the tensor product of these vectors:

$$\vec{e}_{\mu}\vec{e}_{\nu} \equiv \vec{e}_{\mu} \otimes \vec{e}_{\nu}$$

and we should strictly observe the tensor notation. In general, tensor products do not commute.

The tensor $T^{\mu\nu}$ is symmetric if

$$T^{\mu\nu} = T^{\nu\mu} \tag{1.44}$$

and antisymmetric if

$$T^{\mu\nu} = -T^{\nu\mu} \tag{1.45}$$

More generally, a rank k tensor will behave like a product of k vectors:

$$T = T^{\mu\nu\cdots\lambda}\vec{e}_{\mu}\vec{e}_{\nu}\cdots\vec{e}_{\lambda} = T'^{\mu'\nu'\cdots\lambda'}\vec{e}'_{\mu'}\vec{e}'_{\nu'}\cdots\vec{e}'_{\lambda'}$$
$$T^{\mu\nu\cdots\lambda} = T'^{\mu'\nu'\cdots\lambda'}\Lambda^{\mu}_{\mu'}\Lambda^{\nu}_{\nu'}\cdots\Lambda^{\lambda}_{\lambda'} \tag{1.46}$$

1.5 COVARIANT COMPONENTS OF A VECTOR

Starting from the coefficients $g_{\mu\nu}$, which define the metric of the space **E**, we can introduce the coefficients $g^{\mu\nu}$, which define the metric of a dual space, by the relation

$$g_{\mu\lambda}g^{\lambda\nu} = \delta^{\nu}_{\mu} \tag{1.47a}$$

($\delta^{\nu}_{\mu} = 1$ if $\mu = \nu$ and $\delta^{\nu}_{\mu} = 0$ if $\mu \neq \nu$). The dual space $\mathbf{E}^*$ is defined by N basis vectors $\vec{e}^{\mu}$ ($\mu = 0,1,2, \ldots ,N - 1$) formed from the vectors $\vec{e}_{\mu}$:

$$\vec{e}^{\mu} = g^{\mu\nu}\vec{e}_{\nu} \tag{1.47b}$$

It is now easy to see that the dot product of vectors $\vec{e}^{\mu}$ and $\vec{e}^{\nu}$ defines the metric $g^{\mu\nu}$ of the dual space:

$$\vec{e}^{\mu}\cdot\vec{e}^{\nu} = g^{\mu\alpha}\vec{e}_{\alpha}\cdot\vec{e}^{\nu} = g^{\mu\alpha}\vec{e}_{\alpha}\cdot g^{\nu\beta}\vec{e}_{\beta} = g^{\mu\alpha}g^{\nu\beta}g_{\alpha\beta}$$

and with the help of (1.47a),

$$\vec{e}^{\mu}\cdot\vec{e}^{\nu} = g^{\mu\alpha}\delta^{\nu}_{\alpha} = g^{\mu\nu} \tag{1.48}$$

Also,

$$\vec{e}^{\mu}\cdot\vec{e}_{\nu} = g^{\mu\alpha}\vec{e}_{\alpha}\cdot\vec{e}_{\nu} = g^{\mu\alpha}g_{\alpha\nu} = \delta^{\mu}_{\nu} \tag{1.49}$$

The vectors $\vec{e}^{\mu}$ and $\vec{e}_{\nu}$ are thus orthogonal.

By projecting vector V into the dual space, we obtain the contravariant components V^{μ} of this vector. (Figure 2 shows the difference between covariant and contravariant components.)

$$V = V_{\mu}\vec{e}^{\mu} \tag{1.50}$$

The coefficients of the metric enable the transformation of the covariant and contravariant components:

$$\vec{e}^{\mu} = g^{\mu\nu}\vec{e}_{\nu} \qquad \text{[cf. (1.47b)]}$$

and therefore

$$V = V_{\mu}g^{\mu\nu}\vec{e}_{\nu}$$

which is also written as

$$V = V^{\nu}\vec{e}_{\nu}$$

from which, by identification,

$$V^{\nu} = g^{\mu\nu}V_{\mu} \tag{1.51}$$

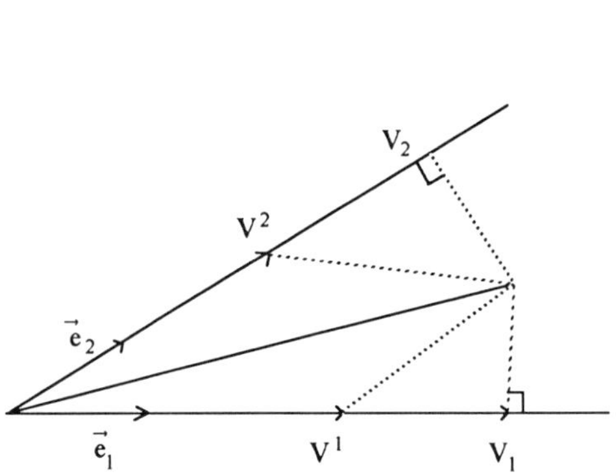

Figure 1.2 Difference between covariant and contravariant components.

But also, according to (1.47b),

$$\vec{e}^{\lambda} = g^{\lambda\nu}\vec{e}_{\nu}$$

and

$$g_{\mu\lambda}\vec{e}^{\lambda} = g_{\mu\lambda}g^{\lambda\nu}\vec{e}_{\nu} = \delta^{\nu}_{\mu}\vec{e}_{\nu}$$

or

$$\vec{e}_{\mu} = g_{\mu\lambda}\vec{e}^{\lambda} \tag{1.52}$$

equivalent to (1.47b). To find a relation similar to (1.51),

$$V = V^{\mu}\vec{e}_{\mu}$$

and with (1.52)

$$V = V^{\mu}g_{\mu\nu}\vec{e}^{\nu}$$

Also,

$$V = V_{\nu}\vec{e}^{\nu}$$

from which

$$V_{\nu} = g_{\mu\nu}V^{\mu} \tag{1.53}$$

From a mathematical point of view the covariant components of a vector behave like the derivative of a scalar function φ.

If we consider the two systems of coordinates

x^{μ} basis: $\{\vec{e}_{\mu}\}$
$x'^{\mu'}$ basis: $\{\vec{e}'_{\mu'}\}$

To say that φ is a scalar function means that we must have

$$\varphi(x^{\mu}) = \varphi(x'^{\mu'})$$

and

$$\frac{\partial\varphi}{\partial x^{\mu}} = \frac{\partial\varphi}{\partial x'^{\mu'}}\frac{\partial x'^{\mu'}}{\partial x^{\mu}} = \frac{\partial\varphi}{\partial x'^{\mu'}}\Lambda^{\mu'}_{\mu}$$

More generally, the covariant components of a vector follow a law of transformation

$$V_{\mu} = V'_{\mu'}\Lambda^{\mu'}_{\mu} = V'_{\mu'}\frac{\partial x'^{\mu'}}{\partial x^{\mu}} \tag{1.54}$$

Note. By using Dirac's notation the scalar product of two vectors V and W is written

$$V\cdot W = \langle V|W\rangle \tag{1.55}$$

where $\langle V|$ is a vector of E^*, said to be bra V, a row vector, and $|W\rangle$ is a column vector of E, said to be ket W. That is,

$$\begin{aligned}
\langle V|W\rangle &= (V_0\vec{e}^0 + V_1\vec{e}^1 + \cdots + V_{N-1}\vec{e}^{N-1}) \\
&\quad \cdot(V^0\vec{e}_0 + \cdots + V^{N-1}\vec{e}_{N-1}) \\
&= (V_0, V_1, \ldots, V_{N-1}) \\
&\quad \times \begin{bmatrix} \vec{e}^0\cdot\vec{e}_0 & & \cdots & \vec{e}^0\cdot\vec{e}_{N-1} \\ \cdot & \cdot & \cdots & \cdot \\ \cdot & \cdot & \cdots & \cdot \\ \cdot & \cdot & \cdots & \cdot \\ \cdot & \cdot & \cdots & \cdot \\ \vec{e}^{N-1}\cdot\vec{e}_0 & \cdot & \cdots & \vec{e}^{N-1}\cdot\vec{e}_{N-1} \end{bmatrix} \begin{bmatrix} V^0 \\ V^1 \\ \cdot \\ \cdot \\ \cdot \\ V^{N-1} \end{bmatrix} \\
&= (V_0, V_1, \ldots, V_{N-1}) \underbrace{\begin{bmatrix} g_0^0 & \cdots & g_{N-1}^0 \\ \cdot & \cdots & \cdot \\ \cdot & \cdots & \cdot \\ \cdot & \cdots & \cdot \\ \cdot & \cdots & \cdot \\ g_0^{N-1} & \cdots & g_{N-1}^{N-1} \end{bmatrix}}_{\text{metric tensor}} \begin{bmatrix} V^0 \\ V^1 \\ \cdot \\ \cdot \\ \cdot \\ V^{N-1} \end{bmatrix}
\end{aligned} \tag{1.56}$$

1.6 COVARIANT COMPONENTS OF A TENSOR

A covariant tensor of rank 1 is a vector:

$$V = V_\mu \vec{e}^\mu$$

Using two vectors V and W such that

$$V = V_\mu \vec{e}^\mu$$
$$W = W_\nu \vec{e}^\nu$$

we can define a new mathematical entity

$$T = V \otimes W$$

the tensorial product of the two covariant vectors V and W such that

$$T = V_\mu \vec{e}^\mu \otimes W_\nu \vec{e}^\nu = V_\mu W_\nu \vec{e}^\mu \otimes \vec{e}^\nu = T_{\mu\nu} \vec{e}^\mu \otimes \vec{e}^\nu$$

or in the form of a table:

$$(T_{\mu\nu}) = ((V \otimes W)_{\mu\nu}) = \begin{bmatrix} V_0W_0 & V_0W_1 & \cdots & V_0W_{N-1} \\ V_1W_0 & V_1W_1 & \cdots & V_1W_{N-1} \\ \vdots & \vdots & & \vdots \\ V_{N-1}W_0 & V_{N-1}W_1 & \cdots & V_{N-1}W_{N-1} \end{bmatrix}$$

Therefore, the covariant tensor of rank 2 transforms itself like a product of two vectors, and the covariant components $T_{\mu\nu}$ of this tensor follow the law of transformation

$$T_{\mu\nu} = V_\mu W_\nu$$

and according to (1.54),

$$T_{\mu\nu} = \Lambda_\mu^{\mu'} V'_{\mu'} \Lambda_\nu^{\nu'} W'_{\nu'} = \Lambda_\mu^{\mu'} \Lambda_\nu^{\nu'} V'_{\mu'} W_{\nu'} = \Lambda_\mu^{\mu'} \Lambda_\nu^{\nu'} T'_{\mu'\nu'}$$

Note that $T_{\mu\nu}$ is twice covariant:

$$T_{\mu\nu} = \Lambda_\mu^{\mu'} \Lambda_\nu^{\nu'} T'_{\mu'\nu'}$$

More generally, a rank k tensor will behave like a product of k vectors:

$$T_{\mu\nu\cdots\lambda} = \Lambda_\mu^{\mu'} \Lambda_\nu^{\nu'} \cdots \Lambda_\lambda^{\lambda'} T'_{\mu'\nu'\cdots\lambda'}$$

1.7 COMPONENTS OF A MIXED TENSOR

The tensor T such that

$$T = T^{\mu\nu}{}_{\lambda\chi} \vec{e}_\mu \vec{e}_\nu \vec{e}^\lambda \vec{e}^\chi$$

is, by definition, a mixed tensor. Its components transform themselves like covariant and contravariant vector products:

$$T^{\mu\nu}{}_{\lambda\chi} = \Lambda_{\mu'}^{\mu} \Lambda_{\nu'}^{\nu} \Lambda_\lambda^{\lambda'} \Lambda_\chi^{\chi'} T'^{\mu'\nu'}{}_{\lambda'\chi'} \tag{1.57}$$

Note. Tensor products do not commute. The index order must be respected.

To find a relation between contravariant, covariant, and mixed components, let us begin with

$$T = T^{\mu\nu} \vec{e}_\mu \vec{e}_\nu$$

According to (1.52),

$$\vec{e}_\mu = g_{\mu\alpha} \vec{e}^\alpha$$

Therefore,

$$T = T^{\mu\nu} g_{\mu\alpha} \vec{e}^\alpha \vec{e}_\nu$$

which we identify with

$$T = T_\alpha{}^\nu \vec{e}^\alpha \vec{e}_\nu$$

a mixed tensor, or

$$T_\alpha{}^\nu = g_{\mu\alpha} T^{\mu\nu}$$

The inverse relation is obtained by multiplying by $g^{\lambda\alpha}$:

$$g^{\lambda\alpha}T_{\alpha}{}^{\nu} = g^{\lambda\alpha}g_{\mu\alpha}T^{\mu\nu} = \delta^{\lambda}_{\mu}T^{\mu\nu} = T^{\lambda\nu}$$

$$T^{\mu\nu} = g^{\mu\alpha}T_{\alpha}{}^{\nu}$$

Equally,

$$T = T^{\mu\nu}\vec{e}_{\mu}\vec{e}_{\nu} = T^{\mu\nu}\vec{e}_{\mu}g_{\nu\alpha}\vec{e}^{\alpha}$$

or

$$T^{\mu}{}_{\alpha} = g_{\nu\alpha}T^{\mu\nu}$$

and by multiplying by $g^{\lambda\alpha}$, we obtain

$$T^{\mu\lambda} = g^{\lambda\alpha}T^{\mu}{}_{\alpha}$$

Finally,

$$T = T^{\mu\nu}\vec{e}_{\mu}\vec{e}_{\nu} = T^{\mu\nu}g_{\mu\alpha}\vec{e}^{\alpha}g_{\nu\beta}\vec{e}^{\beta}$$

or

$$T_{\alpha\beta} = g_{\mu\alpha}g_{\nu\beta}T^{\mu\nu}$$

and by multiplying the two members by $g^{\lambda\alpha}g^{\chi\beta}$, we have

$$T^{\lambda\chi} = g^{\lambda\alpha}g^{\chi\beta}T_{\alpha\beta}$$

The generalization of these two formulas to tensors of rank >2 is easy to achieve.

1.8 CONTRACTION

By using the relation of orthonormalization

$$\vec{e}^{\mu}\cdot\vec{e}_{\nu} = g^{\mu\alpha}\vec{e}_{\alpha}\cdot\vec{e}_{\nu} = g^{\mu\alpha}g_{\alpha\nu} = \delta^{\mu}_{\nu} \tag{1.58}$$

we can write the covariant and contravariant components of a vector or of a tensor as the projection of this vector or tensor onto the basis vectors of space **E** or of the dual space **E***:

$$V = V^{\mu}\vec{e}_{\mu}$$

gives

$$V\cdot\vec{e}^{\nu} = V^{\mu}\vec{e}_{\mu}\cdot\vec{e}^{\nu} = V^{\mu}\delta^{\nu}_{\mu} = V^{\nu}$$

or, in general,

$$V^{\nu} = V\cdot\vec{e}^{\nu} \tag{1.59a}$$

$$V_{\mu} = V\cdot\vec{e}_{\mu} \tag{1.59b}$$

and equally so for the tensors

$$T^{\mu\nu\cdots\lambda} = T\cdot\vec{e}^{\mu}\vec{e}^{\nu}\cdots\vec{e}^{\lambda} \tag{1.60a}$$

and

$$T_{\mu\nu\cdots\lambda} = T\cdot\vec{e}_{\mu}\vec{e}_{\nu}\cdots\vec{e}_{\lambda} \tag{1.60b}$$

By calculating the sum on an index that appears in a superior position (contravariant) and in an inferior position (covariant) we make a contraction that lowers the tensor rank by two:

$$T^{\alpha}{}_{\beta}{}^{\gamma} = T\cdot\vec{e}^{\alpha}\vec{e}_{\beta}\vec{e}^{\gamma} \tag{1.61}$$

and

$$T^{\alpha}{}_{\beta}{}^{\beta} = T\cdot\vec{e}^{\alpha}\vec{e}_{\beta}\vec{e}^{\beta} = T\cdot\vec{e}^{\alpha} = T^{\alpha} \tag{1.62}$$

In fact,

$$\vec{e}_{\beta}\vec{e}^{\beta} = \vec{e}_{\beta}\otimes\vec{e}^{\beta} = |\vec{e}_{\beta}\rangle\langle\vec{e}^{\beta}| \tag{1.63}$$

the closure relation of the basis, or

$$\vec{e}_{\beta}\vec{e}^{\beta} = \hat{1} \tag{1.64}$$

Another example is

$$R^{\alpha}{}_{\beta\gamma\delta} = R\cdot\vec{e}^{\alpha}\vec{e}_{\beta}\vec{e}_{\gamma}\vec{e}_{\delta} \qquad \text{or} \qquad R^{\alpha}{}_{\alpha\gamma\delta} = R\cdot\vec{e}_{\gamma}\vec{e}_{\delta} = R_{\gamma\delta}$$

At the moment of a unitary transformation the mixed tensor T^{μ}_{ν} transforms itself according to the law

$$T^{\mu}{}_{\nu} = \Lambda^{\mu}_{\mu'}\Lambda^{\nu'}_{\nu}T'^{\mu'}{}_{\nu'}$$

and by contraction,

$$T^{\mu}{}_{\mu} = \Lambda^{\mu}_{\mu'}\Lambda^{\nu'}_{\mu}T'^{\mu'}{}_{\nu'} = \delta^{\nu'}_{\mu'}T'^{\mu'}{}_{\nu'} = T'^{\mu'}{}_{\mu'}$$

We thus obtain an invariant under the action of a unitary transformation; the quantity $T^{\mu}{}_{\mu}$ is a scalar.

1.9 DIFFERENTIATION

The gradient whose components are $\partial/\partial x^{\alpha}$ is a covariant vector (cf. Section 1.5). During a change of coordinates the quantity $\partial(T^{\beta\gamma})/\partial x^{\alpha}$ transforms itself into

$$\frac{\partial}{\partial x^\alpha} T^{\beta\gamma} = \Lambda_\alpha^{\alpha'} \frac{\partial}{\partial x'^{\alpha'}} (\Lambda_{\beta'}^{\beta} \Lambda_{\gamma'}^{\gamma} T'^{\beta'\gamma'})$$

$$= \Lambda_\alpha^{\alpha'} \Lambda_{\beta'}^{\beta} \Lambda_{\gamma'}^{\gamma} \frac{\partial}{\partial x'^{\alpha'}} T'^{\beta'\gamma'} + \Lambda_\alpha^{\alpha'} T'^{\beta'\gamma'}$$

$$\times \left(\frac{\partial^2 x^\beta}{\partial x'^{\alpha'}\, \partial x'^{\beta'}} \frac{\partial x^\gamma}{\partial x'^{\gamma'}} + \frac{\partial x^\beta}{\partial x'^{\beta'}} \frac{\partial^2 x^\gamma}{\partial x'^{\alpha'}\, \partial x'^{\gamma'}} \right) \tag{1.65}$$

The differentiation in relation to an index α of a tensor thus increases the rank of this tensor by 1. The quantity $\partial(T^{\beta\gamma})/\partial x^\alpha$ is not a tensor of rank 3. The second term of the right member of (1.65) prevents verification of the rule

$$V_\alpha{}^{\beta\gamma} = \Lambda_\alpha^{\alpha'} \Lambda_{\beta'}^{\beta} \Lambda_{\gamma'}^{\gamma} V'_{\alpha'}{}^{\beta'\gamma'}$$

We write $f, \alpha = \partial f / \partial x^\alpha$ or $\partial_\alpha f = \partial f / \partial x^\alpha$, differentiation with respect to x^α:

$$\frac{\partial}{\partial x^\alpha} T^{\beta\gamma} = \partial_\alpha T^{\beta\gamma} = T^{\beta\gamma}, \alpha \tag{1.66}$$

1.10 CHRISTOFFEL SYMBOLS

1.10.1 Metric Tensor

A space is said to have a *geometric metric* if there is a way to measure the interval separating two infinitely neighboring points. An oriented straight line separating two infinitely neighboring points will be called an infinitesimal vector $d\vec{s}$:

$$d\vec{s} = dx^\mu \, \vec{e}_\mu \tag{1.67}$$

The interval ds between two infinitely neighboring points will therefore be given by

$$ds^2 = d\vec{s} \cdot d\vec{s} = dx^\mu \, \vec{e}_\mu \cdot dx^\nu \, \vec{e}_\nu = g_{\mu\nu} \, dx^\mu \, dx^\nu \tag{1.68}$$

The coefficients $g_{\mu\nu}$ that define the metric of the space are the covariant components of a tensor of rank 2.

The scalar product ds^2 is invariant under a unitary transformation:

$$g_{\mu\nu} \, dx^\mu \, dx^\nu = g'_{\mu'\nu'} \, dx'^{\mu'} \, dx'^{\nu'}$$

with

$$dx'^{\mu'} = \Lambda_\mu^{\mu'} \, dx^\mu$$

and the equivalent for $dx'^{\nu'}$ (cf. Section 1.3), so

$$g_{\mu\nu} = g'_{\mu'\nu'}\Lambda^{\mu'}_{\mu}\Lambda^{\nu'}_{\nu}$$

satisfy the law of transformation of a covariant tensor of rank 2. We can then write

$$g = g_{\mu\nu}\vec{e}^{\,\mu}\vec{e}^{\,\nu}$$

with g the metric tensor.

As by definition [cf. (1.47b)]

$$\vec{e}^{\,\mu} = g^{\mu\alpha}\vec{e}_{\alpha}$$

we see that

$$g = g_{\mu\nu}g^{\mu\alpha}\vec{e}_{\alpha}g^{\nu\beta}\vec{e}_{\beta} = g_{\mu\nu}g^{\mu\alpha}g^{\nu\beta}\vec{e}_{\alpha}\vec{e}_{\beta}$$

and with (1.47a),

$$g = \delta^{\alpha}_{\nu}g^{\nu\beta}\vec{e}_{\alpha}\vec{e}_{\beta} = g^{\nu\beta}\vec{e}_{\nu}\vec{e}_{\beta}$$

If we can find a system of coordinates such that

$$ds^2 = \pm(dx^0)^2 \pm (dx^1)^2 \pm (dx^2)^2 \pm (dx^3)^2 \pm \cdots \tag{1.69}$$

the space is said to be *flat*. In this case

$$g_{\mu\nu} = \begin{cases} \pm 1 & \text{if } \mu = \nu \\ 0 & \text{if } \mu \neq \nu \end{cases} \tag{1.70}$$

If all the signs are identical, the space is said to be *euclidean*. If some signs are + and others −, the space is said to be *pseudoeuclidean*.

The usual geometric space using a trirectangular (cartesian) referential $(O\vec{e}_1, \vec{e}_2, \vec{e}_3)$ is such that

$$d\vec{s} = dx^i\ \vec{e}_i = dx\ \vec{e}_1 + dy\ \vec{e}_2 + dz\ \vec{e}_3$$

and

$$ds^2 = dx^2 + dy^2 + dz^2$$

This is a flat euclidean space.

Minkovski space is defined by an interval such that

$$ds^2 = c^2\,dt^2 - dx^2 - dy^2 - dz^2$$

It is therefore a flat pseudoeuclidean space.

A space is said to be *curved* if there is no system of coordinates in which the metric could be flat, that is, could conform to (1.69). On a small scale any geometry appears flat (except at certain specific points). We will say that the metric is locally flat if at a point of the space

$$g_{\mu\nu} = \begin{cases} \pm 1 & \text{if } \mu = \nu \\ 0 & \text{if } \mu \neq \nu \end{cases} \qquad \text{and} \qquad \frac{\partial g_{\mu\nu}}{\partial x^{\alpha}} = 0 \tag{1.71}$$

1.10.2 Covariant Derivation

Consider a space of dimension N, defined by the basis vectors $\vec{e}_{\mu}$. If the space is flat, the direction of the vectors is always the same regardless of the point of coordinates x^{μ}, and the operation of derivation does not affect the direction of these vectors. If, on the other hand, the space is curved, the direction of vector $\vec{e}_{\mu}$ changes when we go from a point x^{μ} to an infinitely neighboring point $x^{\mu} + dx^{\mu}$ (cf. Figure 3). The vectors $\vec{e}_1'$ and $\vec{e}_2'$ have turned in relation to $\vec{e}_1$ and $\vec{e}_2$.

The derivation of a tensor therefore changes the direction of the basis vectors and means that the derivative of a vector is no longer a vector. An operation called *covariant derivation* enables the reestablishment of this property.

Take the example of a vector written with the help of its contravariant components; it follows the relation of transformation:

$$V^{\mu} = V'^{\mu'} \frac{\partial x^{\mu}}{\partial x'^{\mu'}} \qquad [\text{cf. } (1.35)]$$

The value of vector V in point M' (coordinates $x^{\mu} + dx^{\mu}$), an infinite neighbor of M (coordinates x^{μ}), requires evaluation of the differential dV^{μ}:

$$\begin{aligned} dV^{\mu} &= dV'^{\mu'} \frac{\partial x^{\mu}}{\partial x'^{\mu'}} + V'^{\mu'} d\left(\frac{\partial x^{\mu}}{\partial x'^{\mu'}}\right) \\ &= \underbrace{dV'^{\mu'} \frac{\partial x^{\mu}}{\partial x'^{\mu'}}}_{\substack{\text{contravariant} \\ \text{vector}}} + \underbrace{V'^{\mu'} \frac{\partial^2 x^{\mu}}{\partial x'^{\mu'}\,\partial x'^{\lambda'}} dx'^{\lambda'}}_{\substack{\text{additional term} \\ \text{(noncontravariant)}}} \end{aligned}$$

dV^{μ} does not behave like a vector.

Note. In a flat space

$$\frac{\partial^2 x^{\mu}}{\partial x'^{\mu'}\,\partial x'^{\lambda'}} = 0$$

as the parameters of transformation $\Lambda^{\mu}_{\lambda'}$ do not depend on the coordinates.

Consider a vector with contravariant components V^{μ}:

$$V = V^{\mu}\vec{e}_{\mu}$$

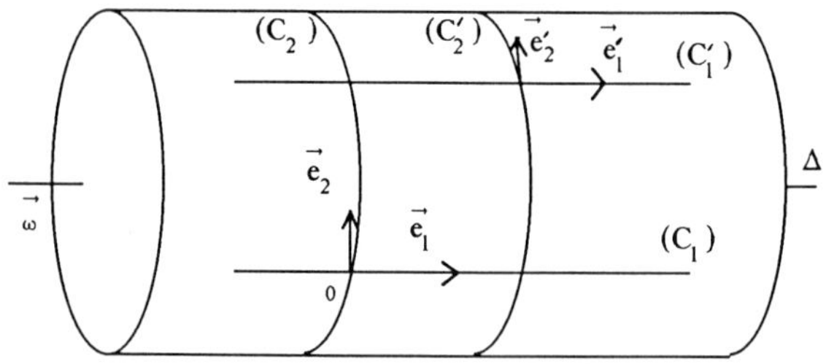

Figure 1.3 Curved space.

By differentiating we obtain

$$dV = dV^{\mu}\, \vec{e}_{\mu} + V^{\mu}\, d\vec{e}_{\mu}$$

with

$$dV^{\mu} = \frac{\partial V^{\mu}}{\partial x^{\nu}}\, dx^{\nu}$$

in the neighborhood of a point x^{ν}. That is,

$$dV = \partial_{\nu} V^{\mu}\, dx^{\nu}\, \vec{e}_{\mu} + V^{\mu}\, d\vec{e}_{\mu} \tag{1.72}$$

The derivative $d\vec{e}_{\mu}$ can be expressed as a linear combination of basis vectors and we write

$$d\vec{e}_{\mu} = \Gamma^{\alpha}_{\mu\nu}\, dx^{\nu}\, \vec{e}_{\alpha} \tag{1.73}$$

which gives

$$dV = (\partial_{\nu} V^{\mu} \vec{e}_{\mu} + V^{\mu} \Gamma^{\alpha}_{\mu\nu} \vec{e}_{\alpha})\, dx^{\nu} = (\partial_{\nu} V^{\mu} + V^{\sigma} \Gamma^{\mu}_{\sigma\nu})\, dx^{\nu}\, \vec{e}_{\mu} \tag{1.74}$$

We call the covariant derivative of vector V the quantity such that

$$\frac{DV^{\mu}}{dx^{\nu}} = \partial_{\nu} V^{\mu} + \Gamma^{\mu}_{\sigma\nu} V^{\sigma} \tag{1.75}$$

We can show that in certain cases

$$\Gamma^{\mu}_{\sigma\nu} = \Gamma^{\mu}_{\nu\sigma} \tag{1.76}$$

and therefore that

$$\frac{DV^{\mu}}{dx^{\nu}} = \partial_{\nu} V^{\mu} + \Gamma^{\mu}_{\nu\sigma} V^{\sigma} \tag{1.77}$$

or

$$dV = DV^{\mu}\vec{e}_{\mu} \tag{1.78}$$

The coefficients $\Gamma^{\mu}_{\nu\sigma}$ are *Christoffel's symbols*; they are not the components of a tensor (cf. Section 1.11.10). We show that the covariant derivative is definitely a tensor (see Section 1.11.13).

For a covariant vector the relation (1.75) is written

$$\frac{DV_{\mu}}{dx^{\nu}} = \partial_{\nu}V_{\mu} - \Gamma^{\sigma}_{\mu\nu}V_{\sigma} \tag{1.79}$$

1.10.3 Covariant Form of Newton's Law

The fundamental principle of dynamics applied to a particle of mass m and speed $\vec{v}$ is written (nonrelativistic case)

$$f = f^i\vec{e}_i = \frac{d}{dt}(mv^i\vec{e}_i)$$

and therefore

$$f^i\vec{e}_i = \left(m\frac{d}{dt}v^i\right)\vec{e}_i + mv^i\frac{d}{dt}\vec{e}_i \tag{1.80}$$

if the mass m is constant in time. As

$$d\vec{e}_i = \Gamma^j_{ik}\,dx^k\,\vec{e}_j$$

(1.80) is written

$$f^i\vec{e}_i = m\frac{d^2x^i}{dt^2}\vec{e}_i + m\frac{dx^i}{dt}\Gamma^j_{ik}\frac{dx^k}{dt}\vec{e}_j$$

We change the notation in the last term to obtain $\vec{e}_i$ instead of $\vec{e}_j$ ($j \to i$ and reciprocally in the second term on the right-hand side):

$$f^i\vec{e}_i = m\frac{d^2x^i}{dt^2}\vec{e}_i + m\frac{dx^j}{dt}\Gamma^i_{jk}\frac{dx^k}{dt}\vec{e}_i$$

or

$$f^i = m\frac{d^2x^i}{dt^2} + m\Gamma^i_{jk}\frac{dx^j}{dt}\frac{dx^k}{dt} \tag{1.81}$$

Multiply this relation by g_{ir} to go from the contravariant components to the covariant component of the force:

$$g_{ir}f^i = f_r = mg_{ir}\frac{d^2x^i}{dt^2} + mg_{ir}\Gamma^i_{jk}\frac{dx^j}{dt}\frac{dx^k}{dt}$$

from which

$$f_r = mg_{ir}\frac{d^2x^i}{dt^2} + m\Gamma_{jk,r}\frac{dx^j}{dt}\frac{dx^k}{dt} \tag{1.82}$$

with, by definition, Christoffel's second-order symbols

$$g_{ir}\Gamma^i_{jk} = \Gamma_{jk,r} \tag{1.83}$$

Equation (1.82) is the covariant form of the fundamental principle (non-relativistic case). By definition, the kinetic energy T is

$$T = \frac{1}{2}m\frac{dx^j}{dt}\vec{e}_j\cdot\frac{dx^k}{dt}\vec{e}_k = \frac{1}{2}mg_{jk}\frac{dx^j}{dt}\frac{dx^k}{dt} = \frac{1}{2}mg_{jk}\dot{x}^j\dot{x}^k \tag{1.84}$$

Evaluate the partial derivative

$$\frac{\partial T}{\partial x^r} = \frac{1}{2}m\dot{x}^j\dot{x}^k\partial_r g_{jk}$$

and

$$\frac{\partial T}{\partial \dot{x}^r} = \frac{1}{2}mg_{jr}\dot{x}^j + \frac{1}{2}mg_{rk}\dot{x}^k = mg_{rk}\dot{x}^k$$

since $g_{ij} = g_{ji}$. Calculate the quantity $d(\partial T/\partial\dot{x}^r)/dt$:

$$\frac{d}{dt}\left(\frac{\partial T}{\partial \dot{x}^r}\right) = mg_{rk}\ddot{x}^k + m\dot{x}^k\frac{dg_{rk}}{dt} = mg_{rk}\ddot{x}^k + m\dot{x}^k(\partial_j g_{rk})\dot{x}^j$$

$$= mg_{rk}\ddot{x}^k + \frac{1}{2}m(\partial_k g_{rj} + \partial_j g_{rk})\dot{x}^k\dot{x}^j$$

so

$$\frac{d}{dt}\left(\frac{\partial T}{\partial \dot{x}^r}\right) - \frac{\partial T}{\partial x^r} = mg_{rk}\ddot{x}^k + \frac{1}{2}m\dot{x}^k\dot{x}^j(\partial_k g_{rj} + \partial_j g_{rk} - \partial_r g_{jk}) \tag{1.85}$$

We show that (cf. Section 1.11.8)

$$\partial_\lambda g_{\mu\nu} + \partial_\mu g_{\lambda\nu} - \partial_\nu g_{\lambda\mu} = 2g_{\rho\nu}\Gamma^\rho_{\lambda\mu} = 2\Gamma_{\lambda\mu,\nu} \tag{1.86}$$

a general formula that in our case gives the relation

$$\partial_j g_{kr} + \partial_k g_{jr} - \partial_r g_{jk} = 2g_{ir}\Gamma^i_{jk}$$

or

$$\partial_k g_{rj} + \partial_j g_{rk} - \partial_r g_{jk} = 2g_{ir}\Gamma^i_{jk}$$

Equation (1.85) then becomes

$$\frac{d}{dt}\left(\frac{\partial T}{\partial \dot{x}^r}\right) - \frac{\partial T}{\partial x^r} = mg_{rk}\ddot{x}^k + mg_{ir}\Gamma^i_{jk}\dot{x}^k\dot{x}^j \tag{1.87}$$

The left-hand side of (1.87) is by definition the component f_r of generalized force. Remember that

$$f_r = \frac{d}{dt}\left(\frac{\partial T}{\partial \dot{x}^r}\right) - \frac{\partial T}{\partial x^r}$$

[cf. A. Charlier, A. Bérard, and M. F. Charlier, *Analytical Mechanics*, Editions Ellipses, Paris, 1989). We have thus [cf. (1.82)]

$$f_r = mg_{rk}\ddot{x}^k + mg_{ir}\Gamma^i_{jk}\dot{x}^k\dot{x}^j$$

In cartesian coordinates we have

$$f_r = m\ddot{x}_r$$

The lagrangian **L** is written

$$\mathbf{L} = T - W = \frac{1}{2}mg_{jk}\frac{dx^j}{dt}\frac{dx^k}{dt} - W(x^i) \tag{1.88}$$

$i = 1, 2, \ldots, g$, where g is the number of degrees of freedom.

1.11 APPLICATIONS

1.11.1 Unitary Space Transformations

A unitary transformation of the space transforms the initial frame $R(0;\vec{i},\vec{j},\vec{k})$ into the frame $R'(0;\vec{i}\,',\vec{j}\,',\vec{k}\,')$. The modulus of vectors is invariant. It is characterized by a matrix (M) of which the terms are the components of the initial basis vectors in the final basis:

$$(M) = \begin{array}{c} \\ \vec{i}\,' \\ \vec{j}\,' \\ \vec{k}\,' \end{array}\!\!\begin{array}{c} \begin{matrix} \vec{i} & \vec{j} & \vec{k} \end{matrix} \\ \begin{bmatrix} x & x & x \\ x & x & x \\ x & x & x \end{bmatrix} \end{array}$$

Example 1. Rotation of the frame around $0z$:

$$(R) = \begin{bmatrix} \cos\alpha & \sin\alpha & 0 \\ -\sin\alpha & \cos\alpha & 0 \\ 0 & 0 & 1 \end{bmatrix}$$

$|R| = +1$, where $|R|$ is the determinant of (R); the frame R' remains direct.

Example 2. Inversion in relation to 0:

$$(I) = \begin{bmatrix} -1 & 0 & 0 \\ 0 & -1 & 0 \\ 0 & 0 & -1 \end{bmatrix}$$

$|I| = -1$; R' is indirect.

This transformation of the frame R leads to a transformation of the components of the vector. Initially,

$$\vec{A} = \begin{bmatrix} a^1 \\ a^2 \\ a^3 \end{bmatrix}_R$$

Then

$$\vec{A} = \begin{bmatrix} a'^{1'} \\ a'^{2'} \\ a'^{3'} \end{bmatrix}_{R'}$$

The transformation being linear,

$$a'^{1'} = \alpha_1^1 a^1 + \alpha_2^1 a^2 + \alpha_3^1 a^3$$

$$a'^{2'} = \alpha_1^2 a^2 + \alpha_2^2 a^2 + \alpha_3^2 a^3$$

$$a'^{3'} = \alpha_1^3 a^1 + \alpha_2^3 a^2 + \alpha_3^3 a^3$$

that is,

$$\begin{bmatrix} a'^{1'} \\ a'^{2'} \\ a'^{3'} \end{bmatrix} = \begin{bmatrix} \alpha_1^1 & \alpha_2^1 & \alpha_3^1 \\ \alpha_1^2 & \alpha_2^2 & \alpha_3^2 \\ \alpha_1^3 & \alpha_2^3 & \alpha_3^3 \end{bmatrix} \begin{bmatrix} a^1 \\ a^2 \\ a^3 \end{bmatrix}$$

we write (T), the transformation matrix, as follows:

If $(T) = (M)$, regardless of the type of transformation, we will say that $\vec{A}$ is a true or polar vector.

If, as is the case for inversion, we obtain $(T) = -(I)$, then $\overset{\hookrightarrow}{A}$ is a pseudovector or axial vector.

Example 1. How does the vector position $\vec{r}$ transform itself during an inversion?

$$\vec{r} = \vec{i}x + \vec{j}y + \vec{k}z = \vec{i}'x' + \vec{j}'y' + \vec{k}'z'$$

$$\begin{bmatrix} -1 \\ 0 \\ 0 \end{bmatrix}_{R'} x + \begin{bmatrix} 0 \\ -1 \\ 0 \end{bmatrix}_{R'} y + \begin{bmatrix} 0 \\ 0 \\ -1 \end{bmatrix}_{R'} z = \begin{bmatrix} x' \\ y' \\ z' \end{bmatrix}_{R'}$$

$$\begin{bmatrix} x' \\ y' \\ z' \end{bmatrix} = \begin{bmatrix} -1 & 0 & 0 \\ 0 & -1 & 0 \\ 0 & 0 & -1 \end{bmatrix} \begin{bmatrix} x \\ y \\ z \end{bmatrix} = (I) \begin{bmatrix} x \\ y \\ z \end{bmatrix}$$

The vector position changes the sign of all its components during the transformation. The components transformation matrix of $\vec{r}$ is equal to (I). The vector position is a true vector.

Example 2. A second example of inversion:

$$\overset{\hookrightarrow}{C} = \vec{A} \wedge \vec{B} = \begin{bmatrix} a^2b^3 - a^3b^2 \\ b^3a^1 - a^3b^1 \\ a^1b^2 - b^1a^2 \end{bmatrix}_R$$

At the moment of inversion the vectors change the sign of all their components. Then

$$\overset{\hookrightarrow}{C} = \begin{bmatrix} c'^{1'} \\ c'^{2'} \\ c'^{3'} \end{bmatrix}_{R'} = \begin{bmatrix} c^1 \\ c^2 \\ c^3 \end{bmatrix}_{R'}$$

Note that $(T) = -(I)$ and not (I): The components of $\overset{\hookrightarrow}{C}$ do not change sign during an inversion. $\overset{\hookrightarrow}{C}$ is not a polar vector but a pseudovector. Examples include the rotation vector and the Poisson vector.

How do scalars transform themselves? Let the scalar be

$$s = \vec{A} \cdot \vec{B} = a^1b^1 + a^2b^2 + a^3b^3$$

After transformation of the space,

$$s = a'^{1'}b'^{1'} + a'^{2'}b'^{2'} + a'^{3'}b'^{3'}$$

For a rotation:

$$\begin{aligned} s &= (\cos \alpha a^1 + \sin \alpha a^2)(\cos \alpha b^1 + \sin \alpha b^2) \\ &\quad + (-\sin \alpha a^1 + \cos \alpha a^2)(-\sin \alpha b^1 + \cos \alpha b^2) + (-a^3)(-b^3) \\ &= a^1b^1 + a^2b^2 + a^3b^3 \end{aligned}$$

s is invariant during the rotation.
For an inversion:

$$
\begin{aligned}
s &= (-a^1)(-b^1) + (-a^2)(-b^2) + (-a^3)(b^3) \\
&= a^1b^1 + a^2b^2 + a^3b^3
\end{aligned}
$$

s is invariant during the inversion.

How do the pseudoscalars transform themselves during an inversion? We can easily check that the pseudoscalar $\vec{A}\cdot\vec{B}$ changes sign during an inversion. The components of the axial vector being invariant and those of the true vector changing sign means that the dot product will change sign. Example: Volume $v = \vec{a}\cdot(\vec{b} \wedge \vec{c})$.

1.11.2 Metric Tensor of Euclidean Space of Dimension 3

Consider the usual geometric space with frame $0xyz$, the unitary vectors on the axes being $\vec{i}$, $\vec{j}$, $\vec{k}$, respectively:

$$d\vec{s} = \vec{i}\,dx + \vec{j}\,dy + \vec{k}\,dz$$

from which

$$ds^2 = dx^2 + dy^2 + dz^2$$

Now, by definition,

$$ds^2 = g_{ij}\,dx^i\,dx^j$$

This leads, by identification, to

$$g_{ij} = 0 \quad \forall i \text{ and } j, \text{ for } i \neq j$$

$$g_{ii} = g_{xx} = g_{yy} = g_{zz} = 1$$

Otherwise,

$$(g_{ij}) = \begin{bmatrix} 1 & 0 & 0 \\ 0 & 1 & 0 \\ 0 & 0 & 1 \end{bmatrix}$$

1.11.3 Metric Tensor of Minkovski Space

We define the interval ds by

$$ds^2 = c^2\,dt^2 - dx^2 - dy^2 - dz^2$$

which we compare with

$$ds^2 = g_{\mu\nu}\,dx^\mu\,dx^\nu$$

from which

$$g_{\mu\nu} = 0 \quad \forall \mu, \nu \qquad \text{and} \qquad \mu \neq \nu$$

$$g_{00} = \eta_{00} = 1$$

$$g_{11} = \eta_{11} = -1$$

$$g_{22} = \eta_{22} = -1$$

$$g_{33} = \eta_{33} = -1$$

or

$$(\eta_{\mu\nu}) = \begin{bmatrix} 1 & 0 & 0 & 0 \\ 0 & -1 & 0 & 0 \\ 0 & 0 & -1 & 0 \\ 0 & 0 & 0 & -1 \end{bmatrix}$$

We can change the *signature* by starting from

$$ds^2 = -c^2\, dt^2 + dx^2 + dy^2 + dz^2$$

in which case

$$(\eta_{\mu\nu}) = \begin{bmatrix} -1 & 0 & 0 & 0 \\ 0 & 1 & 0 & 0 \\ 0 & 0 & 1 & 0 \\ 0 & 0 & 0 & 1 \end{bmatrix}$$

Evaluate the dot product of two vectors:

$$\begin{aligned} A \cdot B &= A_\mu \vec{e}^{\,\mu} \cdot B^\nu \vec{e}_\nu = \eta^\mu_\nu A_\mu B^\nu \\ &= A^\mu \vec{e}_\mu \cdot B^\nu \vec{e}_\nu = \eta_{\mu\nu} A^\mu B^\nu \\ &= A_\mu \vec{e}^{\,\mu} \cdot B_\nu \vec{e}^{\,\nu} = \eta^{\mu\nu} A_\mu B_\nu \end{aligned}$$

or, for example,

$$A \cdot B = -A^0 B^0 + \vec{A} \cdot \vec{B}$$

We also have

$$A \cdot B = A^\mu B_\mu = A^0 B_0 + \vec{A} \cdot \vec{B}$$

since with this last signature

$$A^i = A_i$$

but also, by identification,

$$B^0 = -B_0$$

1.11.4 Metric Tensor of Spherical Coordinates

In spherical coordinates

$$x = r \sin\theta \cos\varphi$$
$$y = r \sin\theta \sin\varphi$$
$$z = r \cos\theta$$

from which the differentials

$$dx = dr \sin\theta \cos\varphi + d\theta\, r \cos\theta \cos\varphi - d\varphi\, r \sin\theta \sin\varphi$$
$$dy = dr \sin\theta \sin\varphi + d\theta\, r \cos\theta \sin\varphi + d\varphi\, r \sin\theta \cos\varphi$$
$$dz = dr \cos\theta - d\theta\, r \sin\theta$$

so

$$ds^2 = dx^2 + dy^2 + dz^2 = (dr)^2 + (r\, d\theta)^2 + (r \sin\theta\, d\varphi)^2$$

which we compare to

$$ds^2 = g_{ij} x^i x^j$$

We see that

$$g_{rr} = 1$$
$$g_{\theta\theta} = r^2$$
$$g_{\varphi\varphi} = r^2 \sin^2\theta$$

all the other elements of metric tensor being zero.

1.11.5 Invariance of Dot Products

For any transformation of coordinates, if A and B are two vectors,

$$A \cdot B = A^\mu B_\mu = \Lambda^\mu_{\mu'} \Lambda^{\nu'}_\mu A'^{\mu'} B'_{\nu'} = \delta^{\nu'}_{\mu'} A'^{\mu'} B'_{\nu'} = A' \cdot B'$$

1.11.6 Metric Tensor

To show that $g_{\mu\nu}$ are the components of a tensor of rank 2 called the *metric tensor*, from the vector

$$d\vec{s} = dx^\mu \vec{e}_\mu$$

we can evaluate ds^2, which measures the square of the distance separating two infinitely neighboring points,

$$ds^2 = dx^\mu\, dx^\nu\, \vec{e}_\mu \cdot \vec{e}_\nu$$

and by definition the $g_{\mu\nu}$:

$$ds^2 = dx^\mu\, dx^\nu\, g_{\mu\nu}$$

As ds^2 is a dot product, it is invariant in any transformation of coordinates

$$dx^\mu\, dx^\nu\, g_{\mu\nu} = dx'^{\mu'}\, dx'^{\nu'}\, g'_{\mu'\nu'}$$

Now

$$x^\alpha = f^\alpha(x'^{\mu'})$$

or

$$dx^\mu = \frac{\partial x^\mu}{\partial x'^{\mu'}}\, dx'^{\mu'} = \Lambda^\mu_{\mu'}\, dx'^{\mu'}$$

and a similar relation for dx^ν, from which

$$g'_{\mu'\nu'} = \Lambda^\mu_{\mu'}\Lambda^\nu_{\nu'} g_{\mu\nu}$$

Otherwise, by multiplying by $\Lambda^{\mu'}_\alpha \Lambda^{\nu'}_\beta$, we obtain

$$g'_{\mu'\nu'}\Lambda^{\mu'}_\alpha \Lambda^{\nu'}_\beta = \delta^\mu_\alpha \delta^\nu_\beta g_{\mu\nu} = g_{\alpha\beta}$$

the law of transformation of a covariant tensor of rank 2 during a change of coordinates.

Starting from

$$ds^2 = dx_\mu\, dx_\nu\, g^{\mu\nu}$$

we arrive at

$$g'^{\mu'\nu'} = \Lambda^{\mu'}_\mu \Lambda^{\nu'}_\nu g^{\mu\nu}$$

a transformation of a contravariant tensor of rank 2. If x, y, and z are the cartesian coordinates and q_1, q_2, and q_3 the curvilinear coordinates (cf. G. B. Arfken, *Mathematical Methods for Physicists*, 3rd ed., Academic Press, New York, 1985)

$$x = x(q_1, q_2, q_3)$$

and similar relations for y and z. We generally define the coefficients h_i and h_{ij} by

$$ds_i = h_i\, dq_i$$

and

$$ds^2 = h_{ij}^2\, dq_i\, dq_j$$

Similarly, for the surface element dS_{ij} and the volume element dV:

$$dS_{ij} = h_i h_j \, dq_i \, dq_j$$

$$dV = h_1 h_2 h_3 \, dq_1 \, dq_2 \, dq_3$$

In cartesian rectilinear coordinates:

$$q_1 = x, \quad q_2 = y, \quad q_3 = z,$$

$$h_{xx} = h_x = 1, \quad h_{yy} = h_y = 1, \quad \text{and} \quad h_{zz} = h_z = 1$$

In spherical coordinates:

$$q_1 = r, \quad q_2 = \theta, \quad q_3 = \varphi,$$

$$h_{rr} = h_r = 1, \quad h_{\theta\theta} = h_\theta = r, \quad \text{and} \quad h_{\varphi\varphi} = h\varphi = r \sin\theta$$

For the two types of preceding coordinates,

$$ds^2 = h_{ij}^2 \, dq_i \, dq_j$$

compared to

$$ds^2 = g_{ij} \, dx^i \, dx^j$$

gives

$$g_{ij} = h_{ij}^2 \delta_{ij}$$

1.11.7 Alternative Formula for Christoffel Symbols

For a free nonrelativistic particle [according to (1.81)],

$$\frac{d^2x^i}{dt^2} + \Gamma^i_{jk} \frac{dx^j}{dt} \frac{dx^k}{dt} = 0 \tag{1}$$

During a transformation of the coordinates,

$$x^i = x^i(x'^{l'}) \qquad \text{with } l' = 1, 2, \ldots \tag{2}$$

and for a free particle the speed is a constant of movement,

$$\frac{d}{dt}\left(\frac{dx^i}{dt}\right) = 0 \tag{3}$$

or

$$\frac{d}{dt}\left(\frac{\partial x^i}{\partial x'^{l'}} \frac{dx'^{l'}}{dt}\right) = 0$$

which is expressed by

$$\frac{\partial x^i}{\partial x'^{l'}} \frac{d^2 x'^{l'}}{dt^2} + \frac{\partial}{\partial x'^{l'}} \left(\frac{dx^i}{dt} \right) \frac{dx'^{l'}}{dt} = 0$$

which is written

$$\frac{\partial x^i}{\partial x'^{l'}} \frac{d^2 x'^{l'}}{dt^2} + \frac{\partial}{\partial x'^{l'}} \left(\frac{\partial x^i}{\partial x'^{k'}} \frac{dx'^{k'}}{dt} \right) \frac{dx'^{l'}}{dt} = 0 \tag{4}$$

or even

$$\frac{\partial x^i}{\partial x'^{l'}} \frac{d^2 x'^{l'}}{dt^2} + \frac{\partial^2 x^i}{\partial x'^{l'}\, \partial x'^{k'}} \frac{dx'^{k'}}{dt} \frac{dx'^{l'}}{dt} = 0 \tag{5}$$

Multiplying by $\partial x'^{m'} / \partial x^i$ yields

$$\frac{d^2 x'^{m'}}{dt^2} + \frac{\partial x'^{m'}}{\partial x^i} \frac{\partial^2 x^i}{\partial x'^{l'} \partial x'^{k'}} \frac{dx'^{k'}}{dt} \frac{dx'^{l'}}{dt} = 0 \tag{6}$$

Comparison with (1) gives

$$\Gamma'^{m'}_{k'l'} = \frac{\partial x'^{m'}}{\partial x^i} \frac{\partial^2 x^i}{\partial x'^{k'}\, \partial x'^{l'}} \tag{7}$$

With the help of the principle of equivalence (cf. E. Elbaz, *Relativité générale et gravitation*, Editions Ellipses, Paris, 1986, p. 124) relation (7) is generalized to

$$\Gamma^{\lambda}_{\mu\nu} = \frac{\partial x^{\lambda}}{\partial \xi^{\alpha}} \frac{\partial^2 \xi^{\alpha}}{\partial x^{\mu}\, \partial x^{\nu}} \tag{8}$$

where $\{\xi^{\alpha}\}$ are the coordinates in a locally flat system and $\{x^{\mu}\}$ are the coordinates in a second, localized system. We note that in formula (8), the Christoffel symbols, so defined, are symmetric under exchange of μ and ν.

1.11.8 Fundamental Relation

According to formula (7) of Section 1.11.7,

$$\Gamma'^{\lambda'}_{\mu'\nu'} = \frac{\partial x'^{\lambda'}}{\partial x^{\alpha}} \frac{\partial^2 x^{\alpha}}{\partial x'^{\mu'}\, \partial x'^{\nu'}} \tag{1}$$

Multiplying this relation by $\Lambda^{\beta}_{\lambda'}$, gives

$$\Lambda^{\beta}_{\lambda'} \Gamma'^{\lambda'}_{\mu'\nu'} = \Lambda^{\beta}_{\lambda'} \Lambda^{\lambda'}_{\alpha} \frac{\partial^2 x^{\alpha}}{\partial x'^{\mu'}\, \partial x'^{\nu'}} = \delta^{\beta}_{\alpha} \frac{\partial^2 x^{\alpha}}{\partial x'^{\mu'}\, \partial x'^{\nu'}}$$

or

$$\frac{\partial}{\partial x'^{\mu'}}(\Lambda^{\beta}_{\nu'}) = \frac{\partial^2 x^{\beta}}{\partial x'^{\mu'}\,\partial x'^{\nu'}} = \Lambda^{\beta}_{\lambda'}\Gamma'^{\lambda'}_{\mu'\nu'} \tag{2}$$

If the x^{α} are the coordinates in a locally flat space,

$$\frac{\partial \eta_{\mu\nu}}{\partial x'^{\lambda'}} = 0$$

but also

$$g'_{\mu'\nu'} = \vec{e}\,'_{\mu'}\cdot\vec{e}\,'_{\nu'} = \Lambda^{\mu}_{\mu'}\Lambda^{\nu}_{\nu'}\vec{e}_{\mu}\cdot\vec{e}_{\nu} = \Lambda^{\mu}_{\mu'}\Lambda^{\nu}_{\nu'}\eta_{\mu\nu} \tag{3}$$

and thus

$$\frac{\partial g'_{\mu'\nu'}}{\partial x'^{\lambda'}} = \eta_{\mu\nu}\Lambda^{\mu}_{\mu'}\frac{\partial \Lambda^{\nu}_{\nu'}}{\partial x'^{\lambda'}} + \eta_{\mu\nu}\Lambda^{\nu}_{\nu'}\frac{\partial \Lambda^{\mu}_{\mu'}}{\partial x'^{\lambda'}}$$

and with the help of (2),

$$\frac{\partial g'_{\mu'\nu'}}{\partial x'^{\lambda'}} = \eta_{\mu\nu}\Lambda^{\mu}_{\mu'}\Lambda^{\nu}_{\alpha'}\Gamma'^{\alpha'}_{\lambda'\nu'} + \eta_{\mu\nu}\Lambda^{\nu}_{\nu'}\Lambda^{\mu}_{\beta'}\Gamma'^{\beta'}_{\lambda'\mu'} \tag{4}$$

Use the definition of the $g'_{\mu'\nu'}$ given by (3):

$$\frac{\partial g'_{\mu'\nu'}}{\partial x'^{\lambda'}} = g'_{\mu'\nu'}\Gamma'^{\alpha'}_{\lambda'\nu'} + g'_{\nu'\beta'}\Gamma'^{\beta'}_{\lambda'\mu'} \tag{5}$$

a relation in which we can also replace the primed quantities by nonprimed quantities. Then do the sum

$$\frac{\partial g_{\mu\nu}}{\partial x^{\lambda}} + \frac{\partial g_{\lambda\nu}}{\partial x^{\mu}} - \frac{\partial g_{\lambda\mu}}{\partial x^{\nu}}$$

We see that

$$2g_{\alpha\nu}\Gamma^{\alpha}_{\mu\lambda} = \frac{\partial g_{\mu\nu}}{\partial x^{\lambda}} + \frac{\partial g_{\lambda\nu}}{\partial x^{\mu}} - \frac{\partial g_{\lambda\mu}}{\partial x^{\nu}} \tag{6}$$

a formula that we remultiply by $g^{\nu\sigma}$, from which

$$\Gamma^{\sigma}_{\mu\lambda} = \frac{1}{2}g^{\nu\sigma}\left(\frac{\partial g_{\mu\nu}}{\partial x^{\lambda}} + \frac{\partial g_{\lambda\nu}}{\partial x^{\mu}} - \frac{\partial g_{\lambda\mu}}{\partial x^{\nu}}\right) \tag{7}$$

This is the *fundamental relation*.

1.11.9 Contraction of Christoffel Symbols

If in formula (7) of Section 1.11.8 we take $\lambda = \sigma$, we obtain

$$\Gamma^{\sigma}_{\mu\sigma} = \frac{1}{2} g^{\nu\sigma} \left(\frac{\partial g_{\mu\nu}}{\partial x^{\sigma}} + \frac{\partial g_{\sigma\nu}}{\partial x^{\mu}} - \frac{\partial g_{\mu\sigma}}{\partial x^{\nu}} \right)$$

Evaluate the expression

$$E = g^{\nu\sigma}(\partial_{\sigma} g_{\mu\nu} - \partial_{\nu} g_{\mu\sigma}) = g^{\nu\sigma}\partial_{\sigma} g_{\mu\nu} - g^{\sigma\nu}\partial_{\nu} g_{\mu\sigma}$$

For the second term on the right-hand side, change the notation as follows:

$$\nu \longrightarrow \sigma$$

$$\sigma \longrightarrow \nu$$

It becomes clear that $E = 0$ and thus that

$$\Gamma^{\sigma}_{\sigma\mu} = \tfrac{1}{2} g^{\nu\sigma}\partial_{\mu} g_{\sigma\nu}$$

1.11.10 Showing That Christoffel Symbols Are Not Tensors

We have seen [cf. formula (7) of Section 1.11.7] that

$$\Gamma'^{\lambda'}_{\mu'\nu'} = \Lambda^{\lambda'}_{\alpha}\partial_{\mu'}\Lambda^{\alpha}_{\nu'} = \frac{\partial x'^{\lambda'}}{\partial x^{\alpha}} \frac{\partial^2 x^{\alpha}}{\partial x'^{\mu'}\, \partial x'^{\nu'}} \tag{1}$$

a formula which means that if we go from one system of coordinates $\{x^{\lambda}\}$ to another $\{x'^{\lambda'}\}$, the connection is made by the Christoffel symbols $\Gamma'^{\lambda'}_{\mu'\nu'}$.

How do the $\Gamma'^{\lambda'}_{\mu'\nu'}$ transform themselves if we change the coordinates from $\{x'^{\lambda'}\}$ to $\{x''^{\lambda''}\}$?

$$\begin{aligned}
\Gamma''^{\lambda''}_{\mu''\nu''} &= \Lambda^{\lambda''}_{\alpha}\partial_{\mu''}\Lambda^{\alpha}_{\nu''} = \frac{\partial x''^{\lambda''}}{\partial x^{\alpha}} \frac{\partial^2 x^{\alpha}}{\partial x''^{\mu''}\, \partial x''^{\nu''}} \\
&= \frac{\partial x''^{\lambda''}}{\partial x'^{\eta'}} \frac{\partial x'^{\eta'}}{\partial x^{\alpha}} \frac{\partial}{\partial x''^{\mu''}} \left(\frac{\partial x^{\alpha}}{\partial x'^{\beta'}} \frac{\partial x'^{\beta'}}{\partial x''^{\nu''}} \right) \\
&= \Lambda^{\lambda''}_{\eta'} \frac{\partial x'^{\eta'}}{\partial x^{\alpha}} \left(\frac{\partial^2 x^{\alpha}}{\partial x''^{\mu''}\, \partial x'^{\beta'}} \frac{\partial x'^{\beta'}}{\partial x''^{\nu''}} + \frac{\partial x^{\alpha}}{\partial x'^{\beta'}} \frac{\partial^2 x'^{\beta'}}{\partial x''^{\mu''}\, \partial x''^{\nu''}} \right) \\
&= \Lambda^{\lambda''}_{\eta'} \frac{\partial x'^{\eta'}}{\partial x^{\alpha}} \left(\frac{\partial^2 x^{\alpha}}{\partial x'^{\gamma'}\, \partial x'^{\beta'}} \frac{\partial x'^{\gamma'}}{\partial x''^{\mu''}} \frac{\partial x'^{\beta'}}{\partial x''^{\nu''}} \right) + \Lambda^{\lambda''}_{\eta'}\delta^{\eta'}_{\beta'} \frac{\partial^2 x'^{\beta'}}{\partial x''^{\mu''}\, \partial x''^{\nu''}} \\
&= \Lambda^{\lambda''}_{\eta'}\Lambda^{\gamma'}_{\mu''}\Lambda^{\beta'}_{\nu''}\Gamma'^{\eta'}_{\gamma'\beta'} + \Lambda^{\lambda''}_{\eta'} \frac{\partial^2 x'^{\eta'}}{\partial x''^{\mu''}\, \partial x''^{\nu''}}
\end{aligned}$$

The last term on the right-hand side prevents the $\Gamma^{\lambda}_{\mu\nu}$ from transforming themselves like a mixed tensor of rank 3.

1.11.11 Christoffel Symbols of Three-Dimensional Euclidean Space

As we have shown,

$$g_{xx} = g_{yy} = g_{zz} = 1$$

The other elements of the metric tensor being zero, according to the fundamental relation (7) of Section 1.11.8,

$$\Gamma^x_{xx} = \Gamma^x_{xy} = \Gamma^x_{xz} = \Gamma^x_{yz} = \Gamma^x_{yy} = \Gamma^x_{zz} = 0$$

as well as all the other symbols obtained by circular permutation of x, y, and z. As the space is flat, these results were inevitable.

1.11.12 Christoffel Symbols for Spherical Coordinates

The only nonzero elements of the metric tensor are

$$g_{rr} = 1 \qquad g_{\theta\theta} = r^2 \qquad g_{\varphi\varphi} = r^2 \sin^2\theta$$

and as

$$g^{ij} g_{i'j} = \delta^i_{i'}$$

we see that

$$g^{rr} = 1 \qquad g^{\theta\theta} = r^{-2} \qquad g_{\varphi\varphi} = (r^2 \sin^2\theta)^{-1}$$

from which, for example,

$$\Gamma^r_{\lambda\mu} = \frac{1}{2} g^{rr}\left(\frac{\partial g_{\mu r}}{\partial x^\lambda} + \frac{\partial g_{\lambda r}}{\partial x^\mu} - \frac{\partial g_{\mu\lambda}}{\partial r}\right)$$

The nonzero symbols are

$$\Gamma^r_{\theta\theta} = -r \qquad \Gamma^\theta_{r\theta} = \frac{1}{r} \qquad \Gamma^\varphi_{\varphi\theta} = \cot\theta$$

$$\Gamma^r_{\varphi\varphi} = -r\sin^2\theta \qquad \Gamma^\theta_{\varphi\varphi} = -\sin\theta\cos\theta \qquad \Gamma^\varphi_{\varphi r} = \frac{1}{r}$$

We can obtain the same results by another method, starting with the lagrangian expression for a free particle [cf. (1.88)]:

$$\mathbf{L} = \frac{1}{2} m g_{jk} \frac{dx^j}{dt}\frac{dx^k}{dt} \tag{1}$$

or in spherical coordinates ($g_{jk} = \delta_{jk}$)

$$\mathbf{L} = \sum_j \frac{1}{2} m \left(\frac{dx^j}{dt}\right)^2 = \frac{1}{2} m(\dot{r}^2 + r^2\dot{\theta}^2 + r^2\dot{\varphi}^2 \sin^2\theta) \tag{2}$$

According to the Lagrange equations, the motion is such that

$$\frac{d}{dt}\left(\frac{\partial \mathbf{L}}{\partial \dot{q}_k}\right) - \frac{\partial \mathbf{L}}{\partial q_k} = 0 \tag{3}$$

For $q_k = r$: $\ddot{r} = r\dot{\theta}^2 + r\dot{\varphi}^2 \sin^2\theta$ (4)

For $q_k = \theta$: $\ddot{\theta} = \dot{\varphi}^2 \sin\theta \cos\theta - \frac{2}{r}\dot{r}\dot{\theta}$ (5)

For $q_k = \varphi$: $\ddot{\varphi} = -2\dot{\varphi}\dot{\theta} \cot\theta - \frac{2}{r}\dot{r}\dot{\varphi}$ (6)

The generalized f_l is such that (cf. Section 1.10.3)

$$f_l = mg_{lk}\ddot{x}^k + g_{il}\Gamma^i_{jk}\dot{x}^k\dot{x}^j \tag{7}$$

and that for a free particle $f_r = 0$, we see that

$$\ddot{x}_l = -\Gamma^l_{jk}\dot{x}^k\dot{x}^j \tag{8}$$

For $x_l = r$,

$$\ddot{r} = -\Gamma^r_{jk}\dot{x}^k\dot{x}^j$$

which when compared with (4) gives

$$\Gamma^r_{\theta\theta} = -r \qquad \Gamma^r_{\varphi\varphi} = -r \sin^2\theta$$

For $x_l = \theta$,

$$\ddot{\theta} = -\Gamma^\theta_{jk}\dot{x}^k\dot{x}^j$$

which when compared with (5) gives

$$\Gamma^\theta_{\varphi\varphi} = -\sin\theta \cos\theta \qquad \Gamma^\theta_{\theta r} = \Gamma^\theta_{r\theta} = \frac{1}{r}$$

For $x_l = \varphi$,

$$\ddot{\varphi} = -\Gamma^\varphi_{jk}\dot{x}^k\dot{x}^j$$

which when compared with (6) gives

$$\Gamma^\varphi_{\varphi r} = \Gamma^\varphi_{r\varphi} = \frac{1}{r} \qquad \Gamma^\varphi_{\theta\varphi} = \Gamma^\varphi_{\varphi\theta} = \cot\theta$$

All the other symbols are zero.

1.11.13 Showing That the Covariant Derivative is a Tensor of Second Rank

According to (1.75),

$$V^{\mu} = \Lambda^{\mu}_{\mu'} V'^{\mu'} \tag{1}$$

Multiply this relation by $\Lambda^{\alpha'}_{\mu}$:

$$\Lambda^{\alpha'}_{\mu} V^{\mu} = \Lambda^{\alpha'}_{\mu} \Lambda^{\mu}_{\mu'} V'^{\mu'} = \delta^{\alpha'}_{\mu'} V'^{\mu'} = V'^{\alpha'} \tag{2}$$

By definition [cf. (1.75)], the covariant derivative is such that

$$\frac{DV'^{\mu'}}{dx'^{\nu'}} = \partial_{\nu'} V'^{\mu'} + \Gamma'^{\mu'}_{\sigma'\nu'} V'^{\sigma'} \tag{3}$$

Thus

$$\partial_{\nu'} V'^{\mu'} = \frac{\partial x^{\alpha}}{\partial x^{\nu'}} \frac{\partial V'^{\mu'}}{\partial x^{\alpha}} = \Lambda^{\alpha}_{\nu'} \frac{\partial V'^{\mu'}}{\partial x^{\alpha}}$$

and with the help of (2),

$$\partial_{\nu'} V'^{\mu'} = \Lambda^{\alpha}_{\nu'} \frac{\partial}{\partial x^{\alpha}} (\Lambda^{\mu'}_{\beta} V^{\beta}) = \Lambda^{\alpha}_{\nu'} \Lambda^{\mu'}_{\beta} \frac{\partial V^{\beta}}{\partial x^{\alpha}} + \Lambda^{\alpha}_{\nu'} V^{\beta} \frac{\partial \Lambda^{\mu'}_{\beta}}{\partial x^{\alpha}} \tag{4}$$

According to Section 1.11.10,

$$\Gamma'^{\mu'}_{\sigma'\nu'} = \Lambda^{\mu'}_{\alpha} \Lambda^{\beta}_{\sigma'} \Lambda^{\gamma}_{\nu'} \Gamma^{\alpha}_{\beta\gamma} + \Lambda^{\mu'}_{\alpha} \frac{\partial^2 x^{\alpha}}{\partial x'^{\sigma'}\, \partial x'^{\nu'}} \tag{5}$$

Thus

$$\delta^{\alpha'}_{\beta'} = \frac{\partial x'^{\alpha'}}{\partial x^{\alpha}} \frac{\partial x^{\alpha}}{\partial x'^{\beta'}}$$

and

$$\frac{\partial \delta^{\alpha'}_{\beta'}}{\partial x'^{\gamma'}} = \frac{\partial^2 x'^{\alpha'}}{\partial x^{\alpha}\, \partial x'^{\gamma'}} \frac{\partial x^{\alpha}}{\partial x'^{\beta'}} + \frac{\partial x'^{\alpha'}}{\partial x^{\alpha}} \frac{\partial^2 x^{\alpha}}{\partial x'^{\gamma'}\, \partial x'^{\beta'}} = 0$$

or

$$\frac{\partial^2 x'^{\alpha'}}{\partial x^{\alpha}\, \partial x^{\rho}} \frac{\partial x^{\rho}}{\partial x'^{\gamma'}} \frac{\partial x^{\alpha}}{\partial x'^{\beta'}} = -\frac{\partial x'^{\alpha'}}{\partial x^{\alpha}} \frac{\partial^2 x^{\alpha}}{\partial x'^{\gamma'}\, \partial x'^{\beta'}} \tag{6}$$

Rewrite relation (5) in terms of (6), with the following change in notation:

$\alpha' \longrightarrow \mu'$
$\gamma' \longrightarrow \sigma'$
$\beta' \longrightarrow \nu'$

$$\Gamma'^{\mu'}_{\sigma'\nu'} = \Lambda^{\mu'}_{\alpha}\Lambda^{\beta}_{\sigma'}\Lambda^{\gamma}_{\nu'}\Gamma^{\alpha}_{\beta\gamma} - \Lambda^{\alpha}_{\nu'}\Lambda^{\rho}_{\sigma'}\frac{\partial^2 x'^{\mu'}}{\partial x^{\alpha}\,\partial x^{\rho}} \tag{7}$$

This enables us to write (3) explicitly as

$$\begin{aligned}
\frac{DV'^{\mu'}}{dx'^{\nu'}} &= \Lambda^{\alpha}_{\nu'}\Lambda^{\mu'}_{\beta}\frac{\partial V^{\beta}}{\partial x^{\alpha}} + \Lambda^{\alpha}_{\nu'}V^{\beta}\frac{\partial \Lambda^{\mu'}_{\beta}}{\partial x^{\alpha}} + \Lambda^{\mu'}_{\alpha}\Lambda^{\beta}_{\sigma'}\Lambda^{\gamma}_{\nu'}\Lambda^{\sigma'}_{\mu}\Gamma^{\alpha}_{\beta\gamma}V^{\mu} \\
&\quad - \Lambda^{\alpha}_{\nu'}\Lambda^{\rho}_{\sigma'}\Lambda^{\sigma'}_{\mu}V^{\mu}\frac{\partial \Lambda^{\mu'}_{\rho}}{\partial x^{\alpha}} \\
&= \Lambda^{\alpha}_{\nu'}\Lambda^{\mu'}_{\beta}\frac{\partial V^{\beta}}{\partial x^{\alpha}} + \Lambda^{\alpha}_{\nu'}V^{\beta}\frac{\partial \Lambda^{\mu'}_{\beta}}{\partial x^{\alpha}} + \Lambda^{\mu'}_{\alpha}\Lambda^{\gamma}_{\nu'}\delta^{\beta}_{\mu}\Gamma^{\alpha}_{\beta\gamma}V^{\mu} \\
&\quad - \Lambda^{\alpha}_{\nu'}\delta^{\rho}_{\mu}V^{\mu}\frac{\partial \Lambda^{\mu'}_{\rho}}{\partial x^{\alpha}} \\
&= \Lambda^{\alpha}_{\nu'}\Lambda^{\mu'}_{\beta}\frac{\partial V^{\beta}}{\partial x^{\alpha}} + \Lambda^{\alpha}_{\nu'}V^{\beta}\frac{\partial \Lambda^{\mu'}_{\beta}}{\partial x^{\alpha}} + \Lambda^{\mu'}_{\alpha}\Lambda^{\gamma}_{\nu'}\Gamma^{\alpha}_{\beta\gamma}V^{\beta} \\
&\quad - \Lambda^{\alpha}_{\nu'}V^{\mu}\frac{\partial \Lambda^{\mu'}_{\mu}}{\partial x^{\alpha}} \\
&= \Lambda^{\alpha}_{\nu'}\Lambda^{\mu'}_{\beta}\frac{\partial V^{\beta}}{\partial x^{\alpha}} + \Lambda^{\mu'}_{\alpha}\Lambda^{\gamma}_{\nu'}\Gamma^{\alpha}_{\beta\gamma}V^{\beta}
\end{aligned}$$

We see clearly the law of transformation of a tensor of rank 2.

1.11.14 Proving Formula (1.79)

If we begin with the vector

$$V = V_{\mu}\vec{e}^{\mu}$$

the differential dV is written

$$dV = dV_{\mu}\vec{e}^{\mu} + V_{\mu}d\vec{e}^{\mu} = \frac{\partial V_{\mu}}{\partial x^{\nu}}dx^{\nu}\vec{e}^{\mu} + V_{\mu}d\vec{e}^{\mu} \tag{1}$$

Thus

$$\begin{aligned}
d(\vec{e}_{\mu}) &= d(g_{\mu\beta}\vec{e}^{\beta}) = g_{\mu\beta}\,d\vec{e}^{\beta} + \vec{e}^{\beta}\,dg_{\mu\beta} \\
&= \Gamma^{\alpha}_{\mu\nu}\,dx^{\nu}\,\vec{e}_{\alpha} \qquad \text{[cf. (1.73)]} \\
&= \Gamma^{\alpha}_{\mu\nu}g_{\alpha\delta}\vec{e}^{\delta}\,dx^{\nu} \qquad \text{[cf. (1.52)]}
\end{aligned}$$

from which we conclude that

$$g_{\mu\beta}\,d\vec{e}^{\beta} = -\vec{e}^{\beta}\,dg_{\mu\beta} + \Gamma^{\alpha}_{\mu\nu}g_{\alpha\delta}\vec{e}^{\delta}\,dx^{\nu} \tag{2}$$

Multiply the two members by $g^{\mu\rho}$:

$$g^{\mu\rho}g_{\mu\beta}\, d\vec{e}^{\beta} = -\vec{e}^{\beta}g^{\mu\rho}\, dg_{\mu\beta} + \Gamma^{\alpha}_{\mu\nu}g^{\mu\rho}g_{\alpha\delta}\vec{e}^{\delta}\, dx^{\nu}$$

Thus

$$\begin{aligned} d\vec{e}^{\rho} &= -\vec{e}^{\beta}g^{\mu\rho}\vec{e}_{\mu}\cdot d\vec{e}_{\beta} - \vec{e}^{\beta}g^{\mu\rho}\, d\vec{e}_{\mu}\cdot\vec{e}_{\beta} + \Gamma^{\alpha}_{\mu\nu}g^{\mu\rho}g_{\alpha\delta}\vec{e}^{\delta}\, dx^{\nu} \\ &= -\vec{e}^{\beta}g^{\mu\rho}\vec{e}_{\mu}\cdot\Gamma^{\alpha}_{\beta\nu}\vec{e}_{\alpha}\, dx^{\nu} - \vec{e}^{\beta}g^{\mu\rho}\Gamma^{\alpha}_{\mu\nu}\vec{e}_{\alpha}\, dx^{\nu}\cdot\vec{e}_{\beta} + \Gamma^{\alpha}_{\mu\nu}g^{\mu\rho}g_{\alpha\delta}\vec{e}^{\delta}\, dx^{\nu} \\ &= -\vec{e}^{\beta}g^{\mu\rho}g_{\mu\alpha}\Gamma^{\alpha}_{\beta\nu}\, dx^{\nu} - \vec{e}^{\beta}g^{\mu\rho}g_{\beta\alpha}\Gamma^{\alpha}_{\mu\nu}\, dx^{\nu} + \Gamma^{\alpha}_{\mu\nu}g^{\mu\rho}g_{\alpha\delta}\vec{e}^{\delta}\, dx^{\nu} \\ &= -\vec{e}^{\beta}g^{\mu\rho}g_{\mu\alpha}\Gamma^{\alpha}_{\beta\nu}\, dx^{\nu} \\ &= -\vec{e}^{\beta}\delta^{\rho}_{\alpha}\Gamma^{\alpha}_{\beta\nu}\, dx^{\nu} = -\vec{e}^{\beta}\Gamma^{\rho}_{\beta\nu}\, dx^{\nu} \end{aligned} \tag{3}$$

After we introduce (3) in (1),

$$dV = \frac{\partial V_{\mu}}{\partial x^{\nu}}\, dx^{\nu}\, \vec{e}^{\mu} - V_{\mu}\vec{e}^{\beta}\Gamma^{\mu}_{\beta\nu}\, dx^{\nu} = \left(\frac{\partial V_{\mu}}{\partial x^{\nu}} - \Gamma^{\alpha}_{\mu\nu}V_{\alpha}\right) dx^{\nu}\, \vec{e}^{\mu}$$

we propose a different but quicker method; note that

$$D_{\mu}V^{\nu} = \frac{DV^{\nu}}{dx^{\mu}} \tag{4}$$

with [cf. (1.75)]

$$D_{\mu}V^{\nu} = \partial_{\mu}V^{\nu} + \Gamma^{\nu}_{\alpha\mu}V^{\alpha} \tag{5}$$

As

$$V^{\nu}V_{\nu} = C \tag{6}$$

where C is a scalar, we should have

$$D_{\mu}(V^{\nu}V_{\nu}) = \partial_{\mu}C = 0 \tag{7}$$

In fact,

$$dC = \frac{\partial C}{\partial x^{\mu}}\, dx^{\mu} = \partial_{\mu}C\, dx^{\mu} = 0 \tag{8}$$

We conclude that

$$\partial_{\mu}(V^{\nu}V_{\nu}) = V^{\nu}\partial_{\mu}V_{\nu} + (\partial_{\mu}V^{\nu})V_{\nu} = 0$$

or

$$(\partial_{\mu}V^{\nu})V_{\nu} = -V^{\nu}\partial_{\mu}V_{\nu} \tag{9}$$

According to (7),

$$(D_{\mu}V^{\nu})V_{\nu} + V^{\nu}D_{\mu}V_{\nu} = 0$$

or

$$(D_\mu V^\nu)V_\nu = -V^\nu D_\mu V_\nu$$

and with (5),

$$V^\nu D_\mu V_\nu = -(\partial_\mu V^\nu)V_\nu - \Gamma^\nu_{\alpha\mu} V^\alpha V_\nu$$

Then with the help of (9),

$$V^\nu D_\mu V_\nu = V^\nu \partial_\mu V_\nu - \Gamma^\nu_{\alpha\mu} V^\alpha V_\nu \tag{10}$$

and by changing for the second term on the right-hand side

$$\alpha \longrightarrow \nu$$
$$\nu \longrightarrow \alpha$$

[relation (10) is valid regardless of ν], we obtain

$$D_\mu V_\nu = \partial_\mu V_\nu - \Gamma^\alpha_{\nu\mu} V_\alpha$$

We see (1.79) again.

1.11.15 Derivation of a Tensor of Second Order

Consider the tensor

$$A = A^{\mu\eta}\vec{e}_\mu \vec{e}_\eta$$

Then

$$dA = dA^{\mu\eta}\, \vec{e}_\mu \vec{e}_\eta + A^{\mu\eta}\, d\vec{e}_\mu\, \vec{e}_\eta + A^{\mu\eta}\vec{e}_\mu\, d\vec{e}_\eta$$

Thus

$$d\vec{e}_\mu = \Gamma^\alpha_{\mu\nu}\, dx^\nu\, \vec{e}_\alpha$$
$$d\vec{e}_\eta = \Gamma^\alpha_{\eta\nu}\, dx^\nu\, \vec{e}_\alpha$$

and

$$dA = dA^{\mu\eta}\, \vec{e}_\mu \vec{e}_\eta + A^{\mu\eta}\Gamma^\alpha_{\mu\nu}\, dx^\nu\, \vec{e}_\alpha \vec{e}_\eta + A^{\mu\eta}\Gamma^\alpha_{\eta\nu}\, dx^\nu\, \vec{e}_\alpha \vec{e}_\mu$$

In the neighborhood of point x^ν, we clearly have

$$dA^{\mu\eta} = \frac{\partial A^{\mu\eta}}{\partial x^\nu}\, dx^\nu$$

from which

$$dA = \left(\frac{\partial A^{\mu\eta}}{\partial x^\nu} \vec{e}_\mu \vec{e}_\eta + A^{\mu\eta}\Gamma^\alpha_{\mu\nu}\vec{e}_\alpha \vec{e}_\eta + A^{\mu\eta}\Gamma^\alpha_{\eta\nu}\vec{e}_\alpha \vec{e}_\mu \right) dx^\nu$$

$$= \left(\frac{\partial A^{\mu\alpha}}{\partial x^\nu} \vec{e}_\mu \vec{e}_\alpha + A^{\mu\alpha}\Gamma^\eta_{\mu\nu}\vec{e}_\eta \vec{e}_\alpha + A^{\mu\eta}\Gamma^\alpha_{\eta\nu}\vec{e}_\alpha \vec{e}_\mu \right) dx^\nu$$

$$= \left(\frac{\partial A^{\mu\alpha}}{\partial x^{\nu'}} \vec{e}_\mu \vec{e}_\alpha + A^{\sigma\alpha}\Gamma^\mu_{\sigma\nu'}\vec{e}_\mu \vec{e}_\alpha + A^{\mu\nu}\Gamma^\alpha_{\nu\nu'}\vec{e}_\alpha \vec{e}_\mu \right) dx^{\nu'}$$

If by generalization of (1.78), we denote by dA the quantity

$$\frac{DA^{\mu\alpha}}{dx^\nu} dx^\nu \vec{e}_\mu \vec{e}_\alpha$$

then

$$\frac{DA^{\mu\alpha}}{dx^\nu} = \frac{\partial A^{\mu\alpha}}{\partial x^\nu} + \Gamma^\mu_{\sigma\nu}A^{\sigma\alpha} + \Gamma^\alpha_{\eta\nu}A^{\mu\eta}$$

In the same way we show that

$$\frac{DA_{\mu\alpha}}{dx^\nu} = \frac{\partial A_{\mu\alpha}}{\partial x^\nu} - \Gamma^\sigma_{\mu\nu}A_{\sigma\alpha} - \Gamma^\eta_{\alpha\nu}A_{\mu\eta}$$

and for a mixed tensor

$$\frac{DA^\mu{}_\alpha}{dx^\nu} = \frac{\partial A^\mu{}_\alpha}{\partial x^\nu} + \Gamma^\mu_{\sigma\nu}A^\sigma{}_\alpha - \Gamma^\beta_{\alpha\nu}A^\mu{}_\beta$$

1.11.16 Curvature and Torsion Tensors

We know that

$$\frac{\partial^2 V}{\partial x^\mu\, \partial x^\nu} = \frac{\partial^2 V}{\partial x^\nu\, \partial x^\mu}$$

that is,

$$\partial_\mu(\partial_\nu V) = \partial_\nu(\partial_\mu V)$$

What are the covariant derivatives? As a general rule,

$$D_\mu D_\nu V \neq D_\nu D_\mu V$$

Denote as the Lie bracket the quantity

$$[D_\mu, D_\nu] = D_\mu D_\nu - D_\nu D_\mu \tag{1}$$

similar to the Poisson bracket and to the commutator of quantum mechanics. According to (1.75),

$$\begin{aligned} D_\mu D_\nu V^\rho &= D_\mu(\partial_\nu V^\rho + \Gamma^\rho_{\sigma\nu} V^\sigma) \\ &= D_\mu(\partial_\nu V^\rho) + D_\mu(\Gamma^\rho_{\sigma\nu} V^\sigma) \\ &= \partial_\mu(\partial_\nu V^\rho) + \Gamma^\rho_{\beta\mu}\,\partial_\nu V^\beta - \Gamma^\gamma_{\mu\nu}\,\partial_\gamma V^\rho + D_\mu(\Gamma^\rho_{\sigma\nu} V^\sigma) \end{aligned}$$

We used result (3) of Section 1.11.13:

$$D_\mu D_\nu V^\rho = D_\mu(S^\rho_\nu) \qquad \text{with } S^\rho_\nu = \partial_\nu V^\rho + \Gamma^\rho_{\sigma\nu} V^\sigma$$

Since $D_\nu V^\rho$ is a tensor,

$$\begin{aligned} D_\mu D_\nu V^\rho &= D_\mu(S^\rho_\nu) = \partial_\mu(S^\rho_\nu) + \Gamma^\rho_{\beta\mu} S^\beta_\nu - \Gamma^\gamma_{\mu\nu} S^\rho_\gamma \\ &= \partial_\mu\,\partial_\nu V^\rho + \partial_\mu(\Gamma^\rho_{\sigma\nu} V^\sigma) + \Gamma^\rho_{\beta\mu}\,\partial_\nu V^\beta + \Gamma^\rho_{\beta\mu}\Gamma^\beta_{\sigma\nu} V^\sigma \\ &\quad - \Gamma^\gamma_{\mu\nu}\,\partial_\gamma V^\rho - \Gamma^\gamma_{\mu\nu}\Gamma^\rho_{\sigma\gamma} V^\sigma \\ D_\nu D_\mu V^\rho &= \partial_\mu\,\partial_\nu V^\rho + \partial_\nu(\Gamma^\rho_{\sigma\mu} V^\sigma) + \Gamma^\rho_{\beta\nu}\,\partial_\mu V^\beta + \Gamma^\rho_{\beta\nu}\Gamma^\beta_{\sigma\mu} V^\sigma \\ &\quad - \Gamma^\gamma_{\nu\mu}\,\partial_\gamma V^\rho - \Gamma^\gamma_{\nu\mu}\Gamma^\rho_{\sigma\gamma} V^\sigma \end{aligned}$$

and thus

$$\begin{aligned} [D_\mu, D_\nu] V^\rho &= \partial_\mu(\Gamma^\rho_{\sigma\nu} V^\sigma) - \partial_\nu(\Gamma^\rho_{\sigma\mu} V^\sigma) + \Gamma^\rho_{\beta\mu}\,\partial_\nu V^\beta - \Gamma^\rho_{\beta\nu}\,\partial_\mu V^\beta \\ &\quad + (\Gamma^\rho_{\beta\mu}\Gamma^\beta_{\sigma\nu} - \Gamma^\rho_{\beta\nu}\Gamma^\beta_{\sigma\mu}) V^\sigma + (\Gamma^\gamma_{\nu\mu} - \Gamma^\gamma_{\mu\nu})\,\partial_\gamma V^\rho \\ &\quad + \Gamma^\rho_{\sigma\gamma}(\Gamma^\gamma_{\nu\mu} - \Gamma^\gamma_{\mu\nu}) V^\sigma \\ &= [(\partial_\mu \Gamma^\rho_{\sigma\nu} - \partial_\nu \Gamma^\rho_{\sigma\mu}) + (\Gamma^\rho_{\beta\mu}\Gamma^\beta_{\sigma\nu} - \Gamma^\rho_{\beta\nu}\Gamma^\beta_{\sigma\mu})] V^\sigma \\ &\quad + (\Gamma^\gamma_{\nu\mu} - \Gamma^\gamma_{\mu\nu})(\partial_\gamma V^\rho + \Gamma^\rho_{\sigma\gamma}) V^\sigma \end{aligned}$$

Define $R^\rho{}_{\sigma\nu\mu}$, as the Riemann curvature tensor, which mixes the components, by

$$R^\rho{}_{\sigma\nu\mu} = \partial_\mu \Gamma^\rho_{\sigma\nu} - \partial_\nu \Gamma^\rho_{\sigma\mu} + \Gamma^\rho_{\beta\mu}\Gamma^\beta_{\sigma\nu} - \Gamma^\rho_{\beta\nu}\Gamma^\beta_{\sigma\mu}$$

Then

$$[D_\mu, D_\nu] V^\rho = R^\rho{}_{\sigma\nu\mu} V^\sigma + (\Gamma^\gamma_{\nu\mu} - \Gamma^\gamma_{\mu\nu})(\partial_\gamma V^\rho + \Gamma^\rho_{\sigma\gamma}) V^\sigma$$

More generally,

$$D_\nu V^\rho = \partial_\nu V^\rho + \Gamma^\rho_{\sigma\nu} V^\sigma$$

and

$$[D_\mu, D_\nu] V^\rho = R^\rho{}_{\sigma\nu\mu} V^\sigma + (\Gamma^\gamma_{\nu\mu} - \Gamma^\gamma_{\mu\nu}) D_\gamma V^\rho$$

We define the torsion tensor $\Omega^\gamma{}_{\nu\mu}$ by

$$\Omega^{\gamma}{}_{\nu\mu} = (\Gamma^{\gamma}_{\nu\mu} - \Gamma^{\gamma}_{\mu\nu}) D_{\gamma} V^{\rho}$$

In a euclidean connection

$$\Gamma^{i}_{jk} = 0 \qquad \Omega^{\gamma}{}_{\nu\mu} = 0 \qquad R^{\rho}{}_{\sigma\nu\mu} = 0$$

When the Christoffel indexes are symmetric, only $\Omega^{\gamma}{}_{\nu\mu} = 0$. By contraction,

$$g^{\nu}_{\gamma} \Omega^{\gamma}{}_{\nu\mu} = \Omega_{\mu}$$

where Ω_{μ} is the torsion vector of classical mechanics.

1.11.17 Useful Formulas Introducing Christoffel Symbols

Write

$$g = |g_{\mu\nu}| = \begin{vmatrix} g_{11} & g_{12} & \cdots & g_{1N} \\ g_{21} & g_{22} & \cdots & g_{2N} \\ \cdot & \cdots & \cdots & \cdot\cdot \\ \cdot & \cdots & \cdots & \cdot\cdot \\ g_{N1} & g_{N2} & \cdots & g_{NN} \end{vmatrix} \tag{1}$$

Let $A_{\mu\nu}$ be the minor associated with the element $g_{\mu\nu}$:

$$g = \sum_{\nu=1}^{N} g_{\mu\nu} A_{\mu\nu} \tag{2}$$

Be careful: Not only is there no summation on μ, but

$$\sum_{\substack{\nu=1 \\ \mu\neq\alpha}}^{N} g_{\mu\nu} A_{\alpha\nu} = 0 \tag{3}$$

We define $g^{\mu\nu}$ by

$$g^{\mu\nu} = \frac{A_{\mu\nu}}{g} \tag{4}$$

a coherent definition since

$$\sum_{\nu} g_{\mu\nu} g^{\alpha\nu} = \frac{1}{g} \sum_{\nu} g_{\mu\nu} A_{\alpha\nu} = \delta_{\mu\alpha} = \delta^{\alpha}_{\mu} \tag{5}$$

According to (6) of Section 1.11.8,

$$2 g_{\alpha\nu} \Gamma^{\alpha}_{\mu\lambda} = \frac{\partial g_{\mu\nu}}{\partial x^{\lambda}} + \frac{\partial g_{\lambda\nu}}{\partial x^{\mu}} - \frac{\partial g_{\lambda\mu}}{\partial x^{\nu}} \tag{6}$$

Multiply by $g^{\lambda\nu}$:

$$2g^{\lambda\nu}g_{\alpha\nu}\Gamma^{\alpha}_{\mu\lambda} = g^{\lambda\nu}\frac{\partial g_{\mu\nu}}{\partial x^{\lambda}} + g^{\lambda\nu}\frac{\partial g_{\lambda\nu}}{\partial x^{\mu}} - g^{\lambda\nu}\frac{\partial g_{\lambda\mu}}{\partial x^{\nu}}$$

$$2\delta^{\lambda}_{\alpha}\Gamma^{\alpha}_{\mu\lambda} = 2\Gamma^{\lambda}_{\mu\lambda} = g^{\lambda\nu}\frac{\partial g_{\lambda\nu}}{\partial x^{\mu}}$$

Since the quantity I, defined by

$$I = g^{\lambda\nu}\frac{\partial g_{\mu\nu}}{\partial x^{\lambda}} - g^{\lambda\nu}\frac{\partial g_{\lambda\mu}}{\partial x^{\nu}}$$

is zero (to be convinced, simply exchange λ and ν in the second term on the right-hand side of I, remembering that $g_{\mu\nu} = g_{\nu\mu}$),

$$\Gamma^{\lambda}_{\mu\lambda} = \frac{1}{2}g^{\lambda\nu}\frac{\partial g_{\lambda\nu}}{\partial x^{\mu}} \tag{7}$$

Also [cf. (4)],

$$\Gamma^{\lambda}_{\mu\lambda} = \frac{A_{\lambda\nu}}{2g}\frac{\partial g_{\lambda\nu}}{\partial x^{\mu}} = \frac{1}{2g}A_{\lambda\nu}\frac{\partial g_{\lambda\nu}}{\partial x^{\mu}} \tag{8}$$

According to (3),

$$\frac{\partial g}{\partial x^{\mu}} = \sum_{\nu=1}^{N}\left(\frac{\partial g_{\alpha\nu}}{\partial x^{\mu}}A_{\alpha\nu} + g_{\alpha\nu}\frac{\partial A_{\alpha\nu}}{\partial x^{\mu}}\right)$$

If we take $\alpha = \mu$,

$$\frac{\partial g}{\partial x^{\mu}} = \sum_{\nu=1}^{N}\left(\frac{\partial g_{\mu\nu}}{\partial x^{\mu}}A_{\mu\nu} + g_{\mu\nu}\frac{\partial A_{\mu\nu}}{\partial x^{\mu}}\right)$$

As $A_{\mu\nu}$ does not contain $g_{\mu\nu}$,

$$\frac{\partial A_{\mu\nu}}{\partial x^{\mu}} = 0$$

There remains

$$\frac{\partial g}{\partial x^{\mu}} = \sum_{\nu=1}^{N}\frac{\partial g_{\mu\nu}}{\partial x^{\mu}}A_{\mu\nu} \tag{9}$$

(note that there is no summation on μ). The summations are written explicitly in (8):

$$\Gamma^{\lambda}_{\mu\lambda} = \frac{1}{2g}\sum_{\lambda}\sum_{\nu}A_{\lambda\nu}\frac{\partial g_{\lambda\nu}}{\partial x^{\mu}} = \frac{1}{2g}\sum_{\nu}A_{\mu\nu}\frac{\partial g_{\mu\nu}}{\partial x^{\mu}}$$

In fact, in the summation on λ, only $\lambda = \mu$ gives a nonzero value for the partial derivative $\partial g_{\lambda\nu}/\partial x^{\mu}$. Taking (2) into consideration, we have

$$\Gamma^{\lambda}_{\mu\lambda} = \frac{1}{2g}\frac{\partial g}{\partial x^{\mu}} \tag{10}$$

Note that

$$\frac{\partial}{\partial x^{\mu}} \ln \sqrt{g} = \frac{1}{2}\frac{\partial}{\partial x^{\mu}} \ln g = \frac{1}{2}\left(\frac{\partial}{\partial g} \ln g\right)\frac{\partial g}{\partial x^{\mu}} = \frac{1}{2}\frac{\partial g}{\partial x^{\mu}}$$

and thus

$$\Gamma^{\lambda}_{\mu\lambda} = \frac{1}{2g}\frac{\partial g}{\partial x^{\mu}} = \frac{\partial}{\partial x^{\mu}} \ln \sqrt{g} \tag{11}$$

but also,

$$\frac{1}{\sqrt{g}}\frac{\partial}{\partial x^{\mu}}\sqrt{g} = \frac{1}{\sqrt{g}}\left(\frac{\partial}{\partial g}\sqrt{g}\right)\frac{\partial g}{\partial x^{\mu}} = \frac{1}{2g}\frac{\partial g}{\partial x^{\mu}}$$

or

$$\Gamma^{\lambda}_{\mu\lambda} = \frac{1}{\sqrt{g}}\frac{\partial}{\partial x^{\mu}}\sqrt{g} \tag{12}$$

APPENDIX 1 DIRAC FORMALISM

A1.1 Ket: Element of Hilbert Space of Dimension *N*

An element of the space will be indicated and recognized by the symbol $|f\rangle$: ket f. However, its description can be made only after having specified the basis $\{|e_i\rangle\}$ by giving the N components of this element in the basis. These components will be listed in a column matrix of N rows:

$$|f\rangle = \begin{bmatrix} f_1 \\ f_2 \\ \vdots \\ f_N \end{bmatrix} = \sum_{i=1}^{N} f_i|e_i\rangle$$

This is only an extension of a vector notation used in classical mechanics, where the symbol $\vec{v}$ enables us to indicate a vector by freeing us from the explicit choice of basis.

A1.2 Bra: Dual Space

To every ket $|f\rangle$, an element of Hilbert space, there corresponds an element in an associated space called *dual space*; this element will be denoted $\langle f|$: bra f. The description of this bra is made on the basis $\{\langle e_i|\}$ of the dual space. The bra is a linear combination of N bras $\langle e_i|$, of which the coefficients are the complex conjugates of the ket $|f\rangle$ components in the basis $\{|e_i\rangle\}$:

$$\langle f| = \sum_{i=1}^{N} f_i^* \langle e_i|$$

The bra components will be listed in a row matrix of N columns:

$$\langle f| = (f_1^*, f_2^*, \ldots, f_N^*) = \sum_{i=1}^{N} f_i^* \langle e_i|$$

A1.3 Scalar Product $S = (|f\rangle, |g\rangle) = \langle f|g\rangle$

The *scalar (dot) product* is a binary mapping that associates a scalar S with two kets $|f\rangle$ and $|g\rangle$ taken in this order:

$$S = \sum_{i=1}^{N} f_i^* g_i$$

Only the multiplication of a row matrix by a column matrix enables us to obtain a scalar, which is the reason the scalar product $(|f\rangle, |g\rangle)$ appears as matrix multiplication of $\langle f|$ by $|g\rangle$, or

$$(|f\rangle, |g\rangle) = \langle f|g\rangle$$

This procedure at once gives us a well-known property of the scalar product:

$$\langle f|g\rangle = \langle g|f\rangle^*$$

since

$$\langle f|g\rangle = \sum_{i=1}^{N} f_i^* g_i = \left(\sum_{i}^{N} f_i g_i^*\right)^* = \langle g|f\rangle^*$$

The orthogonality of two kets is, of course, written $\langle f|g\rangle = 0$, and in the case of two basis kets,

$$\langle e_i|e_j\rangle = \delta_{ij}$$

The ket $|f\rangle$ components in the basis $\{|e_i\rangle\}$ are scalar products:

$$|f\rangle = \sum_{i=1}^{N} f_i |e_i\rangle$$

$$\langle e_j|f\rangle = \sum_{i=1}^{N} f_i \langle e_j|e_i\rangle = \sum_{i=1}^{N} f_i \delta_{ij} = f_j$$

Finally, when the basis is complete, the sum of the projections onto all the basis kets must be the ket itself:

$$|f\rangle = \sum_{i=1}^{N} f_i|e_i\rangle = \sum_{i=1}^{N} \langle e_i|f\rangle|e_i\rangle = \sum_{i=1}^{N} |e_i\rangle\langle e_i|f\rangle$$

This is the closure relation, which must be satisfied by the kets of a basis.

Finally, note that the order in which the terms appear in a formal expression is extremely important; thus $|f\rangle\langle g|$ clearly does not have the same meaning as the expression $\langle f|g\rangle$. Only the scalar product set can be moved at will.

A1.4 The Transformation of Elements: Operator $\hat{O}$

$\hat{O}$ indicates the operator enabling the transformation of $|f\rangle$ into $|g\rangle$ following the symbolic notation

$$\hat{O}|f\rangle = |\hat{O}f\rangle = |g\rangle$$

The description of the kets in a basis $\{|e_i\rangle\}$ equally imposes the description of the operator in this basis. We go from one column matrix of N rows to a matrix of the same type by the intermediary of an $N \times N$ matrix of which the element O_{mn} is found at the intersection of the mth row and the nth column. Applying $\hat{O}$ to the basis ket $|e_1\rangle$ yields

$$\hat{O}|e_1\rangle = |e_1'\rangle = \begin{bmatrix} O_{11} & \cdots & O_{1N} \\ \cdot\cdot & \cdots & \cdot\cdot \\ \cdot\cdot & \cdots & \cdot\cdot \\ O_{N1} & \cdots & O_{NN} \end{bmatrix} \begin{bmatrix} 1 \\ 0 \\ 0 \\ 0 \end{bmatrix} = \begin{bmatrix} O_{11} \\ \vdots \\ O_{N1} \end{bmatrix}$$

This provides O_{i1} as the component of $|e_1'\rangle$ on the basis ket $|e_i\rangle$, or

$$\langle e_i|e_1'\rangle = \langle e_i|\hat{O}|e_1\rangle = O_{i1}$$

More generally,

$$O_{mn} = \langle e_m|\hat{O}|e_n\rangle$$

Note that the relation of closure is a relation that defines an operator

$$\sum_{i=1}^{N} |e_i\rangle\langle e_i|f\rangle = |f\rangle$$

$$\sum_{i=1} |e_i\rangle\langle e_i| = \hat{1}$$

In this form the closure relation proves itself to be very powerful.

Finally, it is useful to know that the product of two operators has such elements as

$$(\hat{A}\hat{B})_{ij} = \langle e_i|\hat{A}\hat{B}|e_j\rangle = \langle e_i|\hat{A}\left(\sum_{k=1}^{N} |e_k\rangle\langle e_k|\right)\hat{B}|e_j\rangle$$

$$= \sum_{i=1}^{N} A_{ik}B_{kj}$$

In the dual space the bra $\langle f|$ is transformed into bra $\langle g|$ by the intermediary of the adjoint operator O^+ working on the left:

$$\langle \hat{O}f| = \langle f|\hat{O}^+ = \langle g|$$

The operator $\hat{O}^+$ is also represented in the basis $\{|e_i\rangle\}$ by an $N \times N$ matrix whose elements satisfy

$$O^+_{mn} = O^*_{nm}$$

A1.5 Eigenvalues and Eigenstates of an Operator

In a space of dimension N there are at least N kets of which the "direction" remains unchanged by application of the operator $\hat{O}$. These kets, called *eigenstates* of $\hat{O}$, satisfy the equation

$$\hat{O}|f_k\rangle = o_k|f_k\rangle$$

The terms o_k are *eigenvalues* of the operator; there are at the very most N different eigenvalues in a space of dimension N. The eigenvalues are obtained by solving the defining equation above in a basis $\{|e_i\rangle\}$, which leads to the secular equation

$$\det|(\hat{O}) - o_k(\hat{1})| = 0$$

An eigenvalue is simple if it appears only once in solutions of the secular equation. So only one eigenstate, ket $|f_k\rangle$, corresponds to it.

An eigenvalue o_m is g_m times degenerate if we find this value g_m times in the spectrum of the eigenvalues of the operator. An infinity of eigenstates, assembled in an eigensubspace ϵ_m of dimension g_m, corresponds to it. This means that among kets of type $|f_m\rangle$ there are some g_m that are orthogonal among themselves and can be identified by denoting them $|f^i_m\rangle$ with $i = 1, 2, \ldots, g_m$. Any ϵ_m ket can be decomposed into

$$|f_m\rangle = \sum_{i=1}^{g_m} \lambda_i|f^i_m\rangle$$

Two eigenstates corresponding to two different eigenvalues are orthogonal since if

$$\hat{O}|f_k\rangle = o_k|f_k\rangle \qquad \text{and} \qquad \hat{O}|f_l\rangle = o_l|f_l\rangle$$

then

$$\langle\hat{O}^+ f_l| = (o_l^*)^*\langle f_l| \qquad \text{and} \qquad \langle f_l|\hat{O} = o_l\langle f_l|$$

In the scalar product

$$\langle f_l|\hat{O}|f_k\rangle = o_k\langle f_l|f_k\rangle = o_l\langle f_l|f_k\rangle$$

the eigenvalues are different. It follows, by hypothesis, that

$$\langle f_l|f_k\rangle = 0$$

In the basis of its eigenstates, an operator is represented by a diagonal matrix. The diagonal elements are the eigenvalues of the operator.

A1.6 Hermitian or Self-adjoint Operator

Hermitian or self-adjoint operators satisfy the equality

$$\hat{O}^+ = \hat{O}$$

The eigenvalues of hermitian operators are real since

$$\hat{O}|f_k\rangle = o_k|f_k\rangle \qquad \text{and} \qquad \langle f_k|\hat{O}^+ = o_k^*\langle f_k|$$

Thus $\hat{O}^+ = \hat{O}$, leading to $o_k = o_k^*$.

These operators satisfy a relation that could prove interesting:

$$\langle f|\hat{O}|g\rangle = \langle g|\hat{O}|f\rangle^*$$

It is proved in the following way:

$$\langle f|\hat{O}|g\rangle = \langle f|\hat{O}g\rangle = \langle\hat{O}g|f\rangle^* = \langle g|\hat{O}^+|f\rangle^* = \langle g|\hat{O}|f\rangle^*$$

APPENDIX 2 TENSOR OF INERTIA

Consider a vector space **E** of dimension 3 for which the basis vectors, denoted $\vec{e}_i$, are linearly independent. The scalar product

$$\vec{A}\cdot\vec{B} = A^iB_i$$

defines the metric g_j^i such that

$$\vec{e}^{\,i}\cdot\vec{e}_j = g_j^i \qquad \text{(A2.1)}$$

Note that x^1, x^2, and x^3 are the contravariant components of the element of mass $m(P)$ found at the point M_P, and G is the center of mass of the

Σ system (see Figure A2.1). The coordinates of the vector $\overrightarrow{GM}_P$ are given by

$$\overrightarrow{GM}_P = x^1(P)\vec{e}_1 + x^2(P)\vec{e}_2 + x^3(P)\vec{e}_3 \tag{A2.2}$$

Looking again at Figure A2.1, the momentum of inertia, in relation to a line Δ going through the center of mass G of the Σ system, is obtained by adding the N moments of inertia $I(P)$ of every particle P of mass $m(P)$ defined by

$$(I_{/\Delta})_{\{\vec{e}_i\}}(P) = m(P)H_PM_P^2 \tag{A2.3}$$

or

$$(I_{/\Delta})_{\{\vec{e}_i\}} = \sum_P m(P)H_PM_P^2 = \sum_P m(P)(GM_P^2 - GH_P^2) \tag{A2.4}$$

Let u_1, u_2, and u_3 be the covariant components of the unitary vector $\vec{u}$ directing line Δ such that

$$\vec{u}\cdot\vec{e}_i = u_i = \cos(\Delta, Gx_i) \tag{A2.5}$$

The vector $\overrightarrow{GH}_P$ is obtained by projecting the vector $\overrightarrow{GM}_P$ onto line Δ:

$$\overrightarrow{GH}_P = \vec{u}(\overrightarrow{GM}_P\cdot\vec{u}) = \vec{u}(x^i(P)\vec{e}_i)\cdot(u_j\vec{e}^{\,j}) = \vec{u}x^i(P)u_i \tag{A2.6}$$

From (A2.2) and (A2.6) we obtain the expression of the momentum of inertia $I_{/\Delta}$ of the system Σ:

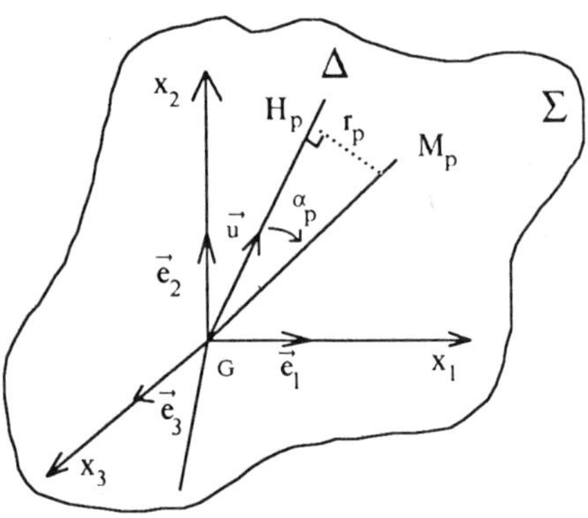

Figure A2.1 Notations.

$$I_{/\Delta} = \sum_{P} m(P)[(x^i(P)\vec{e}_i)^2 - (x^i u_i)^2] \tag{A2.7}$$

remembering that, by definition,

$$\|\vec{u}\|^2 = u_i u^i = 1 \tag{A2.8}$$

Development of the expression $H_P M_P^2$ gives

$$[x^i(P)\vec{e}_i]^2 - (x^i u_i)^2 = x^i(P)x_j(P) - (x^i u_i)(x_j u^j) \tag{A2.9}$$

Now substitute condition (A2.8) in (A2.9), or

$$\begin{aligned} H_P M_P^2 = {} & [x^1(P)x_1(P) + x^2(P)x_2(P) + x_3(P)x^3(P)] \\ & \times (u_1 u^1 + u_2 u^2 + u_3 u^3) - [x^1(P)u_1 + x^2(P)u_2 + x^3(P)u_3] \\ & \times [x_1(P)u^1 + x_2(P)u^2 + x_3(P)u^3] \end{aligned}$$

or

$$\begin{aligned} H_P M_P^2 = {} & [x^2(P)x_2(P) + x^3(P)x_3(P)]u_1 u^1 \\ & + [x^1(P)x_1(P) + x^3(P)x_3(P)]u_2 u^2 \\ & + [x^1(P)x_1(P) + x^2(P)x_2(P)]u_3 u^3 \\ & - x^1(P)x_2(P)u_1 u^2 - x^2(P)x_1(P)u_2 u^1 - x^1(P)x_3(P)u_1 u^3 \\ & - x^3(P)x_1(P)u_3 u^1 - x^2(P)x_3(P)u_2 u^3 - x^3(P)x_2(P)u_3 u^2 \end{aligned} \tag{A2.10}$$

The calculation of $H_P M_P^2$ therefore leads to the establishment of nine coefficients of the quantity $u_i u^j$, obtained by evaluating the product

$$[u_1, u_2, u_3](I) \begin{Bmatrix} u^1 \\ u^2 \\ u^3 \end{Bmatrix} \tag{A2.11}$$

with (I) the tensor of inertia of rank 2. Note: The brackets denote a covariant tensor and the braces a contravariant tensor.

The momentum of inertia $I_{/\Delta}$ is thus obtained by multiplying by the covariant tensor $[u]$, the product of the inertial mixed tensor (I), by the contravariant tensor $\{u\}$:

$$I_{/\Delta} = [u_1, u_2, u_3] \begin{vmatrix} I_1^1 & I_2^1 & I_3^1 \\ I_1^2 & I_2^2 & I_3^2 \\ I_1^3 & I_2^3 & I_3^3 \end{vmatrix} \begin{Bmatrix} u^1 \\ u^2 \\ u^3 \end{Bmatrix} \tag{A2.12}$$

where the nine coefficients I_j^i of the mixed tensor are established from relations (A2.4) and (A2.10). Thus

$$I_1^1 = \sum_P m(P)[x^2(P)x_2(P) + x^3(P)x_3(P)] \tag{A2.13a}$$

$$I_2^2 = \sum_P m(P)[x^1(P)x_1(P) + x^3(P)x_3(P)] \tag{A2.13b}$$

$$I_3^3 = \sum_P m(P)[x^1(P)x_1(P) + x^2(P)x_2(P)] \tag{A2.13c}$$

and

$$I_j^i = -\sum_P m(P)x^i(P)x_j(P) \qquad \text{with } i \neq j \tag{A2.14}$$

To find a general relation between the various terms I_j^i, rewrite the quantities (A2.13) in the following way:

$$I_1^1 = \sum_P m(P)[x^i(P)x_i(P) - x^1(P)x_1(P)] \tag{A2.15}$$

By generalizing to two other relations (A2.13) we eventually write

$$I_i^i = \sum_P m(P)x^i(P)x_i(P) - \sum_P m(P)[x^i(P)]^2 \tag{A2.16}$$

Coming back to the matrix (A2.12), the tensor of inertia (I) is written as a summation of two tensors (I') and (I''):

$$(I) = (I') + (I'') \tag{A2.17}$$

of which only one will have elements on the principal diagonal:

$$(I') = \begin{bmatrix} \sum_P m(P)x^i(P)x_i(P) & 0 & 0 \\ 0 & \sum_P m(P)x^i(P)x_i(P) & 0 \\ 0 & 0 & \sum_P m(P)x^i(P)x_i(P) \end{bmatrix} \tag{A2.18}$$

and

$$
(I'') = \begin{bmatrix}
-\sum_P m(P)x^1(P)x_1(P) & -\sum_P m(P)x^1(P)x_2(P) & -\sum_P m(P)x^1(P)x_3(P) \\
-\sum_P m(P)x^2(P)x_1(P) & -\sum_P m(P)x^2(P)x_2(P) & -\sum_P m(P)x^2(P)x_3(P) \\
-\sum_P m(P)x^3(P)x_1(P) & -\sum_P m(P)x^3(P)x_2(P) & -\sum_P m(P)x^3(P)x_3(P)
\end{bmatrix} \tag{A2.19}
$$

or

$$
I^i_j = \sum_P m(P)[x^l(P)x_l(P)\delta^i_j - x^i(P)x_j(P)] \tag{A2.20}
$$

the mixed component of a tensor of inertia, according to the identities

$$
x^l = x_n g^{nl}
$$

and

$$
I_{mj} = I^i{}_j g_{mi}
$$

$$
I_{mj} = \sum_P m(P)[x_n(P)x_l(P)g^{nl}\delta_{mj} - x_m(P)x_j(P)] \tag{A2.21}
$$

This is the covariant component, and by similar calculations we can show that the contravariant component of the inertia tensor (I) is such that

$$
I^{mj} = \sum_P m(P)[x^n(P)x^l(P)g_{nl}\delta^{mj} - x^m(P)x^j(P)] \tag{A2.22}
$$

APPENDIX 3 TENSORS OF ELASTICITY

In this section we focus on continuous infinitely small and homogeneous deformations.

A3.1 Tensor of Homogeneous Deformation

At the moment of deformation a solid Σ transforms itself into a solid Σ', with the result that a vector $\overrightarrow{AB}$ finds itself in $\overrightarrow{A'B'}$ (Figure A3.1). We have

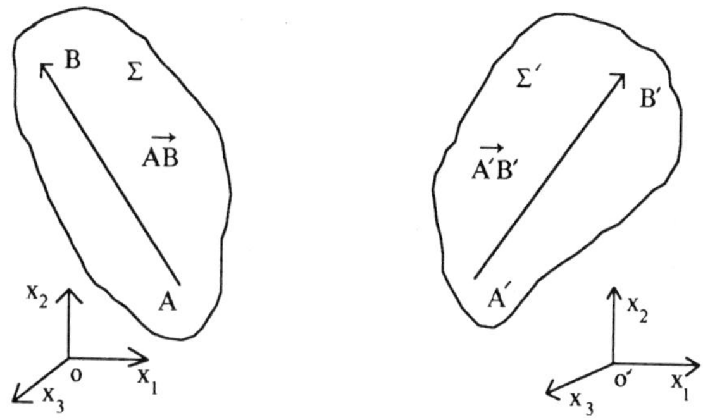

Figure A3.1 Deformation.

$$A(x_1,x_2,x_3)$$
$$B(x_1 + dx_1,\ x_2 + dx_2,\ x_3 + dx_3)$$
$$A'(X_1,X_2,X_3)$$
$$B'(X_1 + dX_1,\ X_2 + dX_2,\ X_3 + dX_3)$$

Assume $\vec{u}$ and $\vec{U}$ to be the respective displacements of the vectors $\overrightarrow{AA'}$ and $\overrightarrow{BB'}$, and u_i and U_i ($i = 1,2,3$) to be the coordinates of the displacement vectors. These coordinates satisfy the following two relations:

$$\begin{aligned} &\overrightarrow{AA'} \longrightarrow X_i - x_i = u_i(x_1,x_2,x_3) \\ &\overrightarrow{BB'} \longrightarrow (X_i + dX_i) - (x_i + dx_i) = U_i(X_1,X_2,X_3) \end{aligned} \tag{A3.1}$$

Otherwise,

$$u_i + \left(\frac{\partial}{\partial x^j}\right) u_i\, dx^j = U_i$$

or

$$U_i = u_i + (\partial_j u_i)\, dx^j \tag{A3.2}$$

Consequently, after deformation the components of $\overrightarrow{AB}$ become $\overrightarrow{A'B'}$ such that

$$dX_i = dx_i + (\partial_j u_i)\, dx^j \tag{A3.3}$$

We now define the general homogeneous deformation tensor, denoted (D), by

$$(D_{ij}) = \begin{bmatrix} \partial_1 u_1 & \partial_2 u_1 & \partial_3 u_1 \\ \partial_1 u_2 & \partial_2 u_2 & \partial_3 u_2 \\ \partial_1 u_3 & \partial_2 u_3 & \partial_3 u_3 \end{bmatrix}$$

This tensor is a tensor of rank 2 which decomposes into a symmetric tensor (a tensor of pure deformation) and an antisymmetric tensor (a tensor of pure rotation). Assume (ϵ_{ij}) and (ω_{ij}) to be the matrix elements of pure deformation and pure rotation, and observe that

$$(\partial_j u_i) = (\epsilon_{ij}) + (\omega_{ij}) \tag{A3.4}$$

For that write

$$\begin{aligned} \partial_j u_i &= \tfrac{1}{2}(\partial_j u_i + \partial_i u_j) + \tfrac{1}{2}(\partial_j u_i - \partial_i u_j) \\ &= (\epsilon_{ij}) + (\omega_{ij}) \end{aligned}$$

with

$$\epsilon_{ij} = \tfrac{1}{2}(\partial_j u_i + \partial_i u_j) \tag{A3.5}$$

$$\omega_{ij} = \tfrac{1}{2}(\partial_j u_i - \partial_i u_j) \tag{A3.6}$$

We have thus constructed two tensors, one symmetric, the other antisymmetric. In conclusion, a deformation is thus the outcome of a pure

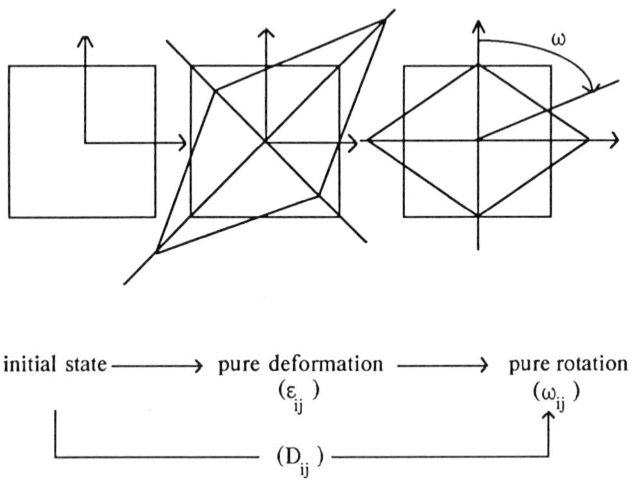

Figure A3.2 Decomposition of the deformation.

deformation and a pure rotation (Figure A3.2). Write the tensors (ϵ_{ij}) and (ω_{ij}) as follows:

$$(\epsilon_{ij}) = \begin{bmatrix} \partial_1 u_1 & \frac{1}{2}(\partial_1 u_2 + \partial_2 u_1) & \frac{1}{2}(\partial_1 u_3 + \partial_3 u_1) \\ \frac{1}{2}(\partial_2 u_1 + \partial_1 u_2) & \partial_2 u_2 & \frac{1}{2}(\partial_2 u_3 + \partial_3 u_2) \\ \frac{1}{2}(\partial_1 u_3 + \partial_3 u_1) & \frac{1}{2}(\partial_2 u_3 + \partial_3 u_2) & \partial_3 u_3 \end{bmatrix} \tag{A3.7}$$

$$(\omega_{ij}) = \begin{bmatrix} 0 & \partial_1 u_2 - \partial_2 u_1 & \partial_1 u_3 - \partial_3 u_1 \\ \partial_2 u_1 - \partial_1 u_2 & 0 & \partial_2 u_3 - \partial_3 u_2 \\ \partial_1 u_3 - \partial_3 u_1 & \partial_2 u_3 - \partial_3 u_2 & 0 \end{bmatrix} \tag{A3.8}$$

A3.2 Trace of the Tensor of Deformations

A scalar called a *trace* (spur) is associated with tensor D of rank 2. The trace, which in this case is called a *coefficient of cubic dilatation*, is invariant regardless of the system of axes and such that

$$\mathrm{tr}(D) = e = \epsilon_i^i \tag{A3.9}$$

A3.3 Tensor of Constraints

Let dF_i be the components of the vector $\overrightarrow{dF}$. The components of the tensor of constraints are defined by the following matrix relation:

$$\begin{bmatrix} dF_1 \\ dF_2 \\ dF_3 \end{bmatrix} = \begin{bmatrix} \sigma_{11} & \sigma_{12} & \sigma_{13} \\ \sigma_{21} & \sigma_{22} & \sigma_{23} \\ \sigma_{31} & \sigma_{32} & \sigma_{33} \end{bmatrix} \begin{bmatrix} adS \\ bdS \\ cdS \end{bmatrix} \tag{A3.10}$$

where the σ_{ij} are the constraint components in M on the faces parallel to the planes of coordinates. They can be represented graphically on the faces of an elementary cube (Figure A3.3) parallel to these planes.

More generally, the component σ_{ij} works on the side perpendicular to $0x_i$ and is directed according to $0x_j$. All the components for which $i = j$ are therefore normal constraints. Sometimes we denote by σ and τ the normal and tangent constraints. Here, to simplify the writing, we are not going to use this notation.

The solid is in equilibrium if the resulting momentum of the forces is zero:

$$\sum_{i,j} \vec{M}_0(\vec{F}_{ij}) = \vec{0}$$

which means that

$$\sigma_{ij} = \sigma_{ji} \tag{A3.11}$$

The tensor is symmetric.

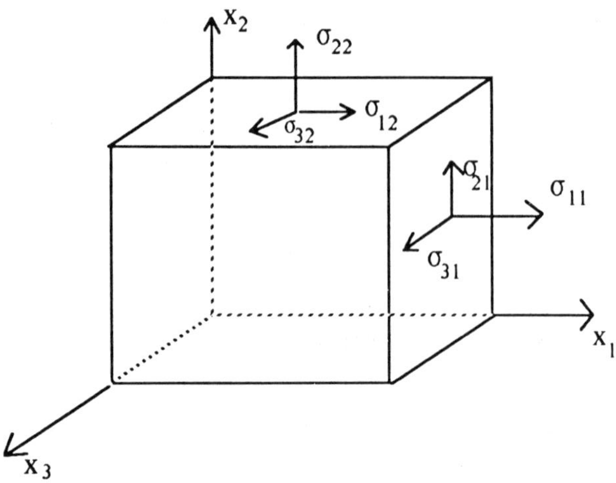

Figure A3.3 Constraints.

A3.4 Relations Between Constraints and Deformations

When an elastic environment is deformed, constraints are created and a tensor of constraints corresponds to each tensor of deformations. The aim of this section is to establish a very general linear relation between the tensors of deformations and those of constraints.

Suppose, first, that the tensor of constraints is a function of that of deformations:

$$\sigma_{ij} = \lambda\epsilon^i_i g_{ij} + 2\mu\epsilon_{ij} \tag{A3.12}$$

with λ and μ Lamé coefficients. For the moment ϵ_{ij} is expressed as a function of σ_{ij}. The matrix form of (A3.12) is written

$$\begin{bmatrix} \sigma_{11} & \sigma_{12} & \sigma_{13} \\ \sigma_{21} & \sigma_{22} & \sigma_{23} \\ \sigma_{31} & \sigma_{32} & \sigma_{33} \end{bmatrix} = \lambda \begin{bmatrix} \epsilon_{11} & 0 & 0 \\ 0 & \epsilon_{22} & 0 \\ 0 & 0 & \epsilon_{33} \end{bmatrix} \begin{bmatrix} g_{11} & g_{12} & g_{13} \\ g_{21} & g_{22} & g_{23} \\ g_{31} & g_{32} & g_{33} \end{bmatrix} + 2\mu \begin{bmatrix} \epsilon_{11} & \epsilon_{12} & \epsilon_{13} \\ \epsilon_{21} & \epsilon_{22} & \epsilon_{23} \\ \epsilon_{31} & \epsilon_{32} & \epsilon_{33} \end{bmatrix} \tag{A3.13}$$

The trace of the tensor (σ_{ij}) is, according to (A3.12) or (A3.13),

$$\sigma^i{}_i = \lambda\epsilon^i{}_i g^i{}_i + 2\mu\epsilon^i{}_i \tag{A3.14}$$

Otherwise,

$$\sigma^{i}_{\ i} = \lambda(3\epsilon^{i}_{\ i}) + 2\mu\epsilon^{i}_{\ i} = (3\lambda + 2\mu)\epsilon^{i}_{\ i} \tag{A3.15}$$

Note that s (respectively, e) is the trace of the tensor σ_{ij} (respectively, ϵ_{ij}):

$$s = (3\lambda + 2\mu)e$$

or

$$e = \frac{s}{3\lambda + 2\mu} \tag{A3.16}$$

We obtain from (A3.12)

$$\epsilon_{ij} = \frac{1}{2\mu}(\sigma_{ij} - \lambda e g_{ij}) = \frac{1}{2\mu}\sigma_{ij} - \frac{\lambda s}{2\mu(3\lambda + 2\mu)} g_{ij} \tag{A3.17}$$

We now write the coefficients E and ν (called *Young's modulus* and *Poisson's coefficient*, respectively) as

$$E = \frac{\mu(3\lambda + 2\mu)}{\lambda + \mu} \tag{A3.18}$$

$$\nu = \frac{\lambda}{2(\lambda + \mu)} \tag{A3.19}$$

We can now put (A3.17) into the classical form

$$\epsilon_{ij} = \frac{1}{E}[(1 + \nu)\sigma_{ij} - \nu s g_{ij}] \tag{A3.20}$$

The total work given by the forces during the deformations is written

$$W = \tfrac{1}{2}\sigma^{ij}\epsilon_{ij} \tag{A3.21}$$

A3.5 Elasticity Constants

Hooke's law enables us to state that the linear components of the deformation tensor ϵ_{kl} are functions of the coefficients σ_{ij} of the constraints tensor. The following two relations are derived from this law:

$$\sigma_{ij} = c_{ij}^{\ kl}\epsilon_{kl} \tag{A3.22}$$

and

$$\epsilon_{ij} = s_{ij}^{\ kl}\sigma_{kl} \tag{A3.23}$$

where $c_{ij}^{\ kl}$ and $s_{ij}^{\ kl}$ are components of the tensor of rank 4, called the tensor of the elasticity modulus and the tensor of the elasticity components, respectively. The matrix for $(c_{ij}{}^{kl})$, for example, is written

$$(c_{ij}{}^{kl}) = \begin{bmatrix} c_{11}{}^{11} & c_{12}{}^{11} & \cdots & c_{33}{}^{11} \\ c_{11}{}^{12} & \cdot & \cdots & c_{33}{}^{12} \\ c_{11}{}^{13} & \cdot & \cdots & c_{33}{}^{13} \\ \cdot & \cdot & \cdots & \cdot \\ \cdot & \cdot & \cdots & \cdot \\ \cdot & \cdot & \cdots & \cdot \\ \cdot & \cdot & \cdots & \cdot \\ \cdot & \cdot & \cdots & \cdot \\ c_{11}{}^{33} & c_{12}{}^{33} & \cdots & c_{33}{}^{33} \end{bmatrix} \tag{A3.24}$$

These tensors thus have 81 coefficients, a good number of which, to be shown shortly, are equal or cancel each other out, due to the symmetries of crystal and the ϵ_{ij}, σ_{ij}.

A3.6 Symmetry of the Coefficients ϵ_{ij} and σ_{ij}

According to definition (A3.5), the coefficients ϵ_{ij} are

$$\epsilon_{ij} = \tfrac{1}{2}(\partial_j u_i + \partial_i u_j) = \tfrac{1}{2}(\partial_i u_j + \partial_j u_i) = \epsilon_{ji} \tag{A3.25}$$

The internal energy of a crystal is defined by

$$U = \tfrac{1}{2} c_{ij}{}^{kl} \epsilon^{ij} \epsilon_{kl} \tag{A3.26}$$

If we exchange the silent indices i and j, we obtain

$$U = \tfrac{1}{2} c_{ji}{}^{kl} \epsilon^{ji} \epsilon_{kl} \tag{A3.27}$$

By identifying (A3.26) and (A3.27) on the basis of (A3.25), we see that

$$c_{ij}{}^{kl} = c_{ji}{}^{kl} \tag{A3.28}$$

Equally, by exchanging the silent indices k and l,

$$c_{ij}{}^{kl} = c_{ij}{}^{lk} \tag{A3.29}$$

Simultaneous exchange of the silent indices i,j and k,l gives

$$c_{ij}{}^{kl} = c_{ji}{}^{lk} \tag{A3.30}$$

We also have, by definition,

$$U = \tfrac{1}{2} c_{ij}{}^{kl} \epsilon^{ij} \epsilon_{kl} = \tfrac{1}{2} c_{ij}{}^{kl} \epsilon_{pq} g^{pi} g^{qj} \epsilon^{rs} g_{rk} g_{sl} = \tfrac{1}{2} c^{pq}{}_{rs} \epsilon^{rs} \epsilon_{pq}$$

or by changing notation ($r \to i$, $s \to j$, $p \to k$, $q \to l$),

$$U = \tfrac{1}{2} c_{ij}{}^{kl} \epsilon^{ij} \epsilon_{kl} = \tfrac{1}{2} c^{kl}{}_{ij} \epsilon^{ij} \epsilon_{kl}$$

or

$$c_{ij}{}^{kl} = c^{kl}{}_{ij} \tag{A3.31}$$

The internal energy U is a real quantity

$$U = U^*$$

This condition leads to the identity

$$\tfrac{1}{2}(c_{pq}{}^{rs}\epsilon^{pq}\epsilon_{rs})^* = \tfrac{1}{2}c_{ij}{}^{kl}\epsilon^{ij}\epsilon_{kl}$$

or

$$c_{pq}{}^{rs}\epsilon^{rs}\epsilon_{pq} = c_{ij}{}^{kl}\epsilon^{ij}\epsilon_{kl}$$

which we write

$$c_{ij}{}^{kl} = c_{kl}{}^{ij} \tag{A3.32}$$

This supplementary condition indicates that the tensor is symmetric with respect to the principal diagonal.

Observe that for the index couples $m = (i,j)$ and $n = (k,l)$ there are only six possibilities:

$$\begin{aligned} 11 &\longrightarrow 1 \\ 22 &\longrightarrow 2 \\ 33 &\longrightarrow 3 \\ 23 = 32 &\longrightarrow 4 \\ 13 = 31 &\longrightarrow 5 \\ 12 = 21 &\longrightarrow 6 \end{aligned} \tag{A3.33}$$

This represents a table of correspondence between

$$\sigma_{ij} = c_{ij}{}^{kl}\epsilon_{kl}$$

and

$$\sigma_m = c_m{}^n\epsilon_n \tag{A3.34}$$

where $m, n = 1, 2, 3, 4, 5, 6$; we are at present working in a space of dimension 6. The indices $m = 1, 2, 3$ may, for example, represent the chosen axes in the study of a crystal.

Similarly, we write

$$\epsilon_m = s_m{}^n\sigma_n \tag{A3.35}$$

and

$$U = \tfrac{1}{2} c_m{}^n \epsilon^m \epsilon_n \tag{A3.36}$$

the internal energy in a new formulation. We have therefore reduced the tensor of rank 4 to a tensor of rank 2 with 21 components:

$$(c_m^n) = \begin{bmatrix} c_1{}^1 & c_1{}^2 & c_1{}^3 & c_1{}^4 & c_1{}^5 & c_1{}^6 \\ c_1{}^2 & c_2{}^2 & c_2{}^3 & c_2{}^4 & c_2{}^5 & c_2{}^6 \\ c_1{}^3 & c_2{}^3 & c_3{}^3 & c_3{}^4 & c_3{}^5 & c_3{}^6 \\ c_1{}^4 & c_2{}^4 & c_3{}^4 & c_4{}^4 & c_4{}^5 & c_4{}^6 \\ c_1{}^5 & c_2{}^5 & c_3{}^5 & c_4{}^5 & c_5{}^5 & c_5{}^6 \\ c_1{}^6 & c_2{}^6 & c_3{}^6 & c_4{}^6 & c_5{}^6 & c_6{}^6 \end{bmatrix} \tag{A3.37}$$

The number of components of this tensor may be reduced even more, thanks to the crystal symmetries studied earlier.

A3.7 Reduction by Symmetry of Tensor of Elasticity Moduli

Consider, by way of example, the case of a crystal of quadratic symmetry (Figure A3.4), the lattice parameters of which are $\beta = \alpha = \gamma = 90°$ and $a = b \neq c$. A rotation of $\pi/2$ around the axis $0z$ results in a transformation of the coordinates x, y, z to x', y', z' such that

$$\begin{bmatrix} x' \\ y' \\ z' \end{bmatrix} = \begin{bmatrix} \cos\frac{\pi}{2} & \sin\frac{\pi}{2} & 0 \\ -\sin\frac{\pi}{2} & \cos\frac{\pi}{2} & 0 \\ 0 & 0 & 1 \end{bmatrix} \begin{bmatrix} x \\ y \\ z \end{bmatrix} = \begin{bmatrix} y \\ -x \\ z \end{bmatrix} \tag{A3.38}$$

The internal energy U is written, in general,

$$\begin{aligned} U = \tfrac{1}{2}[c_1{}^1\epsilon^1\epsilon_1 &+ 2c_1{}^2\epsilon^1\epsilon_2 + 2c_1{}^3\epsilon^1\epsilon_3 + 2c_1{}^4\epsilon^1\epsilon_4 + 2c_1{}^5\epsilon^1\epsilon_5 + 2c_1{}^6\epsilon^1\epsilon_6 \\ &+ c_2{}^2\epsilon^2\epsilon_2 + 2c_2{}^3\epsilon^2\epsilon_3 + 2c_2{}^4\epsilon^2\epsilon_4 + 2c_2{}^5\epsilon^2\epsilon_5 + 2c_2{}^6\epsilon^2\epsilon_6 \\ &+ c_3{}^3\epsilon^3\epsilon_3 + 2c_3{}^4\epsilon^3\epsilon_4 + 2c_3{}^5\epsilon^3\epsilon_5 + 2c_3{}^6\epsilon^3\epsilon_6 \\ &+ c_4{}^4\epsilon^4\epsilon_4 + 2c_4{}^5\epsilon^4\epsilon_5 + 2c_4{}^6\epsilon^4\epsilon_6 \\ &+ c_5{}^5\epsilon^5\epsilon_5 + 2c_5{}^6\epsilon^5\epsilon_6 \\ &+ c_6{}^6\epsilon^6\epsilon_6] \end{aligned}$$

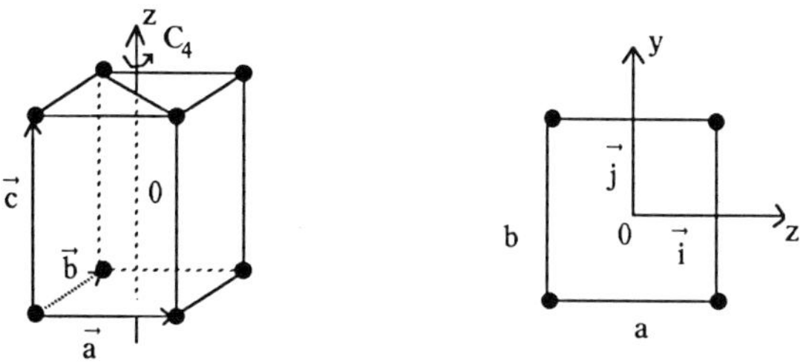

Figure A3.4 Quadratic lattice.

This quantity must remain invariant under the application of any operation of symmetry causing the crystal to have the same volume of geometric space.

Write the transformation of the quantities ϵ_m at the moment of the rotation of $\pi/2$ around the axis $0z$:

$$\epsilon_1 = \epsilon_{xx} \longrightarrow \epsilon_{yy} = \epsilon_2$$

$$\epsilon_2 = \epsilon_{yy} \longrightarrow \epsilon_{(-x)(-x)} = \epsilon_{xx} = \epsilon_1$$

$$\epsilon_3 = \epsilon_{zz} \longrightarrow \epsilon_{zz} = \epsilon_3$$

$$\epsilon_4 = \epsilon_{yz} \longrightarrow \epsilon_{(-x)z} = -\epsilon_{xz} = -\epsilon_5$$

$$\epsilon_5 = \epsilon_{xz} \longrightarrow \epsilon_{yz} = \epsilon_4$$

$$\epsilon_6 = \epsilon_{xy} \longrightarrow \epsilon_{y(-x)} = -\epsilon_{yx} = -\epsilon_6$$

relations that we write symbolically, under action of the rotation C_4, that is, of the angle $\pi/2$ around $0z$:

$$xx \longrightarrow yy$$

$$yy \longrightarrow xx$$

$$zz \longrightarrow zz$$

$$yz \longrightarrow -xz$$

$$xz \longrightarrow yz$$

$$xy \longrightarrow -yx$$

Similarly, we list in a 6×6 table the transformation of the elasticity coefficients during the transformation defined previously.

			n					
		m	1	2	3	4	5	6
2	$yy \leftarrow xx$	1	c_2^2	c_2^1	c_2^3	c_2^{-5}	c_2^4	c_2^{-6}
1	$xx \leftarrow yy$	2	c_1^2	c_1^1	c_1^3	c_1^{-5}	c_1^4	c_1^{-6}
3	$zz \leftarrow zz$	3	c_3^2	c_3^1	c_3^3	c_3^{-5}	c_3^4	c_3^{-6}
−5	$-xz \leftarrow yz$	4	$c^2{}_{-5}$	$c^1{}_{-5}$	$c^3{}_{-5}$	c_{-5}^{-5}	$c^4{}_{-5}$	c_{-5}^{-6}
4	$yz \leftarrow xz$	5	c_4^2	c_4^1	c_4^3	c_4^{-5}	c_4^4	c_4^{-6}
−6	$-yx \leftarrow xy$	6	$c^2{}_{-6}$	$c^1{}_{-6}$	$c^3{}_{-6}$	c_{-6}^{-5}	$c^4{}_{-6}$	c_{-6}^{-6}

The invariance of the energy U under rotation of $\pi/2$ implies the nullity of the coefficients where only one negative sign appears, given that when the product $\epsilon_m\epsilon^n$ produces a minus sign in front of the quantity $c_m^n\epsilon_m\epsilon^n$, and by identification of the coefficients in $\epsilon_m\epsilon^n$ of the two forms of internal energy U, the following equation holds:

$$c_m{}^n\epsilon_n\epsilon^m = -c_m{}^n\epsilon_n\epsilon^m$$

or

$$c_m{}^n = 0$$

Consequently, with the preceding table in mind,

$$c_1{}^4 = c_4{}^1 = c_1{}^6 = c_6{}^1 = c_2{}^4 = c_4{}^2 = c_2{}^6 = c_6{}^2 = 0$$

$$c_3{}^4 = c_4{}^3 = c_3{}^6 = c_6{}^3 = c_4{}^5 = c_5{}^4 = c_5{}^6 = c_6{}^5 = 0$$

Moreover, according to the preceding table,

$$c_1{}^1 = c_2{}^2 \qquad c_1{}^3 = c_2{}^3 \qquad c_5{}^5 = c_4{}^4$$

The simplified expression of the internal energy U of the crystal is

$$\begin{aligned} U = \tfrac{1}{2}[c_1{}^1\epsilon^1\epsilon_1 &+ 2c_1{}^2\epsilon^1\epsilon_2 + 2c_1{}^3\epsilon^1\epsilon_3 + 2c_1{}^5\epsilon^1\epsilon_5 \\ &+ c_1{}^1\epsilon^2\epsilon_2 + 2c_1{}^3\epsilon^2\epsilon_3 + 2c_2{}^5\epsilon^2\epsilon_5 \\ &+ c_3{}^3\epsilon^3\epsilon_3 + 2c_3{}^5\epsilon^3\epsilon_5 \\ &+ c_4{}^4\epsilon^4\epsilon_4 + 2c_4{}^6\epsilon^4\epsilon_6 \\ &+ c_4{}^4\epsilon^5\epsilon_5 \\ &+ c_6{}^6\epsilon^6\epsilon_6] \end{aligned}$$

or

$$U = \tfrac{1}{2}c_1{}^1(\epsilon^1\epsilon_1 + \epsilon^2\epsilon_2) + c_1{}^2\epsilon^1\epsilon_2 + c_1{}^3(\epsilon^1\epsilon_3 + \epsilon^2\epsilon_3) + c_1{}^5\epsilon^1\epsilon_5 + c_2{}^5\epsilon^2\epsilon_5$$
$$+ \tfrac{1}{2}c_3{}^3\epsilon^3\epsilon_3 + c_3{}^5\epsilon^3\epsilon_5 + \tfrac{1}{2}c_4{}^4(\epsilon^4\epsilon_4 + \epsilon^5\epsilon_5) + c_4{}^6\epsilon^4\epsilon_6 + \tfrac{1}{2}c_6{}^6\epsilon^6\epsilon_6$$

The plane $x0z$ being a plane of symmetry,

$$x \longrightarrow x$$
$$y \longrightarrow -y$$
$$z \longrightarrow z$$

must leave U invariant. This is also true for the plane of symmetry $y0z$, characterized by the invariance under the transformation

$$x \longrightarrow -x$$
$$y \longrightarrow y$$
$$z \longrightarrow z$$

Using the same procedure as before for the rotation of $\pi/2$ around the axis $0z$, all the coefficients c_m^n, of which the sum of the indexes contains an odd number of times x (symmetry in relation to the plane $y0z$) or y (symmetry in relation to the plane $x0z$), will be zero:

$$c_1{}^5 = c_5{}^1 = 0$$
$$c_2{}^5 = c_5{}^2 = 0$$
$$c_3{}^5 = c_5{}^3 = 0$$
$$c_4{}^6 = c_6{}^4 = 0$$

from which the final expression of internal energy of a crystal of quadratic symmetry is

$$U = \tfrac{1}{2}c_1{}^1(\epsilon^1\epsilon_1 + \epsilon^2\epsilon_2) + c_1{}^2\epsilon^1\epsilon_2 + c_1{}^3(\epsilon^1\epsilon_3 + \epsilon^2\epsilon_3) + \tfrac{1}{2}c_3{}^3\epsilon^3\epsilon_3$$
$$+ \tfrac{1}{2}c_4{}^4(\epsilon^4\epsilon_4 + \epsilon^5\epsilon_5) + \tfrac{1}{2}c_6{}^6\epsilon^6\epsilon_6$$

In conclusion, the matrix of the coefficients c_m^n is reduced to

$$(c_m^n) = \begin{bmatrix} c_1{}^1 & c_1{}^2 & c_1{}^3 & 0 & 0 \\ c_2{}^1 & c_1{}^1 & c_1{}^3 & 0 & 0 \\ c_1{}^3 & c_1{}^3 & c_3{}^3 & 0 & 0 \\ 0 & 0 & 0 & c_4{}^4 & 0 \\ 0 & 0 & 0 & 0 & c_4{}^4 \\ 0 & 0 & 0 & 0 & 0 \end{bmatrix} \begin{matrix} 0 \\ 0 \\ 0 \\ 0 \\ 0 \\ c_6{}^6 \end{matrix}$$

There are six parameters. For a cubic crystal we easily prove that

$$(c_m^n) = \begin{bmatrix} c_1^{\,1} & c_1^{\,2} & c_1^{\,2} & 0 & 0 & 0 \\ c_1^{\,2} & c_1^{\,1} & c_1^{\,2} & 0 & 0 & 0 \\ c_1^{\,2} & c_1^{\,2} & c_1^{\,1} & 0 & 0 & 0 \\ 0 & 0 & 0 & c_4^{\,4} & 0 & 0 \\ 0 & 0 & 0 & 0 & c_4^{\,4} & 0 \\ 0 & 0 & 0 & 0 & 0 & c_4^{\,4} \end{bmatrix}$$

There are only three parameters.

2
Minkowski Space

2.1 PRINCIPLE OF RELATIVITY

The principle of relativity was discovered by C. Huygens (1629–1695) and A. Einstein (1879–1955). In classical kinematics the space is euclidean independent of the observer. A referential of space $(O,\vec{e}_1,\vec{e}_2,\vec{e}_3)$ and a referential of time, which is defined by an origin and a unity, are associated with an observer O. The square of the distance separating two points M_1 and M_2 (Figure 4) is then written

$$d^2_{M_1M_2} = \overrightarrow{M_1M_2^2} = (\vec{r}_2 - \vec{r}_1)^2 = (x_2 - x_1)^2 + (y_2 - y_1)^2 + (z_2 - z_1)^2$$

This is a positive quadratic form.

For an observer O' in the referential $(O',\vec{e}'_1,\vec{e}'_2,\vec{e}'_3)$

$$d^2_{M_1M_2} = \overrightarrow{M_1M_2^2} = (\vec{r}'_2 - \vec{r}'_1)^2$$
$$= (x'_2 - x'_1)^2 + (y'_2 - y'_1)^2 + (z'_2 - z'_1)^2$$

a quadratic form that is also positive. Classical kinematics postulates that

$$(\vec{r}_2 - \vec{r}_1)^2 = (\vec{r}'_2 - \vec{r}'_1)^2$$

the preservation of the scalar product and therefore of the norm of the vector, which is equivalent to

$$d_{M_1M_2} = d'_{M_1M_2}$$

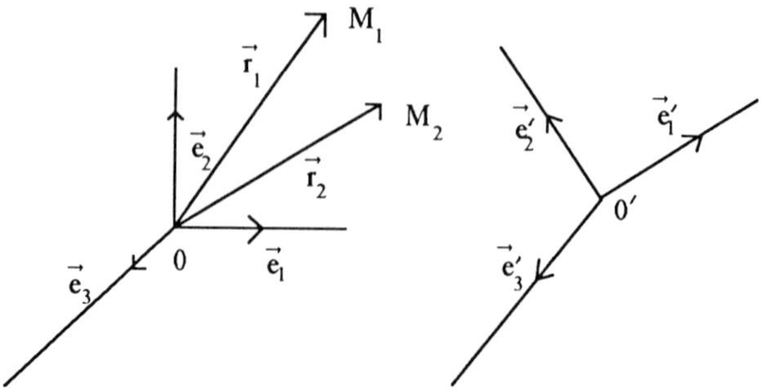

Figure 4 Referentials.

that is, that space and time are independent and, moreover, independent of the observer. We should add

$$t_1 = t_1' \qquad \text{and} \qquad t_2 = t_2'$$

to the previous equation, as the time is absolute as well as independent of the observer. Relativistic kinematics rejects these hypotheses.

To guarantee the validity of a measurement and to obtain a law of physics from such a measurement, of which the measurement is the desired target, the result must be independent of:

The place where the experiment was conducted (homogenity of space)
The positioning of the laboratory on the measurement site (isotropy of space)
The time at which the experiment took place (uniformity of space)

A free particle is a particle on which no force works. A galilean referential is such that movement of a free particle is rectilinear and uniform. The referentials of inertia or of Galileo are such that:

The geometry is euclidean: $ds^2 = dx^2 + dy^2 + dz^2$.
The watches are synchronized.
The positions, directions, or moments are all equivalent for formulation of the equations of movement; there is therefore isotropy, homogenity of space, and uniformity of time.

A family of galilean referentials is created in the following way: The referentials, for which all points are driven by a constant velocity in relation to a referential of inertia, are inertia referentials. The classical mechanics definition given above is supposed to be general and valid in relativistic mechanics:

The geometric space of dimension 3 is euclidean, homogeneous, and isotropic.
The referentials are such that the isolated particles describe rectilinear and uniform movements.

In restricted relativity we are only going to examine galilean referentials: The referentials are driven by one another by the movement of rectilinear and uniform translation. It is obvious that all the referentials of inertia are equivalent for formulation of the equations of movement. Restricted relativity is based on two postulates:

1. All the laws of physics are invariant with a change of galilean referential. This is a generalization of the principle of galilean relativity, which considers only the laws of mechanics; the principle of restricted relativity also considers the laws of electromagnetism.
2. The velocity $\vec{c}$ of light in a vacuum is a universal constant; that is, the speed $\vec{c}$ is invariant with a change in galilean referential (cf. the Michelson–Morley experiment).

The second postulate has serious consequences. Recall the law of composition of velocities:

$$\vec{v}_{\text{absolute}} = \vec{v}_{\text{relative}} + \vec{v}_{\text{drag}}$$

which means that $\vec{c}$ should not have the same value in two galilean frames driving each other with a movement of translation characterized by the speed $\vec{u}$:

$$\vec{v}_{\text{absolute}} = \vec{c} + \vec{u}$$

To reconcile the two preceding postulates, relativistic kinematics considers the notions of time and space as nonindependent. The measurement of time and space is joined to the galilean referential used and thus

$$t_1 \neq t_1' \qquad t_2 \neq t_2' \qquad d_{M_1M_2} \neq d'_{M_1M_2}$$

Time transforms itself into a fourth variable, indicating the coordinates of an event, and the invariant scalar product (such that $\vec{r} \cdot \vec{r} = x^2 + y^2 + z^2$) must be replaced by a new invariant which introduces x, y, z, and t: the scalar product associated with the norm of a four-component vector.

2.2 SPECIAL LORENTZ TRANSFORMATION

In an inertia referential (I) an *event* in a point M at the moment t is characterized by the following four quantities: the three coordinates x, y, z of the point M and the value of the parameter t. They are called *time-space coordinates*. The time is measured by a clock placed in M and at rest (proper time).

Consider the event M with coordinates

x, y, z, t in the referential (I) supposed galilean
x', y', z', t' in the referential (I') supposed galilean

We set $\vec{u}$ as the velocity of (I') in relation to (I). We presume that the origins O and O' of (I) and (I') coincide with the origin of the times, that is, when $t = t' = 0$. Hypothetically, we presume $\vec{u}$ to be parallel to Ox. The referential (I') is a rectilinear and uniform translation at speed $\vec{u}$ in relation to (I).

We intend to link (x',y',z',t') to (x,y,z,t). To do this, note that the transformation must be *linear* as a rectilinear, and uniform movement in (I) must remain rectilinear and uniform in (I'). Nothing is taken from the generality of the problem by choosing with care the direction of the axes. The geometric space is assumed isotropic and homogeneous, and thus the origins O and O' interchanged if $t = t' = 0$. This particular choice (Figure 5) enables simplification of the calculations:

$$x' = \alpha x + \beta t \tag{2.1}$$

$$y' = y \tag{2.2}$$

$$z' = z \tag{2.3}$$

$$t' = \gamma x + \delta t \tag{2.4}$$

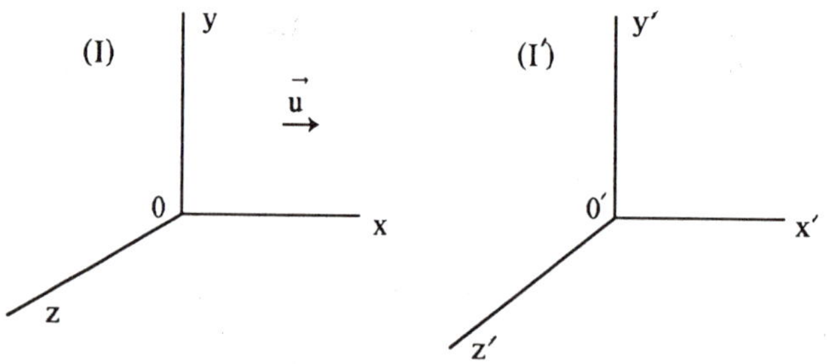

Figure 5 Frames of special Lorentz transformation.

Note that $\alpha = \alpha(\vec{u})$ and similarly for β, γ, and δ.

If O'_t is the position of the origin O' of (I') at a nonzero moment, t, time measured by O, O'_t, has the following coordinates:

In (I): $(x,0,0,t)$ with $x = ut$ as O' moves at the speed u along Ox
In (I'): $(0, 0, 0, t')$

The preceding equations give [cf. (2.1) to (2.4)]

$$0 = \alpha x + \beta t = \alpha ut + \beta t$$

or

$$\beta = -\alpha u \tag{2.5}$$

which enables us to rewrite (2.1) to (2.4) as

$$x' = \alpha(x - ut) \tag{2.6}$$

$$y' = y \tag{2.7}$$

$$z' = z \tag{2.8}$$

$$t' = \gamma x + \delta t \tag{2.9}$$

Equally, if O_t is the position of the origin in (I) at the same nonzero moment t measured by O, O_t has the following coordinates:

In (I): $(0,0,0,t)$
In (I'): $(x',0,0,t')$

with

$$x' = -ut' \tag{2.10}$$

The referential (I) is driven in relation to (I') by a rectilinear and uniform movement of velocity $-\vec{u}$. Equations (2.1) and (2.4) are written

$$x' = -\alpha ut$$

$$t' = \delta t$$

or

$$x' = -\frac{\alpha u}{\delta} t'$$

and by identification with (2.10),

$$\alpha = \delta$$

Substituting in (2.6) to (2.9) yields

$$x' = \alpha(x - ut) \tag{2.11}$$

$$y' = y \tag{2.12}$$

$$z' = z \tag{2.13}$$

$$t' = \gamma x + \alpha t \tag{2.14}$$

To establish the values of α and γ, consider at the moment $t = 0$ a point light source in O (thus also in O' at the moment $t' = 0$) that gives off a luminous vibration. This spherical wave moves in all directions with the same velocity c for observer O in (I) and O' in (I'). Consider the event M to be the reception of this luminous vibration at a subsequent moment.

The coordinates of M prove:

In (I):

$$x^2 + y^2 + z^2 = c^2t^2 \tag{2.15}$$

since the vibration moves at velocity c in (I), and
In (I'):

$$x'^2 + y'^2 + z'^2 = c^2t'^2 \tag{2.16}$$

since the vibration propagates at the velocity c in (I'), according to the second postulate of restricted relativity.

Introduce (2.11) and (2.14) into (2.16):

$$(\alpha^2 - c^2\gamma^2)x^2 + y^2 + z^2 - 2\alpha xt(\gamma c^2 + \alpha u) = \alpha^2\left(1 - \frac{u^2}{c^2}\right)c^2t^2 \tag{2.17}$$

Compare (2.17) with (2.15):

$$\alpha^2 - c^2\gamma^2 = 1 \qquad \gamma = -\frac{\alpha u}{c^2} \qquad \alpha^2\left(1 - \frac{u^2}{c^2}\right) = 1$$

which gives, by taking for α the positive solution,

$$\alpha = \frac{1}{\sqrt{1 - u^2/c^2}} \qquad \gamma = -\frac{u/c^2}{\sqrt{1 - u^2/c^2}}$$

(the coefficient α being real, it is necessary for $u \leqslant c$).

The equations of transformation of the coordinates of time-space for any point M are obtained using (2.11) to (2.14):

$$x' = \frac{x - ut}{\sqrt{1 - u^2/c^2}}$$

$$y' = y$$

$$z' = z$$

$$t' = \frac{t - ux/c^2}{\sqrt{1 - u^2/c^2}}$$

These are the formulas of Lorentz's special transformation. It is said to be special as:

$O = O'$ if $t = t' = 0$.
$\vec{e}_1$, $\vec{e}_1'$, and $\vec{u}$ are parallel and of the same direction and sign.
$\vec{e}_2$ and $\vec{e}_2'$ (respectively, $\vec{e}_3$ and $\vec{e}_3'$) are parallel and of the same direction and sign.

Under this transformation the quantity

$$s^2 = x^2 + y^2 + z^2 - c^2t^2$$

is an invariant. If we take $c \to \infty$, we come back to the galilean transformation.

The inverse relations are obtained by exchanging the primed quantities with the same nonprimed quantities and by replacing u by $-u$:

$$x = \frac{x' + ut'}{\sqrt{1 - u^2/c^2}}$$

$$y = y'$$

$$z = z'$$

$$t = \frac{t' + ux'/c^2}{\sqrt{1 - u^2/c^2}}$$

The best way to check is to replace in the relations expressing x', y', z', t' the latter values of x, y, z, t. For example,

$$x' = \frac{x'(ut') - u(t' + ux'/c^2)}{1 - u^2/c^2} = x'$$

and similarly for t'.

We generally write

$$\Gamma = \left(1 - \frac{u^2}{c^2}\right)^{-1/2} \tag{2.18}$$

from which

$$x' = \Gamma(x - ut)$$
$$y' = y$$
$$z' = z \tag{2.19}$$
$$t' = \Gamma\left(t - \frac{ux}{c^2}\right)$$

For a "general" Lorentz transformation but without rotation, the velocity $\vec{u}$, the relative speed of (I') in relation to (I), is not parallel to Ox. We show that

$$x' = \left[1 + (\Gamma - 1)\frac{u_x^2}{u^2}\right] x + (\Gamma - 1)\frac{u_x u_y}{u^2} y$$
$$+ (\Gamma - 1)\frac{u_x u_z}{u^2} z - u_x \Gamma t$$

$$y' = (\Gamma - 1)\frac{u_y u_x}{u^2} x + \left[1 + (\Gamma - 1)\frac{u_y^2}{u^2}\right] y$$
$$+ (\Gamma - 1)\frac{u_y u_z}{u^2} z - u_y \Gamma t$$

$$z' = (\Gamma - 1)\frac{u_z u_x}{u^2} x + (\Gamma - 1)\frac{u_z u_y}{u^2} y$$
$$+ \left[1 + (\Gamma - 1)\frac{u_z^2}{u^2}\right] z - u_z \Gamma t$$

$$t' = -\Gamma\frac{u_x}{c^2} x - \Gamma\frac{u_y}{c^2} u - \Gamma\frac{u_z}{c^2} z + \Gamma t$$

formulas that are assembled into

$$\vec{r}' = \vec{r} + \vec{u}\left[(\Gamma - 1)\frac{\vec{r}\cdot\vec{u}}{u^2} - \Gamma t\right] \tag{2.20a}$$

$$t' = \Gamma\left(t - \frac{\vec{r}\cdot\vec{u}}{c^2}\right) \tag{2.21a}$$

a generalization of the relations in (2.19).

The inverse relations are clearly obtained by replacing the primed quantities by the same nonprimed quantities and, reciprocally,

$$(x,y,z,t,\vec{u}) \longleftrightarrow (x',y',z',t',\vec{u}')$$

with

$$\vec{u}' = -\vec{u} \tag{2.22}$$

or

$$\vec{r} = \vec{r}' + \vec{u}' \left[(\Gamma - 1) \frac{\vec{r}' \cdot \vec{u}'}{u^2} - \Gamma t' \right] \tag{2.20b}$$

$$t = \Gamma \left(t' - \frac{\vec{r}' \cdot \vec{u}'}{c^2} \right) \tag{2.21b}$$

Using (2.20a) and (2.21a) yields

$$\frac{u}{c} = \tanh \alpha \tag{2.23}$$

$$\vec{u} = u\vec{n} \tag{2.24}$$

so

$$\Gamma = \cosh \alpha$$

We obtain a commonly used equivalent formulation:

$$\vec{r}' = \vec{r} + \vec{u} \left[(\cosh \alpha - 1)\vec{r} \cdot \frac{\vec{u}}{u^2} - (\cosh \alpha)t \right]$$

$$t' = (\cosh \alpha) \left(t - \frac{\vec{r} \cdot \vec{n} u}{c^2} \right)$$

that is,

$$\vec{r}' = \vec{r} + [(\vec{r} \cdot \vec{n})(\cosh \alpha - 1) - ct \sinh \alpha]\vec{n} \tag{2.25}$$

$$ct' = ct \cosh \alpha - (\vec{n} \cdot \vec{r})\sinh \alpha \tag{2.26}$$

relations commonly used (see, e.g., C. Itzykson and J. B. Zuber, *Quantum Field Theory*, McGraw-Hill, New York, 1980).

Note. In the case of general relativity, which is to say if $\vec{u}$ is not parallel to the axes Ox, and especially if the axes of (I') are directed in no particular way in relation to (I), different rotations must first be carried out in order to come back to the preceding case. Use D^{-1}, the rotation matrix of the axes of (I'), in relation to those of (I) in order to come back to the preceding *general transformation*. The rotation D then brings the vector $\vec{r}'$ into such a position as to enable application of the general transformation without rotation. In (2.20b) and (2.21b), $\vec{r}'$ must be replaced by $D\vec{r}'$ and $\vec{u}'$ by $D\vec{u}'$.

The equivalent of (2.22) is

$$D\vec{u}' = -\vec{u} \tag{2.27}$$

with

$$\|\vec{u}'\| = u' = u \tag{2.28}$$

and

$$\vec{r} = D\vec{r}' - \vec{u}\left[(\Gamma - 1)\frac{\vec{r}'\cdot\vec{u}'}{u^2} - \Gamma t'\right] \tag{2.29}$$

$$t = \Gamma\left(t' - \frac{\vec{u}'\cdot\vec{r}'}{c^2}\right) \tag{2.30}$$

Remember that

$$D^{-1} = D^{+}$$

(unitary transformation). Similarly, (2.20a) and (2.21a) are written

$$\vec{r}' = D^{-1}\vec{r} - \vec{u}'\left[(\Gamma - 1)\frac{\vec{r}\cdot\vec{u}}{u^2} - \Gamma t\right] \tag{2.31}$$

(where $u' = -D^{-1}\vec{u}$) and

$$t' = \Gamma\left(t - \frac{\vec{u}\cdot\vec{r}}{c^2}\right) \tag{2.32}$$

Furthermore, the most general Lorentz transformations do not constitute a group; the product of two of them is in general the product of a rotation and a special transformation.

2.3 MINKOWSKI'S UNIVERSE

By writing

$$\beta = \frac{u}{c} \tag{2.33}$$

and by preserving $\Gamma = 1/\sqrt{1 - \beta^2}$, matrix notation can be used to represent the special Lorentz transformation

$$\begin{bmatrix} x' \\ y' \\ z' \\ ct' \end{bmatrix} = \begin{bmatrix} \Gamma & 0 & 0 & -\beta\Gamma \\ 0 & 1 & 0 & 0 \\ 0 & 0 & 1 & 0 \\ -\beta\Gamma & 0 & 0 & \Gamma \end{bmatrix} \begin{bmatrix} x \\ y \\ z \\ ct \end{bmatrix} = (L)\begin{bmatrix} x \\ y \\ z \\ ct \end{bmatrix} \tag{2.34}$$

where (L) is the Lorentz matrix. It can easily be checked that

$$L^+ = \tilde{L}^* = L$$

and that

$$LL^+ = \begin{bmatrix} \Gamma^2(1+\beta^2) & 0 & 0 & -2\beta\Gamma^2 \\ 0 & 1 & 0 & 0 \\ 0 & 0 & 1 & 0 \\ -2\beta\Gamma^2 & 0 & 0 & \Gamma^2(1+\beta^2) \end{bmatrix}$$

which means that $L^+ \neq L^{-1}$ and the matrix associated with the Lorentz transformation is not unitary. Note that the determinant of L is equal to

$$\det(L) = \Gamma^2(1 - \beta^2) = 1$$

In the Minkowski time-space, the metric tensor g is represented by

$$g_{\nu\mu} = \eta_{\mu\nu} = \begin{bmatrix} 1 & 0 & 0 & 0 \\ 0 & -1 & 0 & 0 \\ 0 & 0 & -1 & 0 \\ 0 & 0 & 0 & -1 \end{bmatrix}$$

and it can easily be proved that

$$Lg\tilde{L} = g$$

To obtain the matrix of the inverse transformation, all that is required is the changing of u into $-u$, that is, β into $-\beta$. If $u \ll c$, we find Galileo's transformation and the laws of classical kinematics (especially that of the composition of velocities).

2.3.1 Composition of Velocities

A particle has a speed

$$\vec{v} = \begin{bmatrix} \dfrac{dx}{dt} \\ \dfrac{dy}{dt} \\ \dfrac{dz}{dt} \end{bmatrix} \quad \text{in } (I) \qquad \text{and} \qquad \vec{v}\,' = \begin{bmatrix} \dfrac{dx'}{dt'} \\ \dfrac{dy'}{dt'} \\ \dfrac{dz'}{dt'} \end{bmatrix} \quad \text{in } (I')$$

By differentiating the Lorentz formulas

$$\begin{aligned} dx' &= \Gamma(dx - \beta c\, dt) \\ dy' &= dy \\ dz' &= dz \\ c\, dt' &= \Gamma(-\beta\, dx + c\, dt) \end{aligned} \tag{2.35}$$

we obtain

$$\frac{dx'}{dt'} = \frac{(dx/dt) - \beta c}{1 - \dfrac{\beta}{c}\dfrac{dx}{dt}} = \frac{v_x - \beta c}{1 - \dfrac{\beta}{c} v_x} \tag{2.36}$$

$$\frac{dy'}{dt'} = \frac{dy/dt}{\Gamma\left(1 - \dfrac{\beta}{c}\dfrac{dx}{dt}\right)} = \frac{v_y}{\Gamma\left(1 - \dfrac{\beta}{c} v_x\right)} \tag{2.37}$$

$$\frac{dz'}{dt'} = \frac{dz/dt}{\Gamma\left(1 - \dfrac{\beta}{c}\dfrac{dx}{dt}\right)} = \frac{v_z}{\Gamma\left(1 - \dfrac{\beta}{c} v_x\right)} \tag{2.38}$$

We see clearly the law of composition of velocities:

$$\vec{v}_{\text{absolute}} = \vec{v}_{\text{relative}} + \vec{v}_{\text{drag}}$$

is not proved (explicitly, we should have $\vec{v} = \vec{v}\,' + \vec{u}$). However, if $u \ll c$, then $\beta \to 0$, $\beta c \to u$, $\Gamma \to 1$ and we find the classical case (nonrelativistic). The relations (2.36) to (2.38) then prove the law of the composition of velocities.

2.3.2 Interval

Consider the event M, reception of light vibration given off in $O = O'$ at the time $t = t' = 0$. Remember that O and O' are the respective origins of (I) and (I'). The quantities

$$s_{12}^2 = c^2(t_2 - t_1)^2 - (x_2 - x_1)^2 - (y_2 - y_1)^2 - (z_2 - z_1)^2$$
$$s_{12}'^2 = c^2(t_2' - t_1')^2 - (x_2' - x_1')^2 - (y_2' - y_1')^2 - (z_2' - z_1')^2$$

are such that (use the Lorentz formulas)

$$s_{12}^2 = s_{12}'^2$$

or

$$s_{12} = s_{12}'$$

[The sign is, by necessity, a + sign since $u = 0$, that is, (I) and (I') mixed; we must have $s_{12} = s'_{12}$.] The quantity s_{12} is called the *interval* between the events M_1 and M_2. The interval is invariant under a change of relativistic galilean referential.

The invariant desired for replacing the preservation of

$$\vec{r}\cdot\vec{r} = x^2 + y^2 + z^2$$

in order to go from classical galilean mechanics to relativistic mechanics is thus the quantity s^2:

$$s^2 = c^2t^2 - x^2 - y^2 - z^2$$

We can write in tensor notation

$$s^2 = X^{\mu}X_{\mu}$$

with

$$X^0 = ct \qquad X^1 = x \qquad X^2 = y \qquad X^3 = z$$

$$X_0 = ct \qquad X_1 = -x \qquad X_2 = -y \qquad X_3 = -z$$

Note that s^2 is a pseudoeuclidean quadratic form.

2.3.3 Proper Time

Now consider a moving particle in (I) at the moment t; its coordinates of space are x, y, and z and its speed is $\vec{v}$ with (Figure 6)

$$v^2 = \left(\frac{dx}{dt}\right)^2 + \left(\frac{dy}{dt}\right)^2 + \left(\frac{dz}{dt}\right)^2$$

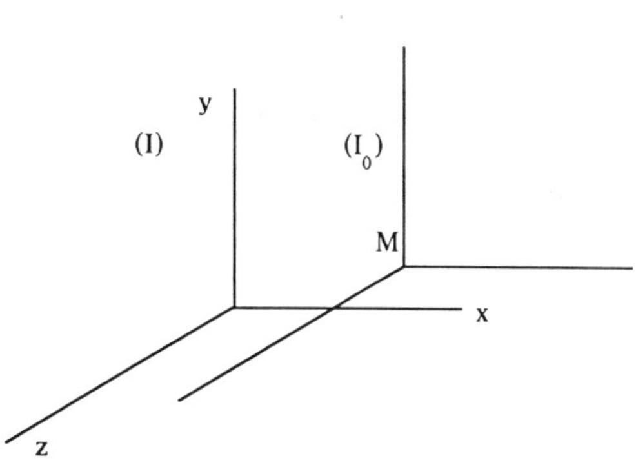

Figure 6 Proper time.

We write

$$\gamma = \frac{1}{\sqrt{1 - v^2/c^2}} \tag{2.39}$$

[not to be confused with $\Gamma = (1 - u^2/c^2)^{-1/2}$].

We now chose the galilean referential I_0, in which the particle is momentarily immobile at the moment t: (I_0) is called a *proper referential*. We could equally choose the axes of (I), the referential of the observer O, and those of (I_0) in order to apply the special Lorentz transformation.

Let there be two neighboring events M and $M + dM$, positions of the particle at the neighboring moments t and $t + dt$ in (I). The square of the elementary interval between these two events is written as follows:

In (I): $ds^2 = c^2\, dt^2 - dx^2 - dy^2 - dz^2$
In (I_0): $ds_0^2 = c^2\, dt_0^2$

Now

$$ds^2 = ds_0^2$$

invariance of the pseudoeuclidean quadratic form, or

$$ds^2 = ds_0^2 = c^2\, dt_0^2$$

The quantity dt_0 is an *interval of proper time* measured in the proper referential (I_0). Recall a well-known result, the phenomenon of dilation of time:

$$(c^2 - v^2)\, dt^2 = c^2\, dt_0^2 \qquad \text{or} \qquad dt^2 = \frac{dt_0^2}{1 - v^2/c^2}$$

or

$$dt = \gamma\, dt_0 \tag{2.40}$$

$(dt > dt_0)$. dt_0 is the variation of time measured with a clock attached to the particle in motion, dt, which is measured by observer O.

Note. The proper time t_0 is often noted τ. We then use either one or the other convention of notation.

2.3.4 Quadrivectors

Event M can be represented by its coordinates (x,y,z,t) in a galilean referential (I). We may also consider the coordinates x, y, z, and ct, which have the advantage of being of the same dimensions and of transforming themselves at the moment of a change in galilean referential by a linear application represented by a Lorentz matrix for a special transformation

$$\begin{bmatrix} x' \\ y' \\ z' \\ ct' \end{bmatrix} = (L) \begin{bmatrix} x \\ y \\ z \\ ct \end{bmatrix}$$

Quadrivector time-space X is called the set of coordinates (x,y,z,ct). It is the vector of a four-dimensional space said to be *time-space* or *Minkowski space*.

The square of the interval between the event of origin (0,0,0,0) and the event (x,y,z,ct) defines the *pseudonorm* of the quadrivector X:

$$s^2 = (X)^2 = c^2t^2 - x^2 - y^2 - z^2 \tag{2.41}$$

Note. s^2 is not necessarily ≥ 0, and Minkowski space is not a euclidean space.

More generally, quadrivector A is called the entire set of the four quantities A_0, A_1, A_2, A_3 of the same dimensions which transform themselves by a Lorentz transformation at the moment of a change of galilean referential. For a special transformation,

$$\begin{bmatrix} A_0' \\ A_1' \\ A_2' \\ A_3' \end{bmatrix}_{(I')} = (L) \begin{bmatrix} A_0 \\ A_1 \\ A_2 \\ A_3 \end{bmatrix}_{(I)}$$

The pseudonorm of (A) is, by definition,

$$(A)^2 = A_0^2 - A_1^2 - A_2^2 - A_3^2 \tag{2.42}$$

We can write that

$$(A)^2 = A \cdot A = A_\mu \vec{e}^{\mu} \cdot A_\nu \vec{e}^{\nu} = A_\mu A_\nu \eta^{\mu\nu} \tag{2.43}$$

where $\eta^{\mu\nu}$ are the components of a metric tensor in Minkowski space.

By identification with (2.42) we see that

$$\eta^{00} = 1 \qquad \eta^{ii} = -1 \quad (i = 1,2,3) \qquad \eta^{\mu\nu} = 0 \quad \text{if } \mu \neq \nu$$

$$(\eta^{\mu\nu}) = \begin{bmatrix} 1 & 0 & 0 & 0 \\ 0 & -1 & 0 & 0 \\ 0 & 0 & -1 & 0 \\ 0 & 0 & 0 & -1 \end{bmatrix}$$

As

$$\vec{e}^{\mu}\cdot\vec{e}_{\nu} = \eta^{\mu\alpha}\vec{e}_{\alpha}\cdot\vec{e}_{\nu} = \eta^{\mu\alpha}\eta_{\alpha\nu} = \delta^{\mu}_{\nu}$$

we must thus have

$$\eta_{\mu\nu} = \eta^{\mu\nu}$$

so

$$(A)^2 = (A^0)^2 - (A^1)^2 - (A^2)^2 - (A^3)^2$$

but also

$$\eta^{\mu}{}_{\nu} = \vec{e}^{\mu}\cdot\vec{e}_{\nu} = \delta^{\mu}_{\nu}$$

$$(\eta^{\mu}{}_{\nu}) = \begin{bmatrix} 1 & 0 & 0 & 0 \\ 0 & 1 & 0 & 0 \\ 0 & 0 & 1 & 0 \\ 0 & 0 & 0 & 1 \end{bmatrix}$$

This means that

$$(A)^2 = A_{\mu}\vec{e}^{\mu}\cdot A^{\nu}\vec{e}_{\nu} = A_{\mu}A^{\nu}\eta^{\mu}{}_{\nu} = A_0A^0 + A_1A^1 + A_2A^2 + A_3A^3$$

By comparing with (2.42) we see that

$$A_0 = A^0 \qquad A_i = -A^i \qquad (i = 1,2,3)$$

In quadrivector notation the interval

$$s^2 = c^2t^2 - x^2 - y^2 - z^2$$

is written

$$s^2 = (X)^2 = X_{\mu}X^{\mu} = X^{\mu}X_{\mu} \tag{2.44}$$

with

$$X^{\mu} = (ct,\vec{r}) = (ct,x,y,z) \tag{2.45a}$$

$$X_{\mu} = (ct,-\vec{r}) = (ct,-x,-y,-z) \tag{2.45b}$$

Equally, by generalizing,

$$A^{\mu} = (A_0,\vec{A}) = (A_0,A_x,A_y,A_z) \tag{2.46a}$$

$$A_{\mu} = (A_0,-\vec{A}) = (A_0,-A_x,-A_y,-A_z) \tag{2.46b}$$

For the scalar product, by definition,

$$R\cdot Q = R^{\mu}Q_{\mu} = R_0Q_0 - R_xQ_x - R_yQ_y - R_zQ_z \tag{2.47}$$

The gradient operator is a covariant vector, by definition; thus we must have

$$\nabla^{\mu} = \left(\frac{1}{c}\frac{\partial}{\partial t}, -\partial_i\right) = \left(\frac{1}{c}\frac{\partial}{\partial t}, -\frac{\partial}{\partial x}, -\frac{\partial}{\partial y}, -\frac{\partial}{\partial z}\right)$$

$$\nabla_{\mu} = \left(\frac{1}{c}\frac{\partial}{\partial t}, \partial_i\right) = \left(\frac{1}{c}\frac{\partial}{\partial t}, \frac{\partial}{\partial x}, \frac{\partial}{\partial y}, \frac{\partial}{\partial z}\right)$$

2.3.5 Quadrivector Velocity

It is defined by

$$(V) = \frac{d}{dt_0}(X) \tag{2.48}$$

where t_0 is the proper time such that $dt_0 = dt/\gamma$. We see that

$$(V) = \gamma \frac{d}{dt}(X) \tag{2.49}$$

or

$$V^{\mu} = \gamma \frac{d}{dt} X^{\mu} = \gamma \frac{d}{dt}(ct, \vec{r}) = (\gamma c, \gamma \vec{v}) \tag{2.50}$$

The pseudonorm of the velocity vector is equal to

$$(V)^2 = \gamma^2(c^2 - v^2) = \frac{1}{1 - v^2/c^2}(c^2 - v^2) = c^2 \tag{2.51}$$

a constant, regardless of the referential.

2.4 RELATIVISTIC MASS

For a nonrelativistic particle the differential equation associated with its motion is the Schrödinger equation:

$$\hat{H}|\Psi\rangle = \hat{E}|\Psi\rangle \tag{2.52}$$

that is,

$$\left(-\frac{\hbar^2}{2m}\nabla^2 + \hat{W}\right)|\Psi\rangle = i\hbar\frac{\partial}{\partial t}|\Psi\rangle \tag{2.53}$$

obtained by applying Ehrenfest's theorem to the preservation of total (classical mechanics) energy (the simplified Jacobi equation $H = E$) with

$$H = T + W = \frac{p^2}{2m} + W$$

Equation (2.53) can be obtained by the operator correspondences

$$\hat{p} \longrightarrow -i\hbar\vec{\nabla} \tag{2.54}$$

$$\hat{E} \longrightarrow i\hbar\,\frac{\partial}{\partial t} \tag{2.55}$$

The principle of relativity postulates that the systems of inertia (those for which the principle of inertia has been proved) are all equivalent; that is, that the laws of mechanics must be identical in these referentials or invariant under a Lorentz transformation going from a system of inertia (I) to system (I').

We assume that the notions of mass and speed have a meaning in relativity: If $u \ll c$, they are classical notions. It has been shown that [cf. (2.31) and (2.32)]

$$\vec{r}\,' = D^{-1}\vec{r} - \vec{u}\,'\left[\frac{\vec{r}\cdot\vec{u}}{u^2}(\Gamma - 1) - \Gamma t\right] \tag{2.56a}$$

$$t' = \Gamma\left(t - \frac{\vec{u}\cdot\vec{r}}{c^2}\right) \tag{2.57a}$$

with

$$\Gamma = \frac{1}{\sqrt{1 - u^2/c^2}}$$

Remember that for a special transformation, $D^{-1} = D = E$ and $\vec{u}\,' = -\vec{u}$. Differentiate the preceding relations:

$$d\vec{r}\,' = d\vec{r} + \vec{u}\left[\frac{\vec{dr}\cdot\vec{u}}{u^2}(\Gamma - 1) - \Gamma\,dt\right]$$

$$dt' = \Gamma\left(dt - \frac{\vec{u}\cdot\vec{dr}}{c^2}\right) = \Gamma\,dt\left(1 - \frac{\vec{u}\cdot\vec{v}}{c^2}\right)$$

from which the velocity

$$\vec{v}\,' = \frac{(1 - u^2/c^2)^{1/2}\vec{v} + [(1 - \sqrt{1 - u^2/c^2})(\vec{v}\cdot\vec{u})/u^2 - 1]\vec{u}}{1 - (\vec{u}\cdot\vec{v})/c^2} \tag{2.58a}$$

Similarly, relations (2.29) and (2.30),

$$\vec{r} = D\vec{r}' - \vec{u}\left[\frac{\vec{r}'\cdot\vec{u}'}{u^2}(\Gamma - 1) - \Gamma t'\right] \tag{2.56b}$$

$$t = \Gamma\left(t' - \frac{\vec{u}'\cdot\vec{r}'}{c^2}\right) \tag{2.57b}$$

give

$$\vec{v} = \frac{(1 - u^2/c^2)^{1/2}\vec{v}' + [(1 - \sqrt{1 - u^2/c^2})(\vec{v}'\cdot\vec{u})/u^2 + 1]\vec{u}}{1 + (\vec{u}\cdot\vec{v}')/c^2} \tag{2.58b}$$

Relations (2.58a) and (2.58b) reduce to $\vec{v}' = \vec{v} - \vec{u}$ if $c \to \infty$.

Consider that in (I) two particles have the speeds $\vec{v}_1$ and $\vec{v}_2$, respectively. The frame (I') is driven in relation to (I) by a velocity $\vec{u}$; imagine an elastic collision between two identical particles.

According to the laws of conservation of impulse in (I),

$$m_1\vec{v}_1 + m_2\vec{v}_2 = m_1\vec{V}_1 + m_2\vec{V}_2 \tag{2.59}$$

(the vectors $\vec{v}_1$ and $\vec{v}_2$ are the speeds of the identical particles 1 and 2 before the collision, and $\vec{V}_1$ and $\vec{V}_2$ the respective speeds after the elastic shock) and of the conservation of total kinetic energy,

$$\tfrac{1}{2}m_1v_1^2 + \tfrac{1}{2}m_2v_2^2 = \tfrac{1}{2}m_1V_1^2 + \tfrac{1}{2}m_2V_2^2 \tag{2.60}$$

where $m_1 = m_2$, the preceding equations lead to the condition

$$\vec{v}_1\cdot\vec{v}_2 = \vec{V}_1\cdot\vec{V}_2 \tag{2.61}$$

If movement takes place along the Ox axis, for example, and after the collision particle 1 begins to travel in the opposite direction,

$$\vec{V}_1 = -\alpha\vec{v}_1 \tag{2.62}$$

and according to (2.61),

$$\vec{V}_2 = -\frac{1}{\alpha}\vec{v}_2 \tag{2.63}$$

where α is a positive number.

Choose frames (I) and (I') before collision in such a way that

$$\vec{v}_2' = -\vec{v}_1 \tag{2.64a}$$

According to (2.58a) and (2.58b),

$$\vec{v}_2' = \frac{(1 - u^2/c^2)^{1/2}\vec{v}_2 + [(1 - \sqrt{1 - u^2/c^2})(\vec{v}_2 \cdot \vec{u})/u^2 - 1]\vec{u}}{1 - (\vec{u} \cdot \vec{v}_2)/c^2}$$

$$\vec{v}_1 = \frac{(1 - u^2/c^2)^{1/2}\vec{v}_1' + [(1 - \sqrt{1 - u^2/c^2})(\vec{v}_1' \cdot \vec{u})/u^2 + 1]\vec{u}}{1 + (\vec{u} \cdot \vec{v}_1')/c^2}$$

By clarifying the identity (2.64) we obtain a relation that cannot be true until

$$\vec{v}_1' = -\vec{v}_2 \qquad (2.65a)$$

We are following the 1909 Lewis and Tolman demonstration. The theorem of conservation of impulse if true in (I) must be true in (I'). We thus choose the collision such that after the shock the particles would have the velocities $\vec{V}_1$ and $\vec{V}_2$ in (I) and $\vec{V}_1'$ and $\vec{V}_2'$ in (I') such that [cf. (2.64a) and (2.65a)]

$$\vec{V}_2' = -\vec{V}_1 \qquad (2.64b)$$

$$\vec{V}_1' = -\vec{V}_2 \qquad (2.65b)$$

A particle of speed $\vec{v}$ in relation to an inertia frame (I) will have an impulse

$$\vec{p} = m\vec{v} \qquad (2.66)$$

By analogy with the classical case, m is a factor of proportionality called the mass of the particle.

Imagine m a nonconstant but universal function $f(v)$ with $v = \|\vec{v}\|$, that is,

$$m = m(v) = f(v) \qquad (2.67)$$

(remember that m is a scalar).

If the speed of the particle is $\vec{v}'$ in frame (I'),

$$\vec{p}' = m'\vec{v}'$$

with

$$m' = m'(v') = f(v') \qquad (2.68)$$

It is the same function f as for m. All the referentials are *identical*; that is, they must be treated as being *equivalent*—f is a universal function.

Rewrite (2.59), the law of preservation of the total impulse, as

$$f(v_1)\vec{v}_1 + f(v_2)\vec{v}_2 = f(V_1)\vec{V}_1 + f(V_2)\vec{V}_2 \tag{2.69}$$

By adding a simplifying condition

$$\vec{v}_1 \cdot \vec{u} = 0 \tag{2.70}$$

the velocity of particle 1 is considered perpendicular to $\vec{u}$. The result of (2.70) is [cf. (2.64a)]

$$\vec{v}_2' \cdot \vec{u} = -\vec{v}_1 \cdot \vec{u} = 0 \tag{2.71}$$

and [cf. (2.62)]

$$\vec{V}_1 \cdot \vec{u} = 0 \tag{2.72}$$

With (2.64b) and (2.72),

$$\vec{V}_2' \cdot \vec{u} = 0 \tag{2.73}$$

In these conditions (2.58b) enables us to write that

$$\vec{v}_2 = \sqrt{1 - \frac{u^2}{c^2}}\, \vec{v}_2' + \vec{u} \tag{2.74a}$$

and with (2.64a),

$$\vec{v}_2 = -\sqrt{1 - \frac{u^2}{c^2}}\, \vec{v}_1 + \vec{u} \tag{2.75a}$$

Thus

$$v_2^2 = \left(1 - \frac{u^2}{c^2}\right) v_1^2 + u^2 \tag{2.76a}$$

In the same way, starting from

$$\vec{V}_2 = \sqrt{1 - \frac{u^2}{c^2}}\, \vec{V}_2' + \vec{u} \tag{2.74b}$$

and with (2.64b),

$$\vec{V}_2 = -\sqrt{1 - \frac{u^2}{c^2}}\, \vec{V}_1 + \vec{u} \tag{2.75b}$$

then

$$V_2^2 = \left(1 - \frac{u^2}{c^2}\right) V_1^2 + u^2 \tag{2.76b}$$

Introduce (2.75a)–(2.75b), and (2.76a)–(2.76b) in (2.69):

$$\begin{aligned} f(v_1)\vec{v}_1 &+ \left(-\sqrt{1 - \frac{u^2}{c^2}}\,\vec{v}_1 + \vec{u}\right) f\left(\sqrt{v_1^2\left(1 - \frac{u^2}{c^2}\right) + u^2}\right) \\ &= f(V_1)\vec{V}_1 + \left(-\sqrt{1 - \frac{u^2}{c^2}}\,\vec{V}_1 + \vec{u}\right) \\ &\quad \times f\left(\sqrt{V_1^2\left(1 - \frac{u^2}{c^2}\right) + u^2}\right) \end{aligned} \tag{2.77}$$

Scalar multiply the two members of (2.77) by $\vec{u}$ and bear in mind (2.70) and (2.72):

$$f\left(\sqrt{\left(1 - \frac{u^2}{c^2}\right) v_1^2 + u^2}\right) = f\left(\sqrt{\left(1 - \frac{u^2}{c^2}\right) V_1^2 + u^2}\right) \tag{2.78}$$

if we assume that the function f is monotone, (2.80) is equivalent to

$$v_1 = V_1 \tag{2.79}$$

and thus, according to (2.62),

$$\alpha = 1 \tag{2.80}$$

Rewrite (2.77) in terms of (2.79):

$$(\vec{V}_1 - \vec{v}_1)\left[f(\vec{v}_1) - \sqrt{1 - \frac{u^2}{c^2}}\, f\left(\sqrt{v_1^2\left(1 - \frac{u^2}{c^2}\right) + u^2}\right)\right] = \vec{0} \tag{2.81}$$

Note that (2.62) with $\alpha = 1$ means that

$$\vec{V}_1 = -\vec{v}_1 \tag{2.82}$$

so

$$\vec{V}_1 - \vec{v}_1 = -2\vec{v}_1 \neq \vec{0} \tag{2.83}$$

It follows from relation (2.77) that

$$\frac{f(v_1)}{\sqrt{1 - u^2/c^2}} = f\left(\sqrt{v_1^2\left(1 - \frac{u^2}{c^2}\right) + u^2}\right) \tag{2.84}$$

which must be proved regardless of the speeds v_1 and u; choose the case $v_1 = 0$, from which

$$f(u) = \frac{f(0)}{\sqrt{1 - u^2/c^2}} \tag{2.85}$$

or

$$m(u) = \frac{m(0)}{\sqrt{1 - u^2/c^2}}$$

The expression (2.85) is a general solution for (2.84), as can easily be shown. Take $u \to v_1$ in (2.85):

$$f(v_1) = \frac{f(0)}{\sqrt{1 - v_1^2/c^2}} \tag{2.86}$$

Then we substitute in (2.84). We thus write more generally

$$m(v) = \frac{m(0)}{\sqrt{1 - v^2/c^2}} = \frac{m_0}{\sqrt{1 - v^2/c^2}} \tag{2.87}$$

by writing $m(0) = m_0$, the mass at rest, and

$$\vec{p} = m\vec{v} = \frac{m_0\vec{v}}{\sqrt{1 - v^2/c^2}} \tag{2.88}$$

2.5 QUADRIVECTOR ENERGY–MOMENTUM

The quantity of motion of the particle is the vector

$$\vec{p} = \gamma m_0 \vec{v} \tag{2.89}$$

(momentum of the particle) with

$$\gamma = \left(1 - \frac{v^2}{c^2}\right)^{-1/2} \tag{2.90}$$

m_0 is the mass at rest, a scalar invariant with a change of galilean referential. We call energy the quantity

$$E = \gamma m_0 c^2 = mc^2 \tag{2.91}$$

The quadrivector *energy–momentum* is defined by

$$(P) = m_0(V) \tag{2.92}$$

Now

$$V^\mu = (\gamma c, \gamma \vec{v})$$

so

$$P^\mu = (\gamma m_0 c, \gamma m_0 \vec{v}) = \left(\frac{E}{c}, \vec{p}\right) \tag{2.93}$$

The pseudonorm of this quadrivector is equal to

$$(P)^2 = P_\mu P^\mu = \frac{E^2}{c^2} - p^2 = \gamma^2 m_0^2 (c^2 - v^2) = m_0^2 c^2$$

from which

$$E^2 = m_0^2 c^4 + p^2 c^2 \tag{2.94}$$

When the speed is zero, $E_0 = m_0 c^2$, it is the energy at rest. From this we have the definition of kinetic energy is relativity:

$$T = E - E_0 = \gamma m_0 c^2 - m_0 c^2 = (\gamma - 1) m_0 c^2 \tag{2.95}$$

T is the work that must be done to lead the particle at rest to the speed $\vec{v}$. Note that for $v \ll c$ we recall all the classical mechanics definitions: $\vec{p} = m_0 \vec{v}$ and $T = \frac{1}{2} m_0 v^2$.

2.6 QUADRIVECTOR WAVE

For a photon of mass zero

$$P^\mu = \left(\frac{E}{c}, \vec{p}\right) \tag{2.96}$$

with

$$E = h\nu = \hbar\omega \tag{2.97}$$

$$\vec{p} = \hbar\vec{k} \tag{2.98}$$

We often write

$$\vec{k} = k\vec{n} \tag{2.99}$$

With (2.97) and (2.98) the expression for P^μ transforms itself into

$$P^\mu = \left(\frac{\hbar\omega}{c}, \hbar\vec{k}\right) = \hbar K^\mu \tag{2.100}$$

if, by definition, the quadrivector wave K^μ is given by

$$K^\mu = \left(\frac{\omega}{c}, \vec{k}\right) \tag{2.101}$$

Note

$$X \cdot K = X^\mu K_\mu = \omega t - \vec{k}\cdot\vec{r} \tag{2.102}$$

which is an invariant under any Lorentz transformation (property of a scalar product), called phase φ of the wave that propagates:

$$\varphi = \omega t - \vec{k}\cdot\vec{r}$$

The phase velocity $\vec{v}$ is given by

$$\frac{d\varphi}{dt} = \omega - \vec{k}\cdot\vec{v} = 0$$

and if the propagation is made according to axis Oz,

$$v_z = \frac{\omega}{k_z} \tag{2.103}$$

2.7 QUADRIVECTOR FORCE

By definition,

$$(F) = \frac{d}{dt_0}(P) = \gamma\frac{d}{dt}(P) \tag{2.104}$$

where t_0 is the proper time, or

$$F^\mu = \gamma\frac{d}{dt}\left(\frac{E}{c}, \vec{p}\right) = \left(\frac{\gamma}{c}\frac{dE}{dt}, \gamma\vec{f}\right) \tag{2.105}$$

where $\vec{f}$ is the force that works on the particle:

$$\vec{f} = \frac{d\vec{p}}{dt} \qquad \text{with} \quad \vec{p} = \gamma m_0 \vec{v}$$

Note that

$$(F)\cdot(V) = F^{\mu}V_{\mu} = \gamma^2 \left(\frac{dE}{dt} - \vec{f}\cdot\vec{v} \right)$$

as

$$(F) = \frac{d}{d\tau}(P) = \gamma \frac{d}{dt}(P) = \gamma \frac{d}{dt}(m_0 V) = m_0\gamma \frac{d}{dt}(V)$$

We can therefore evaluate the same expression in a second way:

$$(F)\cdot(V) = m_0\gamma \left[\frac{d}{dt}(V) \right]\cdot(V)$$

Now

$$\frac{d}{dt}(V)^2 = 2(V)\cdot\frac{d}{dt}(V)$$

from which

$$(F)\cdot(V) = \frac{1}{2} m_0\gamma \frac{d}{dt}(V^2)$$

But also, according to (2.51),

$$(V)^2 = c^2$$

from which

$$(F)\cdot(V) = 0 \tag{2.106}$$

and

$$\frac{dE}{dt} = \vec{f}\cdot\vec{v} \tag{2.107}$$

This is the theorem of kinetic energy in the galilean frame (I). We can therefore write

$$F^{\mu} = \left(\frac{\gamma}{c} \vec{f}\cdot\vec{v}, \gamma\vec{f} \right) \tag{2.108}$$

2.8 LAGRANGIAN RELATIVISTIC MECHANICS

Classically,

$$f_i = \frac{d}{dt} p_i = \frac{d}{dt}(m_i \dot{x}_i) = \frac{d}{dt}\left(\frac{\partial T}{\partial \dot{x}_i}\right) \tag{2.109}$$

where T is the kinetic energy, while in relativity [cf. (2.105)]

$$F^1 = \gamma f_x = \gamma \frac{d}{dt}(\gamma m_0 v_x) \tag{2.110}$$

(remember that $\vec{p} = \gamma m_0 \vec{v}$) or

$$f_x = \frac{d}{dt}(\gamma m_0 \dot{x}) \tag{2.111}$$

If we write

$$R = -m_0 c^2 \left(1 - \frac{v^2}{c^2}\right)^{1/2} \tag{2.112}$$

a function of the speed of the particle but which is not the relativistic kinetic energy, we have

$$\frac{\partial R}{\partial \dot{x}} = -m_0 c^2 \left(\frac{1}{2}\right)\left(-\frac{1}{c^2}\right)\left(1 - \frac{v^2}{c^2}\right)^{-1/2} \frac{\partial v^2}{\partial \dot{x}} = \frac{m_0 \dot{x}}{\sqrt{1 - v^2/c^2}} = m_0 \gamma \dot{x}$$

and

$$f_x = \frac{d}{dt}\left(\frac{\partial R}{\partial \dot{x}}\right) \tag{2.113}$$

and identical formulas in y and z. Thus more generally,

$$f_i = \frac{d}{dt}\left(\frac{\partial R}{\partial \dot{q}_i}\right) - \frac{\partial R}{\partial q_i} \tag{2.114}$$

The function R plays the role of kinetic energy T.

If the forces applied come from a potential energy W, we can define a lagrangian

$$L = R - W = -m_0 c^2 \left(1 - \frac{v^2}{c^2}\right)^{1/2} - W \tag{2.115}$$

This is clearly not the equivalent of a classical lagrangian ($L = T - W$).

2.9 RELATIVISTIC HAMILTONIAN

Starting from the lagrangian expression

$$L = -m_0c^2\left(1 - \frac{v^2}{c^2}\right)^{1/2} - W \tag{2.116}$$

according to [cf. A. Charlier, A. Bérard, and M. F. Charlier, *Analytical Mechanics*, Editions Ellipses, Paris, 1989],

$$p_i = \frac{\partial L}{\partial \dot{q}_i}$$

and

$$H = \sum_i p_i\dot{q}_i - L$$

or

$$H = \sum_i \frac{\partial L}{\partial \dot{q}_i}\dot{q}_i - L$$

and if

$$x_j = x_j(q_k)$$

then the x_k are the cartesian coordinates of the system and the q_k are the generalized coordinates, and thus independent, so

$$\dot{x}_j = \sum_l \frac{\partial x_j}{\partial q_l}\dot{q}_l$$

and

$$\frac{\partial \dot{x}_j}{\partial \dot{q}_l} = \frac{\partial x_j}{\partial q_l}$$

which leads to

$$\frac{\partial L}{\partial \dot{q}_i} = \sum_j \frac{\partial L}{\partial \dot{x}_j}\frac{\partial \dot{x}_j}{\partial \dot{q}_i} = \sum_j \frac{\partial L}{\partial \dot{x}_j}\frac{\partial x_j}{\partial q_i}$$

Then

$$H = \sum_{i,j} \frac{\partial L}{\partial \dot{x}_j}\frac{\partial x_j}{\partial q_i}\dot{q}_i - L$$

and if W does not depend on the $\dot{x}_k$,

$$\frac{\partial L}{\partial \dot{x}_j} = \frac{\partial}{\partial \dot{x}_j}[R - W(q_k)] = m_0\gamma\dot{x}_j$$

from which

$$H = \sum_{i,j} m_0\gamma\dot{x}_j \frac{\partial x_j}{\partial q_i}\dot{q}_i - L = \sum_{i,j} m_0\gamma \frac{dx_j}{dt}\frac{\partial x_j}{\partial q_i}\frac{dq_i}{dt}$$

$$- L = \sum_j m_0\gamma\dot{x}_j^2 - L$$

$$H = m_0\gamma v^2 - L \tag{2.117}$$

or explicitly,

$$H = m_0\gamma v^2 + m_0c^2\left(1 - \frac{v^2}{c^2}\right)^{1/2}$$

$$+ W = m_0\gamma\left[v^2 + c^2\left(1 - \frac{v^2}{c^2}\right)\right] + W$$

$$H = m_0\gamma c^2 + W = mc^2 + W \tag{2.118}$$

We find the equivalent of the classical hamiltonian since $T = (m - m_0)c^2$.

2.10 RELATIVISTIC ACTION OF JACOBI

By definition w the action of Jacobi is written in the ordinary space (N = 3)

$$w = \int_{t_1}^{t_2} L(q_k,\dot{q}_k;t)\,dt \tag{2.119}$$

a formula that is generalized into

$$w = \int_{\tau_1}^{\tau_2} L\left(X^\mu, \frac{\partial X^\mu}{\partial \tau}; \tau\right) d\tau \tag{2.120}$$

in Minkowski space ($N = 4$) and in

$$w = \int_{\lambda_1}^{\lambda_2} L\left(X^\mu, \frac{\partial X^\mu}{\partial \lambda}; \lambda\right) d\lambda \tag{2.121}$$

where λ is the *natural parameter* of the trajectory in a hyperspace of dimension N.

We know from Section 2.8 that for a free particle,

$$L = R = -m_0c^2\left(1 - \frac{v^2}{c^2}\right)^{1/2} \tag{2.122}$$

and thus

$$w = -\int_{t_1}^{t_2} m_0c^2\sqrt{1 - \frac{v^2}{c^2}}\,dt \tag{2.123}$$

Now

$$dt = \gamma d\tau = \frac{1}{\sqrt{1 - v^2/c^2}}\,d\tau$$

by definition of the proper time τ, from which

$$w = -\int_{\tau_1}^{\tau_2} m_0c^2\,d\tau \tag{2.124}$$

2.11 APPLICATIONS

2.11.1 Invariance of s^2 Under a Lorentz Transformation

In the referential (I'), the square of the interval between two events M_1 and M_2 is written

$$s_{12}'^2 = c^2(t_2' - t_1')^2 - (x_2' - x_1')^2 - (y_2' - y_1')^2 - (z_2' - z_1')^2$$

a relation in which we replace x', y', z', and t' by

$$x' = \Gamma(x - \beta ct)$$
$$y' = y$$
$$z' = z$$
$$ct' = \Gamma(ct - \beta x)$$

and we obtain

$$s_{12}'^2 = s_{12}^2$$

2.11.2 Lorentz Contraction

According to formula (2.20a), valid for a special Lorentz transformation

$$\vec{r}\,' = \vec{r} + \vec{u}\left[(\Gamma - 1)\frac{\vec{r}\cdot\vec{u}}{u^2} - \Gamma t\right] \tag{1}$$

by decomposing the vectors $\vec{r}$ and $\vec{r}'$ into components parallel to $\vec{u}$ $(\vec{r}_{\parallel}, \vec{r}'_{\parallel})$ and perpendicular to $\vec{u}(\vec{r}_{\perp}, \vec{r}'_{\perp})$:

$$\vec{r} = \vec{r}_{\parallel} + \vec{r}_{\perp}$$

$$\vec{r}' = \vec{r}'_{\parallel} + \vec{r}'_{\perp}$$

The relation (1) gives us the identities

$$\vec{r}'_{\perp} = \vec{r}_{\perp} \tag{2}$$

$$\vec{r}'_{\parallel} = \Gamma \vec{r}_{\parallel} - \vec{u}\Gamma t \tag{3}$$

For two points M_1 and M_2 of the same solid,

$$(\vec{r}'_2 - \vec{r}'_1)_{\perp} = (\vec{r}_2 - \vec{r}_1)_{\perp}$$

$$(\vec{r}'_2 - \vec{r}'_1)_{\parallel} = \Gamma(\vec{r}_2 - \vec{r}_1)_{\parallel} = \frac{1}{\sqrt{1 - u^2/c^2}}(\vec{r}_2 - \vec{r}_1)_{\parallel}$$

Only the components parallel to the speed $\vec{u}$ undergo Lorentz contraction: As the mobile referential goes to the fixed referential, the component parallel to $\vec{u}$, of length separating two points on the same solid, diminishes.

2.11.3 Angular Contraction

The referential (I') is joined to the event M, which causes a circular movement around O (the origin of frame I). The speed $\vec{u}$ that drives (I') in relation to (I) is brought into the circle by the tangent to M. While M travels the arc (Figure 7)

$$ds = R\, d\theta$$

if M' is an infinitely neighboring point to M, we have

$$ds_{\parallel} = R\, d\theta$$

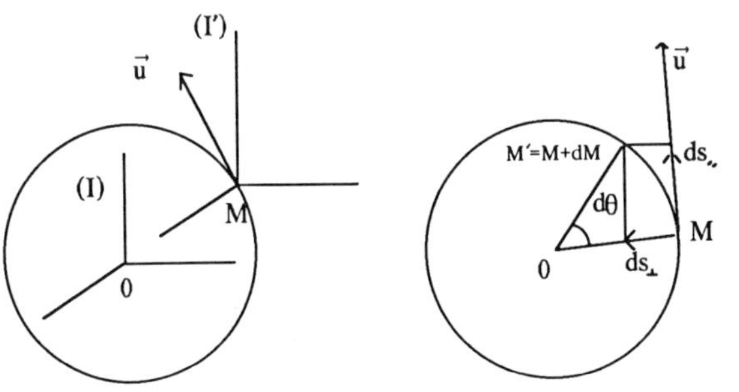

Figure 7 Angular contraction.

and so, according to the preceding exercise,

$$ds'_{\parallel} = \Gamma\, ds_{\parallel} = \Gamma R\, d\theta = R(\Gamma\, d\theta)$$

as R has remained constant (R corresponds to a distance measured perpendicular to $\vec{u}$).

Under the action of the Lorentz transformation, $d\theta$ becomes $\Gamma\, d\theta$. By supposing that M no longer describes the infinitesimal arc $\widehat{MM'}$ but the entire length L of the circle through a sequence of Lorentz infinitesimal transformations, we obtain

$$L' = \Gamma L = 2\pi\Gamma R$$

2.11.4 Proving Formulas (2.20a) and (2.21a)

In the transformation envisaged, vector $\vec{u}$ is an ordinary vector. The event M is located in the referentials (I) and (I') by (x,y,z,t) and (x',y',z',t'), respectively (cf. Figure 8). The transformation must be linear:

$$x' = a_{11}x + a_{12}y + a_{13}z + a_{14}t \tag{1}$$

$$y' = a_{21}x + a_{22}y + a_{23}z + a_{24}t \tag{2}$$

$$z' = a_{31}x + a_{32}y + a_{33}z + a_{34}t \tag{3}$$

$$t' = a_{41}x + a_{42}y + a_{43}z + a_{44}t \tag{4}$$

Now $\overrightarrow{OO'} = \vec{u}t$, which is written as follows:

In (I): $O'(u_xt,u_yt,u_zt,t)$
In (I'): $O'(0,0,0,t')$

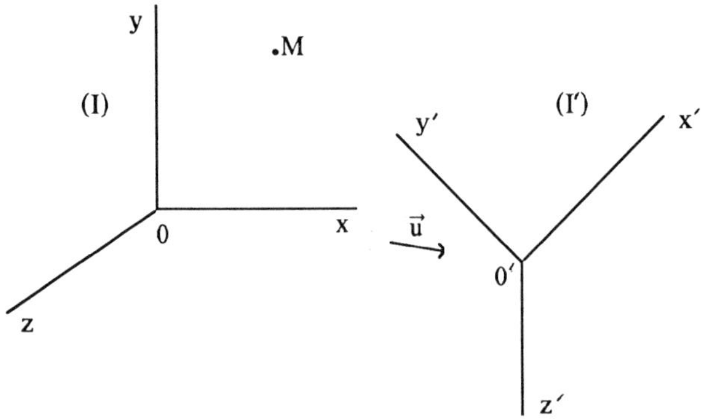

Figure 8 Referentials.

from which we get

$$a_{i4} = -(a_{i1}u_x + a_{i2}u_y + a_{i3}u_z)$$

with $i = 1, 2, 3$; similarly:

In (I): $O(0,0,0,t)$
In (I'): $O(-u_x t', -u_y t', -u_z t', t')$

which gives

$$t' = a_{44}t$$

The invariance

$$x^2 + y^2 + z^2 + c^2t^2 = x'^2 + y'^2 + z'^2 + c^2t'^2$$

provides the system of equations to be solved:

$$a_{11}^2 + a_{21}^2 + a_{31}^2 - c^2a_{41}^2 = 1$$

$$a_{12}^2 + a_{22}^2 + a_{32}^2 - c^2a_{42}^2 = 1$$

$$a_{13}^2 + a_{23}^2 + a_{33}^2 - c^2a_{43}^2 = 1$$

$$-c^2 = (u^2 - c^2)a_{44}^2$$

$$a_{11}a_{12} + a_{21}a_{22} + a_{31}a_{32} - c^2a_{41}a_{42} = 0$$

$$a_{11}a_{13} + a_{21}a_{23} + a_{31}a_{33} - c^2a_{41}a_{43} = 0$$

$$a_{12}a_{13} + a_{22}a_{23} + a_{32}a_{33} - c^2a_{42}a_{43} = 0$$

$$a_{11}u_x + a_{21}u_y + a_{31}u_z + c^2a_{41} = 0$$

$$a_{12}u_x + a_{22}u_y + a_{32}u_z + c^2a_{42} = 0$$

$$a_{13}u_x + a_{23}u_y + a_{33}u_z + c^2a_{43} = 0$$

It can be seen that

$$a_{44}^2 = \frac{1}{1 - u^2/c^2} = \Gamma^2 \qquad a_{12} = a_{21} \qquad a_{13} = a_{31} \qquad -u_x a_{44} = c^2a_{41}$$

$$a_{23} = a_{32} \qquad -u_y a_{44} = c^2a_{43}$$

$$a_{41} = -\frac{u_x}{c^2}\Gamma \qquad a_{42} = -\frac{u_y}{c^2}\Gamma \qquad a_{43} = -\frac{u_z}{c^2}\Gamma$$

and thus

$$t' = \Gamma\left(t - \frac{\vec{u}\cdot\vec{r}}{c^2}\right)$$

$$a_{14} = -u_x\Gamma \qquad a_{24} = -u_y\Gamma \qquad a_{34} = -u_z\Gamma$$

$$a_{12} = (\Gamma - 1)\,\frac{u_x u_y}{u^2}$$

$$a_{13} = (\Gamma - 1)\,\frac{u_x u_z}{u^2}$$

$$a_{23} = (\Gamma - 1)\,\frac{u_y u_z}{u^2}$$

and

$$a_{11} = 1 + \frac{\Gamma - 1}{u^2}\,u_x^2$$

$$a_{22} = 1 + \frac{\Gamma - 1}{u^2}\,u_y^2$$

$$a_{33} = 1 + \frac{\Gamma - 1}{u^2}\,u_z^2$$

or

$$x' = x + u_x\left(\frac{\Gamma - 1}{u^2}\,\vec{u}\cdot\vec{r} - \Gamma t\right)$$

and the identical relations for y' and z'.

2.11.5 Variational Principle

Imagine that the particle being examined moves along the axis Ox. The transformation G is defined such that

$$(x,t) \xrightarrow{G} (q,\tau)$$

where τ is the proper time:

$$x = q \cosh \lambda + \tau c \sinh \lambda$$

$$ct = q \sinh \lambda + \tau c \cosh \lambda$$

with

$$\cosh \lambda = \left(1 - \frac{v^2}{c^2}\right)^{-1/2}$$

$$\sinh \lambda = \frac{v}{c}\left(1 - \frac{v^2}{c^2}\right)^{-1/2}$$

which clearly proves the relation

$$\cosh^2\lambda - \sinh^2\lambda = 1$$

The functional determinant

$$D = \frac{\partial(x,t)}{\partial(q,\tau)} = \begin{vmatrix} \dfrac{\partial x}{\partial q} & \dfrac{\partial t}{\partial q} \\ \dfrac{\partial x}{\partial \tau} & \dfrac{\partial t}{\partial \tau} \end{vmatrix} = \begin{vmatrix} \cosh\lambda & \dfrac{1}{c}\sinh\lambda \\ c\sinh\lambda & \cosh\lambda \end{vmatrix} = 1$$

being equal to 1, the transformation is canonical.

The variational principle

$$\delta w = 0$$

indicates among the set of virtual paths, compatible with the imposed constraints, the one that is taken by the particle (i.e., the real path). This condition is written, if

$$\mathbf{L} = \mathbf{L}(x,\dot{x};t) \qquad \dot{x} = \frac{dx}{dt}$$

$$L = L(q,\dot{q};\tau) \qquad \dot{q} = \frac{dq}{d\tau}$$

which are, respectively, the old and new Lagrange functions of the system

$$\delta\int_{t_1}^{t_2} \mathbf{L}\,dt = \delta\int_{\tau_1}^{\tau_2} L\,d\tau = 0$$

As

$$\frac{dx}{dt} = c\,\frac{dq\cosh\lambda + d\tau\,c\sinh\lambda}{dq\sinh\lambda + d\tau\,c\cosh\lambda}$$

and

$$c\,\frac{dt}{d\tau} = \frac{dq}{d\tau}\sinh\lambda + c\cosh\lambda$$

we obtain

$$\begin{aligned} L = \mathbf{L}\,\frac{dt}{d\tau} &= -m_0c^2\sqrt{1 - \frac{1}{c^2}\left(\frac{dx}{dt}\right)^2}\,\frac{dt}{d\tau} \\ &= -m_0c^2\sqrt{1 - \left(\frac{dq}{c\,d\tau}\right)^2} \\ &= -m_0c^2\sqrt{1 - \frac{1}{c^2}\,\dot{q}^2} \end{aligned}$$

The form of L is identical to that of $\mathbf{L}$: The lagrangian retains its form under the proposed transformation, which is the special Lorentz transformation.

2.11.6 Geodesic

According to (1.82), the force applied to a nonrelativistic particle is

$$f_r = mg_{rk}\ddot{x}^k + mg_{ir}\Gamma^i_{jk}\dot{x}^j\dot{x}^k \tag{1}$$

For a free particle, $f_r = 0$, or

$$g_{rk}\ddot{x}^k + g_{ir}\Gamma^i_{jk}\dot{x}^j\dot{x}^k = 0 \tag{2}$$

Otherwise, by multiplying (2) by g^{nr}, we obtain

$$\ddot{x}^n + \Gamma^n_{jk}\dot{x}^j\dot{x}^k = 0 \tag{3}$$

In an inertia referential the g_{ij} are constants, so [cf. (7) of Section 1.11.8]

$$\Gamma^n_{jk} = 0 \tag{4}$$

and (3) simply becomes

$$\ddot{x}_n = 0 \tag{5}$$

a differential equation that integrates to give

$$x^n = \dot{x}^n t + (x^n)_{t=0} \tag{6}$$

with $\dot{x}^n$ a constant. The movement is uniform and rectilinear. With the shortest way to go from one point to another being a straight line, equation (3) is the equation of the *geodesic* in our euclidean space.

In Minkowski time-space, by generalization, if τ is the proper time, (3) is written ($\nu = 0,1,2,3$)

$$\frac{d^2x^\nu}{d\tau^2} + \Gamma^\nu_{\alpha\beta}\frac{dx^\alpha}{d\tau}\frac{dx^\beta}{d\tau} = 0 \tag{7}$$

For a space of dimension N ($\nu = 0,1,2, \ldots, N - 1$), by noting λ as the main or principal parameter,

$$\frac{d^2x^\nu}{d\lambda^2} + \Gamma^\nu_{\alpha\beta}\frac{dx^\alpha}{d\lambda}\frac{dx^\beta}{d\lambda} = 0 \tag{8}$$

2.11.7 Generalized Lagrange Equation

In our discussion of analytical mechanics it was shown that if a system has g degrees of freedom ($x^1,x^2, \ldots, x^g$) the lagrangian L is a function of the generalized coordinates $x^k = \{x^1, x^2, x^3, \ldots, x^g\}$ and of the speeds $\{\dot{x}^1, \dot{x}^2, \ldots, \dot{x}^g\}$:

$$L = L(x^k, \dot{x}^k; t) \tag{1}$$

such that the equations of motion are the solutions for the generalized Lagrange equation

$$\frac{d}{dt}\left(\frac{\partial L}{\partial \dot{x}^k}\right) - \frac{\partial L}{\partial x^k} = 0 \tag{2}$$

$k = 0, 1, \ldots, g$. In time-space, and in tensor notation, (2) becomes

$$\frac{d}{d\tau}\left(\frac{\partial L}{\partial(\partial x^\mu/\partial\tau)}\right) - \frac{\partial L}{\partial x^\mu} = 0 \tag{3}$$

where $\mu = 0, 1, 2, 3$, and to N dimensions,

$$\frac{d}{d\lambda}\left(\frac{\partial L}{\partial(\partial x^\mu/\partial\lambda)}\right) - \frac{\partial L}{\partial x^\mu} = 0 \tag{4}$$

where $\mu = 0, 1, 2, \ldots, N - 1$. To solve (3) the particular lagrangian is considered in Minkowski time-space:

$$L = -m_0 \sqrt{\eta_{\alpha\beta} \frac{dx^\alpha}{d\tau} \frac{dx^\beta}{d\tau}} \tag{5}$$

Evaluate the terms in (3):

$$\frac{\partial L}{\partial x^\mu} = -\frac{m_0}{2} \frac{\dfrac{\partial \eta_{\alpha\beta}}{\partial x^\mu} \dfrac{dx^\alpha}{d\tau} \dfrac{dx^\beta}{d\tau}}{\sqrt{\eta_{\alpha\beta} \dfrac{dx^\alpha}{d\tau} \dfrac{dx^\beta}{d\tau}}}$$

$$\frac{\partial L}{\partial(dx^\mu/d\tau)} = -m_0 \frac{\eta_{\alpha\mu}(dx^\alpha/d\tau)}{\sqrt{\eta_{\alpha\beta} \dfrac{dx^\alpha}{d\tau} \dfrac{dx^\beta}{d\tau}}}$$

$$\frac{d}{d\tau}\left(\frac{\partial L}{\partial(dx^\mu/d\tau)}\right) = -m_0 \frac{\eta_{\alpha\mu} \dfrac{d^2x^\alpha}{d\tau^2} + \dfrac{\partial \eta_{\alpha\mu}}{\partial x^\beta} \dfrac{dx^\alpha}{d\tau} \dfrac{dx^\beta}{d\tau}}{\sqrt{\eta_{\alpha\beta} \dfrac{dx^\alpha}{d\tau} \dfrac{dx^\beta}{d\tau}}}$$

Replace in (3):

$$\eta_{\alpha\mu}\frac{d^2x^\alpha}{d\tau^2} + \frac{\partial\eta_{\alpha\mu}}{\partial x^\beta}\frac{dx^\beta}{d\tau}\frac{dx^\alpha}{d\tau} - \frac{1}{2}\frac{\partial\eta_{\alpha\beta}}{\partial x^\mu}\frac{dx^\alpha}{d\tau}\frac{dx^\beta}{d\tau} = 0 \tag{6}$$

Now

$$\frac{\partial\eta_{\alpha\mu}}{\partial x^\beta}\frac{dx^\beta}{d\tau}\frac{dx^\alpha}{d\tau} = \frac{1}{2}\left(\frac{\partial\eta_{\alpha\mu}}{\partial x_\beta}\frac{dx^\beta}{d\tau}\frac{dx^\alpha}{d\tau} + \frac{\partial\eta_{\beta\mu}}{\partial x^\alpha}\frac{dx^\beta}{d\tau}\frac{dx^\alpha}{d\tau}\right)$$

and (6) becomes

$$\eta_{\alpha\mu}\frac{d^2x^\alpha}{d\tau^2} + \frac{1}{2}\left(\frac{\partial\eta_{\alpha\mu}}{\partial x^\beta} + \frac{\partial\eta_{\beta\mu}}{\partial x^\alpha} - \frac{\partial\eta_{\alpha\beta}}{\partial x^\mu}\right)\frac{dx^\alpha}{d\tau}\frac{dx^\beta}{d\tau} = 0$$

or, multiplying by $\eta^{\mu\gamma}$, we obtain

$$\frac{d^2x^\gamma}{d\tau^2} + \frac{1}{2}\eta^{\mu\gamma}\left(\frac{\partial\eta_{\alpha\mu}}{\partial x^\beta} + \frac{\partial\eta_{\beta\mu}}{\partial x^\alpha} - \frac{\partial\eta_{\alpha\beta}}{\partial x^\mu}\right)\frac{dx^\alpha}{d\tau}\frac{dx^\beta}{d\tau} = 0$$

By using the fundamental relation

$$\tfrac{1}{2}(\partial_\lambda g_{\mu\nu} + \partial_\mu g_{\lambda\nu} - \partial_\nu g_{\lambda\mu}) = \Gamma_{\lambda\mu,\nu} = g_{\nu\sigma}\Gamma^\sigma_{\lambda\mu}$$

we see that

$$\frac{d^2x^\gamma}{d\tau^2} + \eta^{\mu\gamma}\Gamma_{\alpha\beta,\mu}\frac{dx^\alpha}{d\tau}\frac{dx^\beta}{d\tau} = 0$$

or even

$$\frac{d^2x^\gamma}{d\tau^2} + \eta^{\mu\gamma}\Gamma^\gamma_{\alpha\beta}\frac{dx^\alpha}{d\tau}\frac{dx^\beta}{d\tau} = 0$$

the equation of the geodesic in Minkowski space.

2.11.8 First Integral of Geodesic Equation

Our demonstration is based on equation (3) of Section 2.11.6:

$$\frac{d^2x^n}{dt^2} + \Gamma^n_{jk}\dot{x}^j\dot{x}^k = 0 \tag{1}$$

This choice simplifies the calculations, but the result can be generalized for equations (7) or (8) of Section 2.11.6. Remember that (1) corresponds to the movement of a free particle, so if s is the curvilinear abscissa,

$$\frac{dx^n}{dt} = \frac{ds}{dt}\frac{dx^n}{ds}$$

and

$$\frac{d^2x^n}{dt^2} = \left(\frac{ds}{dt}\right)^2 \frac{d^2x^n}{ds^2}$$

Equation (1) transforms itself to

$$\frac{d^2x^n}{ds^2} + \Gamma^n_{jk}\frac{dx^j}{ds}\frac{dx^k}{ds} = 0 \tag{2}$$

which is multiplied by $g_{nl}\,(dx^l/ds)$:

$$g_{nl}\frac{dx^l}{ds}\frac{d^2x^n}{ds^2} + g_{nl}\Gamma^n_{jk}\frac{dx^j}{ds}\frac{dx^k}{ds}\frac{dx^l}{ds} = 0$$

and with the help of Section 1.11.8 we have

$$2g_{nl}\Gamma^n_{jk} = \frac{\partial g_{jl}}{\partial x_k} + \frac{\partial g_{kl}}{\partial x_j} - \frac{\partial g_{kj}}{\partial x_l} \tag{3}$$

or

$$g_{nl}\frac{dx^l}{ds}\frac{d^2x^n}{ds^2} + \frac{1}{2}\left(\frac{\partial g_{jl}}{\partial x_k} + \frac{\partial g_{kl}}{\partial x_j} - \frac{\partial g_{kj}}{\partial x_l}\right)\frac{dx^j}{ds}\frac{dx^k}{ds}\frac{dx^l}{ds} = 0$$

Now

$$\frac{\partial g_{jl}}{\partial x^k}\frac{dx^j}{ds}\frac{dx^k}{ds}\frac{dx^l}{ds} = \frac{\partial g_{kj}}{\partial x^l}\frac{dx^j}{ds}\frac{dx^k}{ds}\frac{dx^l}{ds}$$

which gives us

$$g_{nl}\frac{dx^l}{ds}\frac{d^2x^n}{ds^2} + \frac{1}{2}\frac{\partial g_{kl}}{\partial x^j}\frac{dx^j}{ds}\frac{dx^k}{ds}\frac{dx^l}{ds} = 0 \tag{4}$$

Bearing in mind that

$$\frac{dg_{kl}}{ds} = \frac{\partial g_{kl}}{\partial x^m}\frac{dx^m}{ds} \tag{5}$$

equation (4) gives

$$g_{nl}\frac{dx^l}{ds}\frac{d^2x^n}{ds^2} + \frac{1}{2}\frac{dg_{kl}}{ds}\frac{dx^k}{ds}\frac{dx^l}{ds} = 0 \tag{6}$$

which we recognize as

$$\frac{d}{ds}\left(g_{kl}\frac{dx^k}{ds}\frac{dx^l}{ds}\right) = 0 \tag{7}$$

where the term in parentheses is a constant in relation to s. The scalar product representing the square of the speed enables us to specify the value of this constant:

$$\vec{v}\cdot\vec{v} = \frac{dx^k}{dt}\vec{e}_k\cdot\frac{dx^l}{dt}\vec{e}_l = \frac{dx^k}{ds}\frac{dx^l}{ds}g_{kl}\left(\frac{ds}{dt}\right)^2$$

Now

$$\vec{v} = \vec{u}\,\frac{ds}{dt}$$

$\vec{u}$ being a unitary vector carried by the tangent to the trajectory and directed in a forward direction; thus

$$g_{kl}\frac{dx^k}{ds}\frac{dx^l}{ds} = 1$$

a relation which shows that

$$ds^2 = g_{kl}\,dx^k\,dx^l$$

is a first integral of the geodesic equation.

2.11.9 Transformation of Energy–Momentum Quadrivector

We have already shown that [cf. (2.36) to (2.38)]

$$v'_x = \frac{v_x - u}{1 - uv_x/c^2}$$

$$v'_y = \frac{\sqrt{1 - u^2/c^2}v_y}{1 - uv_x/c^2}$$

$$v'_z = \frac{\sqrt{1 - u^2/c^2}v_z}{1 - uv_x/c^2}$$

We can therefore evaluate the quantity v'^2/c^2:

$$\frac{v'^2}{c^2} = \frac{(v^2/c^2)(1 - u^2/c^2) + (1 - uv_x/c^2)^2 - (1 - u^2/c^2)}{(1 - uv_x/c^2)^2}$$

and

$$1 - \frac{v'^2}{c^2} = \frac{(1 - u^2/c^2)(1 - v^2/c^2)}{(1 - uv_x/c^2)^2}$$

or

$$\frac{1}{\sqrt{1 - v'^2/c^2}} = \frac{\Gamma(1 - uv_x/c^2)}{\sqrt{1 - v^2/c^2}} \tag{1}$$

and by making $v' \rightarrow v$, $v \rightarrow v'$, and $u \rightarrow -u$, we have

$$\frac{1}{\sqrt{1 - v^2/c^2}} = \frac{\Gamma(1 + uv_x'/c^2)}{\sqrt{1 - v'^2/c^2}} \tag{2}$$

The components of the energy–momentum quadrivector are

$$P^\mu = \left(\frac{E}{c^2}, \vec{p}\right)$$

which means that

$$\frac{E'}{c^2} = \frac{m_0}{\sqrt{1 - v'^2/c^2}} = \frac{m_0\Gamma(1 - uv_x/c^2)}{\sqrt{1 - v^2/c^2}} \tag{3}$$

$$p_x' = \frac{m_0 v_x'}{\sqrt{1 - v'^2/c^2}} = \frac{m_0\Gamma(v_x - u)}{\sqrt{1 - v^2/c^2}} \tag{4}$$

$$p_y' = \frac{m_0 v_y'}{\sqrt{1 - v'^2/c^2}} = \frac{m_0 v_y}{\sqrt{1 - v^2/c^2}} = p_y \tag{5}$$

$$p_z' = \frac{m_0 v_z}{\sqrt{1 - v^2/c^2}} = p_z \tag{6}$$

Now let's go back to relation (4):

$$p_x' = \frac{m_0\Gamma(v_x - u)}{\sqrt{1 - v^2/c^2}} = \frac{mv_x - mu}{\sqrt{1 - u^2/c^2}} = \frac{p_x - uE/c^2}{\sqrt{1 - u^2/c^2}} \tag{7}$$

With the help of (4), equation (3) gives

$$\frac{E'}{c^2} = \frac{m - umv_x/c^2}{\sqrt{1 - u^2/c^2}} = \frac{E/c^2 - up_x/c^2}{\sqrt{1 - u^2/c^2}} \tag{8}$$

which means that the components of P^μ transform themselves, as do those of a quadrivector.

2.11.10 Transformation of Force Components

According to the preceding section,

$$dp_x' = \frac{dp_x - u\,dE/c^2}{\sqrt{1 - u^2/c^2}}$$

$$dp'_y = dp_y$$

$$dp'_z = dp_z$$

$$dt' = dt\,\Gamma\left(1 - \frac{uv_x}{c^2}\right)$$

or

$$F'_x = \frac{dp'_x}{dt'} = \frac{dp_x/dt - (u/c^2)(dE/dt)}{\Gamma(1 - uv_x/c^2)\sqrt{1 - u^2/c^2}} = \frac{1}{1 - uv_x/c^2}\left(F_x - \frac{u}{c^2}\,\vec{F}\cdot\vec{v}\right)$$

$$F'_y = \frac{dp'_y}{dt'} = \frac{dp_y/dt}{\Gamma(1 - uv_x/c^2)} = \frac{F_y\sqrt{1 - u^2/c^2}}{1 - uv_x/c^2}$$

$$F'_z = \frac{dp'_z}{dt'} = \frac{F_z\sqrt{1 - u^2/c^2}}{1 - uv_x/c^2}$$

which are classic formulas.

2.11.11 Orthogonal Groups O(3)

Let E be the euclidean space of dimension 3 with a metric

$$g_{ij} = \delta_{ij} \tag{1}$$

and $\hat{R}$ the operator associated with a linear transformation that preserves the scalar product $X\cdot X$, where X is the position vector:

$$(X)^2 = X\cdot X = X^i\vec{e}_i\cdot X^j\vec{e}_j = X^iX^j\delta_{ij} = x^2 + y^2 + z^2 \tag{2}$$

and by definition of

$$\hat{R}\colon X \xrightarrow{\hat{R}} X'$$

with

$$(X')^2 = \hat{R}X\cdot\hat{R}X = X^2$$

or, in Dirac's notation,

$$\langle \hat{R}X|\hat{R}X\rangle = \langle X|X\rangle \tag{3}$$

or

$$\langle X|\hat{R}^+\hat{R}X\rangle = \langle X|X\rangle$$

By identification,

$$\hat{R}^+\hat{R} = \hat{1} \tag{4}$$

or even

$$\langle \vec{e}_i | \hat{R}^{+} \hat{R} | \vec{e}_j \rangle = \delta_{ij}$$

equivalent to (if I is the 3×3 identity matrix)

$$(R^{+}R)_{ij} = \delta_{ij} I \tag{5}$$

and to

$$R^{+}_{ik} R_{kj} = \delta_{ij} I \tag{6}$$

The coefficients being real,

$$\tilde{R}_{ik} R_{kj} = \delta_{ij} I$$

or, by definition of the transposed matrix,

$$R_{ki} R_{kj} = \delta_{ij} I \tag{7}$$

The transformation studied earlier is unitary:

$$R^{+} = \tilde{R} = R^{-1} \tag{8}$$

and

$$\det(RR^{-1}) = 1 \tag{9}$$

According to relation (8),

$$\tilde{R} = R^{-1}$$

so

$$R\tilde{R} = I \tag{10}$$

As

$$\det(R) = \det(\tilde{R})$$

(the exchange of the rows and columns does not change the value of the determinant), we can deduce from (10) that

$$\det(R) = \pm 1$$

The operators associated with the matrices of determinant equal to $+1$ constitute a group denoted SO(3), the special orthogonal group of dimension 3.

The operations associated with the matrices of determinant equal to -1 do not constitute a group (the product of two matrices will have a determinant equal to $+1$). The union of SO(3) and the "mapping" of improper rotations constitutes the group O(3), an orthogonal group of dimension 3.

2.11.12 Infinitesimal Rotations

The operator associated with an infinitesimal rotation is defined by

$$\hat{R}_\epsilon = \hat{1} + \hat{\epsilon} + \hat{o}(\epsilon^2) \tag{1}$$

where ϵ is, for example, the operator associated with an infinitesimal rotation of angle $d\theta$ around an axis Δ. The matrix element

$$(R_\epsilon)_{ij} = \langle \vec{e}_i | \hat{R}_\epsilon | \vec{e}_j \rangle = \delta_{ij} + \epsilon_{ij} \tag{2}$$

and

$$\begin{aligned}(R_\epsilon \tilde{R}_\epsilon)_{ij} &= \langle \vec{e}_i | \hat{R}_\epsilon \hat{\tilde{R}}_\epsilon | \vec{e}_j \rangle = \sum_l \langle \vec{e}_i | \hat{R}_\epsilon | \vec{e}_l \rangle \langle \vec{e}_l | \hat{\tilde{R}}_\epsilon | \vec{e}_j \rangle \\ &= \sum_l (R_\epsilon)_{il} (\tilde{R}_\epsilon)_{lj} = \sum_l (R_\epsilon)_{il} (R_\epsilon)_{jl} \\ &= \sum_l (\delta_{il} + \epsilon_{il})(\delta_{jl} + \epsilon_{jl})\end{aligned}$$

and by disregarding the terms of second order,

$$(R_\epsilon \tilde{R}_\epsilon)_{ij} = \delta_{ij} + (\epsilon_{ij} + \epsilon_{ji})$$

or

$$R\tilde{R} = I$$

so

$$(R_\epsilon \tilde{R}_\epsilon)_{ij} = \delta_{ij}$$

which means that

$$\epsilon_{ij} = -\epsilon_{ji}$$

The matrix associated with $\hat{\epsilon}$ is antisymmetrical; thus it depends on only three parameters (cf. Section 1.1).

2.11.13 Lorentz Group

In Minkowski time-space the scalar (dot) product of two position quadrivectors X and Y is an invariant under any transformation of the coordinates:

$$X^\mu Y_\mu = X'^{\mu'} Y'_{\mu'} \tag{1}$$

If the coefficients $l^{\mu}_{\mu'}$ define the elements of the matrix of the Lorentz transformation, as X and Y are vectors, they comply with

$$X'^{\mu'} = l^{\mu'}_{\mu} X^\mu \tag{2}$$

$$Y'_{\mu'} = l^{\mu}_{\mu'} Y_{\mu} \tag{3}$$

while

$$X^{\mu} Y_{\mu} = X^{\mu} \eta_{\mu\nu} Y^{\nu} \tag{4}$$

and according to (1),

$$X^{\mu} \eta_{\mu\nu} Y^{\nu} = X'^{\mu'} \eta'_{\mu'\nu'} Y'^{\nu'} = l^{\mu'}_{\alpha} X^{\alpha} \eta'_{\mu'\nu'} l^{\nu'}_{\beta} Y^{\beta} \tag{5}$$

By exchanging $\alpha \rightarrow \mu$ and $\beta \rightarrow \nu$, we obtain

$$X^{\mu} \eta_{\mu\nu} Y^{\nu} = X^{\mu} l^{\mu'}_{\mu} \eta'_{\mu'\nu'} l^{\nu'}_{\nu} Y^{\nu} \tag{6}$$

or

$$\eta_{\mu\nu} = l^{\mu'}_{\mu} \eta'_{\mu'\nu'} l^{\nu'}_{\nu} \tag{7}$$

which is written

$$\eta_{\mu\nu} = l^{\mu'}_{\mu} \eta'_{\mu'\nu'} \tilde{l}^{\nu}_{\nu'} \tag{8}$$

Equation (8) is equivalent to

$$g = Lg\tilde{L}$$

of Section 2.3. We can deduce from this that

$$\det(g) = \det(L) \det(g) \det(L)$$

where

$$[\det(L)]^2 = 1$$

or

$$\det(L) = \pm 1$$

The special Lorentz transformation is, as we saw in Section 2.3, such that

$$\det(L) = 1$$

The Special Lorentz Transformations Form a Group

By definition, for a special Lorentz transformation

$$(x,y,z,t) \xrightarrow{L_1} (x_1,y_1,z_1,t_1)$$

with

$$\begin{bmatrix} x_1 \\ y_1 \\ z_1 \\ ct_1 \end{bmatrix} = \begin{bmatrix} \Gamma_1 & 0 & 0 & -\beta_1\Gamma_1 \\ 0 & 1 & 0 & 0 \\ 0 & 0 & 1 & 0 \\ -\beta_1\Gamma_1 & 0 & 0 & \Gamma_1 \end{bmatrix} \begin{bmatrix} x \\ y \\ z \\ ct \end{bmatrix}$$

The matrix is symmetrical, and for this reason, if

$$(x_1,y_1,z_1,t_1) \xrightarrow{L_2} (x_2,y_2,z_2,t_2)$$

then

$$\begin{bmatrix} x_2 \\ y_2 \\ z_2 \\ ct_2 \end{bmatrix} = \begin{bmatrix} \Gamma & 0 & 0 & -\beta\Gamma \\ 0 & 1 & 0 & 0 \\ 0 & 0 & 1 & 0 \\ -\beta\Gamma & 0 & 0 & \Gamma \end{bmatrix} \begin{bmatrix} x \\ y \\ z \\ ct \end{bmatrix}$$

with, as is shown by the explicit calculation of the matrices associated with L_1 and L_2,

$$\Gamma = (1 + \beta_1\beta_2)\Gamma_2\Gamma_2$$

and

$$\beta\Gamma = (\beta_1 + \beta_2)\Gamma_1\Gamma_2$$

These two relations clearly show that

$$\Gamma^2 = \frac{1}{1 - \beta^2}$$

The neutral element ($\vec{u} = \vec{0}$) and the inverse element ($\vec{u} \to -\vec{u}$) of a determinant equal to 1 do exist. We clearly have a group.

Lorentz Sets

It was shown earlier that for the Lorentz transformations

$$\det(L) = \pm 1$$

The special Lorentz transformations for which $\det(L) = +1$ constitute the first "set," and those of determinant equal to -1 make up the second set. In actual fact, it is preferable to classify the Lorentz transformations in four distinct sets, taking for a second criterion a condition on the element l_0^0 of the Lorentz matrix.

According to formula (7),

$$\eta_{00} = l_0^{\mu'}\eta'_{\mu'\nu'}\tilde{l}_{\nu'}^{0}$$

with

$$\eta_{00} = 1$$

The preceding condition is rewritten

$$1 = \sum_{\mu'} l_0^{\mu'}\eta'_{\mu'\mu'}\tilde{l}_{\mu'}^{0}$$

(a single summation), or

$$1 = (l_0^0)^2 - (l_0^i)^2$$

a relation which implies that

$$(l_0^0)^2 \geq 1$$

from which the four sets are as follows:

Symbol	Determinant	l_0^0	Name
$L_+^{\Uparrow}$	1	≥ 1	Proper Lorentz group, said to be orthochrone
$L_+^{\Downarrow}$	1	≤ -1	Proper heterochrone set
$L_-^{\Uparrow}$	-1	≥ 1	Improper orthochrone set
$L_-^{\Downarrow}$	-1	≤ -1	Improper heterochrone set

Note that

$L_+^{\Uparrow}$ also contains the three-dimensional rotations.
$L_+^{\Downarrow}$ contains the four-dimensional symmetries.
$L_-^{\Uparrow}$ contains the three-dimensional symmetries.
$L_-^{\Downarrow}$ contains the symmetries in relation to time.

2.11.14 Past–Present–Future

In Minkowski space the pseudonorm of a vector is the quantity

$$(X)^2 = X_0^2 - X_x^2 - X_y^2 - X_z^2 = c^2t^2 - \vec{r}^{\,2}$$

The vectors can thus be classified into three types, characterized by the norm:

If $(X)^2 > 0$, the quadriposition vector is said to be of time type. An event associated with this vector moves with a velocity $v < c$.
If $(X)^2 < 0$, the quadrivector is said to be of space type ($v > c$).
If $(X)^2 = 0$, we have a light vector or isotrope ($v = c$).

To show that any vector of time X_t is orthogonal to a vector of space X_e, according to the preceding definitions,

$$(X_t)^2 = c^2t^2 - \vec{r}^{\,2} > 0 \tag{1}$$

Let Y be a vector orthogonal to X_t:

$$X_t \cdot Y = 0 \tag{2}$$

or explicitly,

$$ctY_0 - \vec{r}\cdot\vec{Y} = 0 \tag{3}$$

By using Schwarz's inequality, we have

$$(\vec{r}\cdot\vec{Y})^2 \leqslant \vec{r}^{\,2}\vec{Y}^2 \tag{4}$$

and in terms of (3),

$$c^2t^2Y_0^2 \leqslant \vec{r}^{\,2}\vec{Y}^2$$

Now

$$c^2t^2 > \vec{r}^{\,2}$$

so

$$Y_0^2 < \vec{Y}^2$$

The vector Y complying with (1) is such that

$$Y^2 = Y_0^2 - \vec{Y}^2 < 0$$

and is thus of space type.

We also show that any vector orthogonal to a light vector X_l is either a space vector or a collinear vector (thus a light vector). Indeed, by definition of a light vector,

$$(X_l)^2 = c^2t^2 - \vec{r}^{\,2} = 0$$

We imagine Y to be a vector orthogonal to X_l:

$$X_l\cdot Y = 0$$

or

$$ctY_0 - \vec{r}\cdot\vec{Y} = 0$$

and with Schwarz,

$$Y_0^2 \leqslant \vec{Y}^2$$

from which we conclude that:

If $Y_0^2 < \vec{Y}^2$, then Y is a space vector.
If $Y_0^2 = \vec{Y}^2$, then Y is a light vector proportional to X_l.

Indeed,

$$0 = X_l\cdot Y = ctY_0 - \vec{r}\cdot\vec{Y} = ctY_0 - rY\cos\theta$$

as

$$Y_0 = \pm Y$$

and

$$r = \pm ct$$

That is,

$$\cos\theta = \pm 1$$

It is seen that

$$\theta = \{0, \pi\}$$

or

$$\vec{Y} = \alpha \vec{X}_l \tag{5}$$

Now

$$ctY_0 = \vec{r}\cdot\vec{Y} = \alpha\vec{r}\cdot\vec{X}_l = \alpha r^2 = \alpha c^2t^2$$

from which

$$Y_0 = \alpha ct \tag{6}$$

By comparing (5) and (6) it can be seen that

$$Y = \alpha X_l \tag{7}$$

If we want a geometric representation of these results, we can go back to a space of dimension 3 (one for time and two for space; Figure 9). For a light vector

$$X_l^2 = c^2t^2 - x^2 - y^2 = 0$$

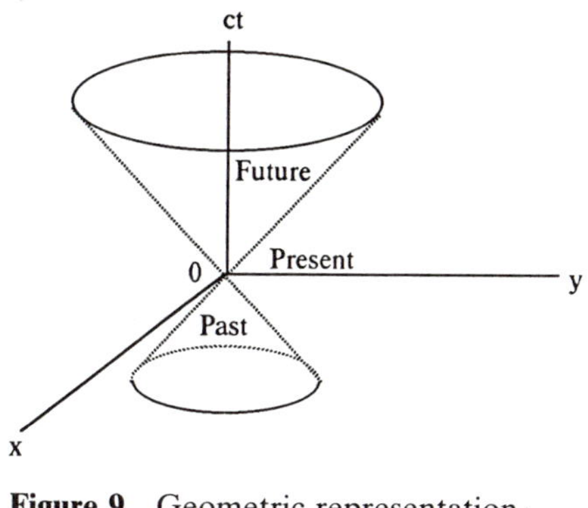

Figure 9 Geometric representation.

or

$$ct = \pm\sqrt{x^2 + y^2}$$

a relation that defines the envelope of two opposing cones by their vertex. For the time vectors

$$X_t^2 = c^2t^2 - (x^2 + y^2) > 0$$

or

$$c^2t^2 > x^2 + y^2$$

which means that X_t is inside the cone. Similarly, for a space vector X_e we end up with

$$c^2t^2 < x^2 + y^2$$

X_e is outside the cone.

APPENDIX 4 CONFORMAL TRANSFORMATION

A4.1 Definition

A *conformal transformation* is defined by

$$g'^{\mu'\nu'}(\vec{X}) = \Omega(\vec{X})g^{\mu'\nu'}(\vec{X}) \tag{A4.1}$$

where $\Omega(\vec{X})$ preserves the angles (the components of the metric tensor are not mixed). Equation (A4.1) causes one zero vector to correspond to another zero vector; the light cone is preserved. From the preceding formula we can deduce that

$$\begin{aligned} ds'^2 &= g'^{\mu'\nu'}X'_{\mu'}X'_{\nu'} = \Omega(\vec{X})g^{\mu'\nu'}(\vec{X})X'_{\mu'}X'_{\nu'} \\ &= \Omega(\vec{X})g^{\mu'\nu'}(\vec{X})\Lambda^{\alpha}_{\mu'}\Lambda^{\beta}_{\nu'}X_\alpha X_\beta = \Omega(\vec{X})g^{\alpha\beta}X_\alpha X_\beta \\ &= \Omega(\vec{X})\, ds^2 \end{aligned} \tag{A4.2}$$

For an infinitesimal transformation

$$X'^\mu = X^\mu + \delta X^\mu \tag{A4.3}$$

$$\Omega(\vec{X}) = \hat{1} + \delta\hat{\Omega} \tag{A4.4}$$

where

$$\delta\hat{\Omega} = -2\hat{\sigma}(\vec{X}) \tag{A4.5}$$

a choice justified afer the calculations. According to (A4.3),

$$g'_{\mu'\nu'}(\vec{X}') = g'_{\mu'\nu'}(\vec{X} + \overrightarrow{\delta X}) = g'_{\mu'\nu'}(\vec{X}) + \partial_\rho[g_{\mu'\nu'}(\vec{X})]\,\delta X^\rho \tag{A4.6}$$

We write by definition

$$\delta g_{\mu'\nu'}(\vec{X}) = g'_{\mu'\nu'}(\vec{X}) - g_{\mu'\nu'}(\vec{X}) \tag{A4.7}$$

or with the help of (A4.6),

$$\delta g_{\mu'\nu'}(\vec{X}) = g'_{\mu'\nu'}(X') - \partial_\rho g_{\mu'\nu'}(\vec{X})\,\delta X^\rho - g_{\mu'\nu'}(\vec{X}) \tag{A4.8}$$

As

$$g'_{\mu'\nu'}(\vec{X}) = \Lambda^\alpha_{\mu'}\Lambda^\beta_{\nu'}g_{\alpha\beta}(\vec{X}) \tag{A4.9}$$

(a relation of transformation of a rank 2 tensor) and as according to (A4.3),

$$\Lambda^\rho_{\mu'} = \frac{\partial X^\rho}{\partial X'^{\mu'}} = \delta^\rho_{\mu'} - \partial_{\mu'}(\delta X^\rho) \tag{A4.10}$$

relation (A4.8) is written

$$\begin{aligned}\delta g_{\mu'\nu'} &= [\delta^\alpha_{\mu'} - \partial_{\mu'}(\delta X^\alpha)][\delta^\beta_{\nu'} - \partial_{\nu'}(\delta X^\beta)]g_{\alpha\beta}(\vec{X})\\ &\quad - \partial_\rho g_{\mu'\nu'}(\vec{X})\,\delta X^\rho - g_{\mu'\nu'}(\vec{X})\end{aligned}$$

or in first order,

$$\begin{aligned}\delta g_{\mu'\nu'} &= -[\partial_{\mu'}(\delta X^\alpha)g_{\alpha\nu'}(\vec{X}) + \partial_{\nu'}(\delta X^\beta)g_{\mu'\beta}(\vec{X})\\ &\quad + \partial_\rho g_{\mu'\nu'}(\vec{X})(\delta X^\rho)\end{aligned} \tag{A4.11}$$

A4.2 Killing-Cartan Equation

The preceding result is applied to the particular case of a flat metric:

$$\eta_{\mu'\nu'} = \begin{bmatrix}1 & 0 & 0 & 0\\ 0 & -1 & 0 & 0\\ 0 & 0 & -1 & 0\\ 0 & 0 & 0 & -1\end{bmatrix} \tag{A4.12}$$

We have chosen that of Minkowski. According to (A4.1), (A4.4), and (A4.5),

$$\eta'_{\mu'\nu'} = [\hat{1} - 2\hat{\sigma}(\vec{X})]\eta_{\mu'\nu'} \tag{A4.13}$$

Relation (A4.11) could be used if so wished:

$$\delta g_{\mu'\nu'}\,(\vec{X}) = -2\hat{\sigma}(\vec{X})\eta_{\mu'\nu'} \tag{A4.14}$$

or

$$\begin{aligned}2\hat{\sigma}(\vec{X})\eta_{\mu'\nu'} &= \partial_{\mu'}(\delta X^\alpha)\eta_{\alpha\nu'} + \partial_{\nu'}(\delta X^\beta)\eta_{\mu'\beta}\\ &= \partial_{\mu'}(\delta X^{\nu'}) + \partial_{\nu'}(\delta X^{\mu'})\end{aligned} \tag{A4.15}$$

Note that the dimension d of the space is given by

$$\eta^{\mu\nu}\eta_{\mu\nu} = d$$

Then, multiplying equation (A4.15) by $\eta^{\mu'\nu'}$, we see that

$$\sigma(\vec{X}) = \frac{1}{d}\,\partial_{\mu'}(\delta X^{\mu'}) \qquad \text{(A4.16)}$$

a contraction. The equation we are looking for, due to Killing and Cartan, is obtained by using this last result with (A4.15):

$$\partial_{\mu'}(\delta X^{\nu'}) + \partial_{\nu'}(\delta X^{\mu'}) - \frac{2}{d}\,\partial_{\rho'}(\delta X^{\rho'})\eta_{\mu'\nu'} = 0 \qquad \text{(A4.17)}$$

For the moment, we are trying to integrate (A4.17). The contraction by $\partial^{\mu'}\,\partial_{\nu'}$ gives

$$(1 - d)\,\Box\partial_{\mu'}(\partial X^{\mu'}) = 0 \qquad \text{(A4.18)}$$

If $d \neq 1$,

$$\Box\partial_{\mu'}(\delta X^{\mu'}) = 0 \qquad \text{(A4.19)}$$

Contract (A4.17) by $\partial^{\lambda'}$:

$$\partial^{\lambda'}\partial_{\mu'}(\delta X^{\nu'}) + \partial^{\lambda'}\partial_{\nu'}(\delta X^{\mu'}) - \frac{2}{d}\,\partial^{\lambda'}\partial_{\rho'}(\delta X^{\rho'})\eta_{\mu'\nu'} = 0 \qquad \text{(A4.20)}$$

Now [cf. (A4.17)] we also have

$$\partial_{\lambda'}(\delta X^{\nu'}) = -\partial_{\nu'}(\delta X^{\lambda'}) + \frac{2}{d}\,\partial_{\rho'}(\delta X^{\rho'})\eta_{\lambda'\nu'} \qquad \text{(A4.21)}$$

and (A4.20) becomes

$$\partial_{\lambda'}\partial_{\nu'}(\delta X_{\mu'}) - \partial_{\mu'}\partial_{\nu'}(\delta X_{\lambda'}) + \frac{2}{d}\,\partial_{\mu'}\partial_{\rho'}(\delta X^{\rho'})\eta_{\lambda'\nu'}$$
$$- \frac{2}{d}\,\partial_{\lambda'}\partial_{\rho'}(\delta X_{\rho'})\eta_{\mu'\nu'} = 0 \qquad \text{(A4.22)}$$

By contracting by $\partial_{\lambda'}$

$$\Box\partial_{\nu'}(\delta X_{\mu'}) - \partial_{\lambda'}\partial_{\mu'}\partial_{\nu'}(\delta X_{\lambda'}) + \frac{2}{d}\,\partial_{\mu'}\partial_{\rho'}\partial_{\nu'}(\delta X_{\rho'})$$
$$- \frac{2}{d}\,\Box\partial_{\rho'}(\delta X_{\rho'})\eta_{\mu'\nu'} = 0$$

and thanks to (A4.19),

$$\Box\partial_{\nu'}(\delta X_{\mu'}) + \left(\frac{2}{d} - 1\right)\partial_{\mu'}\partial_{\nu'}\partial_{\lambda'}(\delta X_{\lambda'}) = 0 \tag{A4.23}$$

As [cf. (A4.17)]

$$\Box\partial_{\nu'}(\delta X_{\mu'}) = \tfrac{1}{2}\Box[\partial_{\nu'}(\delta X_{\mu'}) + \partial_{\mu'}(\delta X_{\nu'})]$$

and according to (A4.17),

$$\Box\partial_{\nu'}(\delta X_{\mu'}) = -\frac{1}{2}\Box\partial_{\rho'}(\delta X_{\rho'})\eta_{\mu'\nu'}$$

or with (A4.19),

$$\Box\partial_{\nu'}(\delta X_{\mu'}) = 0 \tag{A4.24}$$

from which, for (A4.23), if $d \neq 2$,

$$\partial_{\mu'}\partial_{\nu'}[\partial_{\lambda'}(\delta X_{\lambda'})] = 0 \tag{A4.25}$$

Integrating for the first time yields

$$\partial_{\nu'}[\partial_{\lambda'}(\delta X_{\lambda'})] = 2dC_{\nu'} \tag{A4.26}$$

(the factor $2d$ is introduced for later convenience). A second integration gives

$$\partial_{\lambda'}(\delta X_{\lambda'}) = d[2C_{\nu'}X_{\nu'} + A] \tag{A4.27}$$

Equation (A4.17) multiplied by $\partial_{\lambda'}$ is written

$$\partial_{\lambda'}\partial_{\mu'}(\delta X_{\nu'}) + \partial_{\lambda'}\partial_{\nu'}(\delta X_{\mu'}) = \frac{2}{d}\partial_{\lambda'}\partial_{\rho'}(\delta X_{\rho'})\eta_{\mu'\nu'} \tag{A4.28}$$

In this equation we use the relations

$$\lambda' \longrightarrow \mu'$$
$$\mu' \longrightarrow \nu'$$
$$\nu' \longrightarrow \lambda'$$

$$\partial_{\mu'}\partial_{\lambda'}(\delta X_{\nu'}) + \partial_{\mu'}\partial_{\nu'}(\delta X_{\lambda'}) = \frac{2}{d}\partial_{\mu'}\partial_{\rho'}(\delta X_{\rho'})\eta_{\nu'\lambda'} \tag{A4.29}$$

Substracting (A4.28)–(A4.29) gives

$$\partial_{\nu'}[\partial_{\lambda'}(\delta X_{\mu'}) - \partial_{\mu'}(\delta X_{\lambda'})] = \frac{2}{d}[\partial_{\lambda'}\partial_{\rho'}(\delta X_{\rho'})\eta_{\mu'\nu'} - \partial_{\mu'}\partial_{\rho'}(\delta X_{\rho'})\eta_{\nu'\lambda'}] \tag{A4.30}$$

and with (A4.27),

$$\partial_{\nu'}[\partial_{\lambda'}(\delta X_{\mu'}) - \partial_{\mu'}(\delta X_{\lambda'})] = \frac{2}{d}\{\partial_{\lambda'}[d(2c_{\alpha'}X_{\alpha'} + A)\eta_{\mu'\nu'}] - \partial_{\mu'}[d(2C_{\alpha'}X_{\alpha'} + A)\eta_{\nu'\lambda'}]\}$$

or

$$\begin{aligned}\partial_{\nu'}[\partial_{\lambda'}(\delta X_{\mu'}) - \partial_{\mu'}(\delta X_{\lambda'})] &= 4[\partial_{\lambda'}(C_{\alpha'}X_{\alpha'}\eta_{\mu'\nu'}) - \partial_{\mu'}(C_{\alpha'}X_{\alpha'}\eta_{\nu'\lambda'})] \\ &= 4(C_{\alpha'}\delta_{\lambda'\alpha'}\eta_{\mu'\nu'} - C_{\alpha'}\delta_{\mu'\alpha'}\eta_{\nu'\lambda'}) \\ &= 4(C_{\lambda'}\eta_{\mu'\nu'} - C_{\mu'}\eta_{\nu'\lambda'}) \qquad \text{(A4.31)}\end{aligned}$$

The last relation is then integrated:

$$\begin{aligned}\partial_{\lambda'}(\delta X_{\mu'}) - \partial_{\mu'}(\delta X_{\lambda'}) &= 4(C_{\lambda'}\eta_{\mu'\nu'}X_{\nu'} - C_{\mu'}\eta_{\nu'\lambda'}X_{\nu'}) + 2D_{\lambda'\mu'} \\ &= 4(C_{\lambda'}X_{\mu'} - C_{\mu'}X_{\lambda'}) + 2D_{\lambda'\mu'} \qquad \text{(A4.32)}\end{aligned}$$

Observe that

$$D_{\lambda'\mu'} = -D_{\mu'\lambda'} \qquad \text{(A4.33)}$$

since the left-hand term of the identity (A4.32) changes sign if the roles of λ' and μ' are exchanged, as does the first term on the right-hand side.

We replace $\partial_{\mu'}(\delta X_{\lambda'})$ of (A4.32) with the help of (A4.17):

$$\partial_{\lambda'}(\delta X_{\mu'}\ \partial_{\lambda'}(\delta X_{\mu'}) - \frac{2}{d}\partial_{\rho'}(\delta X_{\rho'})\eta_{\mu'\lambda'} = 4(C_{\lambda'}X_{\mu'} - C_{\mu'}X_{\lambda'}) + 2D_{\lambda'\mu'}$$

Then, based on (A4.27),

$$2\partial_{\lambda'}(\delta X_{\mu'}) - 2C_{\nu'}X_{\nu'}\eta_{\mu'\lambda'} - 2A\eta_{\mu'\lambda'} = 4(C_{\lambda'}X_{\mu'} - C_{\mu'}X_{\lambda'}) + 2D_{\lambda'\mu'}$$

We integrate this last relation:

$$\delta X_{\mu'} = C_{\nu'}\eta_{\mu'\lambda'}X_{\nu'}X_{\lambda'} + AX_{\lambda'}\eta_{\mu'\lambda'} + 2\left(C_{\lambda'}X_{\mu'}\frac{X_{\lambda'}}{2} - C_{\mu'}\frac{X_{\lambda'}^2}{2}\right) + D_{\lambda'\mu'}X_{\lambda'} + E_{\mu'}$$

which is written

$$\delta X_{\mu'} = E_{\mu'} + D_{\lambda'\mu'}X_{\lambda'} + AX_{\mu'} + 2C_{\lambda'}X_{\mu'}X_{\lambda'} - C_{\mu'}X_{\lambda'}^2 \qquad \text{(A4.34)}$$

an equation of a conformal *and* general infinitesimal transformation.

Notes

1. The $E_{\mu'}$ are the four parameters of the group of translations in a space of dimension 4 (Minkowski group).

2. The $D_{\lambda'\mu'}$ are the six parameters of the Lorentz group. They are antisymmetrical.
3. The $E_{\mu'}$ and $D_{\lambda'\mu'}$ are the 10 parameters of the Poincaré group.
4. A is the parameter of the dilation group.
5. A special conformal transformation is defined by

$$X_\mu \longrightarrow \frac{X_\mu - C_\mu X^2}{1 - 2CX + C^2X^2}$$

and if C is considered to be small,

$$X_\mu \longrightarrow (X_\mu - C_\mu X^2)(1 + 2CX - \cdots)$$

or

$$X_\mu \longrightarrow (X_\mu + 2CXX_\mu - C_\mu X^2 + \cdots)$$

We recognize

$$\delta X_\mu = 2CXX_\mu - C_\mu X^2$$

that is, the two last terms on the right-hand side of (A4.34). The inversions are the elements of this group, a group that is defined by the four parameters C_μ. The general conformal transformation requires the data of 15 parameters.

APPENDIX 5 RELATIVISTIC STRING

The representation used (Figure A5.1) is of dimension 3; X_0 clearly represents the time. Points A and B are the extremities of the string at the time $X_0 = 0$. When X_0 is altered, the string describes a Riemann surface of dimension 2. Note that τ and σ are the two parameters enabling the

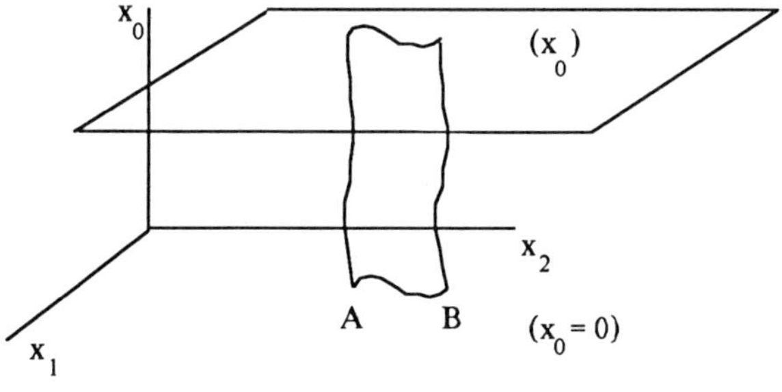

Figure A5.1 String.

definition of a surface point (τ is the proper time and σ the curvilinear abscissa).

The string is imagined to be of finite length. We impose

$$\sigma \in [0,\pi] \tag{A5.1}$$

a choice to be justified later; it leads to an obvious simplification of the formulas.

We are first trying to find the equivalent of the results obtained for a free particle, that is, the equation of the geodesic (Section 2.11.6):

$$\text{(A5.2)} \quad \ddot{X}^{\mu} + \Gamma^{\mu}_{\nu\rho}\dot{X}^{\nu}\dot{X}^{\rho} = 0$$

which is a result of the condition

$$\delta S = 0 \tag{A5.3}$$

with

$$w = -m_0 \int_{\lambda_1}^{\lambda_2} \sqrt{g_{\mu\nu}(X)\dot{X}^{\mu}\dot{X}^{\nu}}\, d\lambda \tag{A5.4}$$

where λ is the principal parameter. In a metric supposed flat,

$$g_{\mu\nu} = \eta_{\mu\nu} \tag{A5.5}$$

the equivalent of the geodesic will be a surface Σ (two parameters, τ and σ, instead of one, λ). A surface element Σ is written

$$ds^2 = dX^{\mu}(\tau,\sigma)\, dX_{\mu}(\tau,\sigma) = \frac{\partial X^{\mu}}{\partial \lambda_i}\, d\lambda_i\, \frac{\partial X_{\mu}}{\partial \lambda_j}\, d\lambda_j \tag{A5.6}$$

with $i, j = 1, 2$:

$$\begin{aligned} \lambda_1 &= \tau \\ \lambda_2 &= \sigma \end{aligned} \tag{A5.7}$$

We write, by definition of the h_{ij}, a metric induced onto the surface,

$$ds^2 = h_{ij}\, d\lambda_i\, d\lambda_j \tag{A5.8}$$

We then write

$$h = \det(h_{ij}) = \begin{vmatrix} h_{\tau\tau} & h_{\tau\sigma} \\ h\sigma\tau & h_{\sigma\sigma} \end{vmatrix} \tag{A5.9}$$

or

$$\begin{aligned} h &= \left(\frac{\partial X^{\mu}}{\partial \tau}\frac{\partial X_{\mu}}{\partial \tau}\right)\left(\frac{\partial X^{\nu}}{\partial \sigma}\frac{\partial X_{\nu}}{\partial \sigma}\right) - \left(\frac{\partial X^{\mu}}{\partial \tau}\frac{\partial X_{\mu}}{\partial \sigma}\right)\left(\frac{\partial X^{\nu}}{\partial \sigma}\frac{\partial X_{\nu}}{\partial \tau}\right) \\ &= \dot{X}^2 X'^2 - (\dot{X}\cdot X')^2 \end{aligned} \tag{A5.10}$$

It can be seen that the first two terms in parentheses are of opposing signs, as for the form h_{00} multiplied by h_{11}. This means that $h \leqslant 0$. We can now show that in even the most general cases

$$d\Sigma^{i_1 \cdots i_n} = \frac{1}{n!} \sqrt{-g}\, \epsilon^{i_1}{}_{i_2 \cdots i_n}\, dX^{i_2} \wedge dX^{i_3} \wedge \cdots \wedge dX^{i_n} \tag{A5.11}$$

which in our case is reduced to

$$d\Sigma = \tfrac{1}{2} \sqrt{-h}\, (d\sigma \wedge d\tau - d\tau \wedge d\sigma) \tag{A5.12}$$

written

$$d\Sigma = \sqrt{-h}\, d\sigma\, d\tau \tag{A5.13}$$

The surface covered by the string is

$$\Sigma = \int_{\sigma=0}^{\pi} \int_{\tau_1}^{\tau_2} \sqrt{h_{\sigma\tau}^2 - h_{\sigma\sigma} h_{\tau\tau}}\, d\sigma\, d\tau \tag{A5.14}$$

which is the quantity that must be made extreme. By analogy with (A5.3) and (A5.4), Jacobi's action can be deduced from this:

$$w = -\frac{1}{2\pi\alpha} \int_{\sigma=0}^{\pi} \int_{\tau_1}^{\tau_2} \sqrt{(\dot{X}X')^2 - \dot{X}^2 X'^2}\, d\sigma\, d\tau \tag{A5.15}$$

α is called Regge's constant. As

$$w = \int_{\tau_1}^{\tau} \mathbf{L}\, d\tau$$

by definition of the Jacobi action w, we see for the lagrangian

$$\mathbf{L} = -\frac{1}{2\pi\alpha} \int_{\sigma=0}^{\pi} \sqrt{(\dot{X}X')^2 - \dot{X}^2 X'^2}\, d\sigma \tag{A5.16}$$

or a lagrangian density $l = l(\dot{X}, X')$ such that

$$l = -\frac{1}{2\pi\alpha} \sqrt{(\dot{X}X')^2 - \dot{X}^2 X'^2} \tag{A5.17}$$

It is important for all the *virtual surfaces* δX to have the same extreme values for the variable X_0:

$$\delta X^{\mu}(\tau = \tau_1, \sigma) = \delta X^{\mu}(\tau = \tau_2, \sigma) = 0 \tag{A5.18}$$

The condition of extremum

$$\delta w = 0$$

means that

$$\delta w = \delta \int_{\tau_1}^{\tau_2} \mathbf{L}\, d\tau = \delta \int_{\tau_1}^{\tau_2} \int_{\sigma=0}^{\pi} l\, d\sigma\, d\tau$$

$$= \int_{\tau_1}^{T_2} \int_{\sigma=0}^{\pi} [l(\dot{X}^\mu + \delta\dot{X}^\mu, X'^\mu + \delta X'^\mu) - l(\dot{X}^u, X'^\mu)]\, d\sigma\, d\tau$$

$$= \int_{\tau_1}^{\tau_2} \int_{\sigma=0}^{\pi} \left[\frac{\partial l}{\partial \dot{X}^\mu}\, \delta(d\dot{X}^\mu) + \frac{\partial l}{\partial X'^\mu}\, \delta(dX'^\mu)\right] d\tau\, d\sigma = 0 \tag{A5.19}$$

The constraint is called an *edge condition* [cf. (A5.18)],

$$\left(\frac{\partial l}{\partial X'^\mu}\, \delta X^\mu\right)_{\sigma=0} = \left(\frac{\partial l}{\partial X'^\mu}\, \delta X^\mu\right)_{\sigma=\pi} = 0$$

and integration by parts leads to the identities

$$\int_{\tau_1}^{\tau_2} \frac{\partial l}{\partial \dot{X}^\mu}\, d(\delta\dot{X}^\mu)\, d\tau = -\int_{\tau_1}^{\tau_2} \delta X^\mu \frac{\partial}{\partial \tau}\left(\frac{\partial l}{\partial \dot{X}^\mu}\right) d\tau$$

$$\int_0^{\pi} \frac{\partial l}{\partial X'^\mu}\, d(\partial X'^\mu)\, d\sigma = -\int_0^{\pi} \delta X^\mu \frac{\partial}{\partial \sigma}\left(\frac{\partial l}{\partial X'^\mu}\right) d\sigma$$

or

$$\delta w = -\int_{\tau_1}^{\tau_2} \int_0^{\pi} \left[\frac{\partial}{\partial \tau}\left(\frac{\partial l}{\partial \dot{X}^\mu}\right) + \frac{\partial}{\partial \sigma}\left(\frac{\partial l}{\partial X'^\mu}\right) \delta X^\mu\, d\tau\, d\sigma\right.$$

Now the surface Σ is minimal if $\delta w = 0$, that is, if

$$\frac{\partial}{\partial \tau}\left(\frac{\partial l}{\partial \dot{X}^\mu}\right) + \frac{\partial}{\partial \sigma}\left(\frac{\partial l}{\partial X'^\mu}\right) = 0 \tag{A5.20}$$

with l given by (A5.17). With the help of the usual definitions, the densities of the generalized momentums of the system are evaluated:

$$p_\tau^\mu = \frac{\partial l}{\partial \dot{X}^\mu} = \frac{1}{2\pi\alpha} \frac{X'^2\dot{X}^\mu - (\dot{X}\cdot X')X'^\mu}{\sqrt{(\dot{X}\cdot X')^2 - \dot{X}^2X'^2}} \tag{A5.21}$$

$$p_\sigma^\mu = \frac{\partial l}{\partial X'^\mu} = \frac{1}{2\pi\alpha} \frac{\dot{X}^2X'^\mu - (\dot{X}\cdot X')\dot{X}^\mu}{\sqrt{(\dot{X}\cdot X')^2 - \dot{X}^2X'^2}} \tag{A5.22}$$

from which we deduce that

$$p_\tau^\mu X'_\mu = 0 \tag{A5.23}$$

$$p_\sigma^\mu \dot{X}_\mu = 0 \tag{A5.24}$$

$$p_\sigma^2 = -\frac{(\dot{X})^2}{(2\pi\alpha)^2} \tag{A5.25}$$

$$p_\tau^2 = -\frac{(X')^2}{(2\pi\alpha)^2} \tag{A5.26}$$

The equations of motion deduced from (A5.20) are

$$\frac{\partial}{\partial\tau}(p_\tau^\mu) + \frac{\partial}{\partial\sigma}(p_\sigma^\mu) = 0 \tag{A5.27}$$

with (edge conditions)

$$(p_\sigma^\mu)_{\sigma=0} = (p_\sigma^\mu)_{\sigma=\pi} = 0 \tag{A5.28}$$

and for the extremities

$$p_\sigma^2 = 0 \quad \text{when } \dot{X}^2 = 0$$

which means that the speed s of the edges is equal to the speed of the propagation of light, since

$$\dot{X}^2 = c^2\tau^2 - \sigma^2$$

The generalized momentum P^μ is such that

$$P^\mu = \int_{\sigma=0}^{\pi} p_\tau^\mu \, d\sigma \tag{A5.29}$$

and with (A5.27),

$$\frac{\partial p^\mu}{\partial\tau} = -\int_{\sigma=0}^{\pi} \frac{\partial}{\partial\sigma}(p_\sigma^\mu)\, d\sigma = -[p_\sigma^\mu]_{\sigma=0}^{\pi} = 0$$

bearing in mind the edge conditions (A5.28). There is an invariance of the energy–momentum quadrivector. Similarly, the generalized momentum is preserved. It is defined by

$$M^{\mu\nu} = \int_{\sigma=0}^{\sigma=\pi} (X^\mu p_\tau^\nu - X^\nu p_\tau^\mu)\, d\sigma = \int_0^\pi M_\tau^{\mu\nu}\, d\sigma \tag{A5.30}$$

and complies with

$$\frac{\partial}{\partial\sigma} M_\sigma^{\mu\nu} + \frac{\partial}{\partial\tau} M_\tau^{\mu\nu} = 0 \tag{A5.31}$$

with

$$M_\sigma^{\mu\nu} = X^\mu p_\sigma^\nu - X^\nu p_\sigma^\mu \tag{A5.32}$$

Indeed, with the help of (A5.27),

$$\frac{\partial}{\partial\sigma} M_\sigma^{\mu\nu} + \frac{\partial}{\partial\tau} M_\tau^{\mu\nu} = (\partial_\sigma X^\mu)p_\sigma^\nu + (\partial_\tau X^\mu)p_\tau^\nu - (\partial_\sigma X^\nu)p_\sigma^\mu - (\partial_\tau X^\nu)p_\tau^\mu$$

and bearing in mind (A5.21) and that, for example,

$$\partial_\sigma X^\mu = X'^\mu \qquad \partial_\tau X^\mu = \dot{X}^\mu$$

the identity (A5.31) is confirmed. The edge conditions being identical for the impulses and the densities of angular (kinetic) momentum, we see that

$$\frac{\partial}{\partial\tau} M^{\mu\nu} = 0 \tag{A5.33}$$

We are now trying to identify the solutions of the equation of motion (A5.27). The space in which we operate is of dimension 4. The surface covered by the string is now examined. The *conformal gauge conditions* are said to add to (A5.27) two equations of constraints, enabling us to have a unique choice of τ and σ:

$$e_\sigma = \partial_\sigma X \tag{A5.34}$$

$$e_\tau = \partial_\tau X \tag{A5.35}$$

that is,

$$h_{ij} = \begin{bmatrix} h_{\tau\tau} & 0 \\ 0 & h_{\sigma\sigma} \end{bmatrix} = \begin{bmatrix} \dot{X}^2 & 0 \\ 0 & X'^2 \end{bmatrix} \tag{A5.36}$$

we see that

$$h = \det(h_{ij}) = \dot{X}^2 X'^2 \tag{A5.37}$$

which when compared with (A5.10) means that

$$X \cdot \dot{X}' = 0 \tag{A5.38}$$

By the choice of gauge we impose (in analogy with the Minkowski gauge)

$$h_{ij} = \begin{bmatrix} -1 & 0 \\ 0 & 1 \end{bmatrix} \tag{A5.39}$$

It follows from this that

$$\dot{X}^2 + X'^2 = 0 \tag{A5.40}$$

by identification with (A5.39). Equations (A5.21), (A5.22), and (A5.32) are then written

$$p_\tau^\mu = \frac{1}{2\pi\alpha} \frac{X'^2 \dot{X}^\mu}{\sqrt{-\dot{X}^2 X'^2}} = -\frac{1}{2\pi\alpha} \dot{X}^\mu \tag{A5.41}$$

$$p_\sigma^\mu = \frac{1}{2\pi\alpha} \frac{\dot{X}^2 X'^\mu}{\sqrt{\dot{X}^2 X'^2}} = \frac{1}{2\pi\alpha} X'^\mu \tag{A5.42}$$

$$M_\sigma^{\mu\nu} = -\frac{1}{2\pi\alpha} (X^\mu X'^\nu - X^\nu X'^\mu) \tag{A5.43}$$

$$M_\tau^{\mu\nu} = -\frac{1}{2\pi\alpha} (X^\mu \dot{X}^\nu - X^\nu \dot{X}^\mu) \tag{A5.44}$$

The lagrangian density l is evaluated based on (A5.17):

$$l = -\frac{1}{2\pi\alpha} \sqrt{-\dot{X}^2 X'^2} = -\frac{1}{2\pi\alpha} \dot{X}^2 \tag{A5.45}$$

and according to (A5.20),

$$-\frac{\partial}{\partial\tau} \dot{X}^\mu + \frac{\partial}{\partial\sigma} X'^\mu = 0$$

or

$$\ddot{X}^\mu - X''^\mu = 0$$

a well-known equation of propagation.

APPENDIX 6 DIFFERENTIAL FORMS

A6.1 Notion of Differential Form

Let $dx\ dy$ be the symbolic notation of a surface element of the plane (Oxy). The change of variables

$$(\vec{x}, \vec{y}) \longrightarrow (\vec{u}, \vec{v}) \tag{A6.1}$$

leads to

$$du\ dv = |J|\ dx\ dy \tag{A6.2}$$

where $du\ dv$ is the surface element of plane $(\vec{u}, \vec{v})$ and J the jacobian of the transformation defined by (A6.1).

We take two linear forms, f_1 and f_2 with two variables:

$$f_1 = P_1\ dx + Q_1\ dy \tag{A6.3}$$

$$f_2 = P_2\ dx + Q_2\ dy \tag{A6.4}$$

The exterior product $f_1 \wedge f_2$ is defined from the two linear forms by

$$f_1 \wedge f_2 = -f_1 \wedge f_2 \tag{A6.5}$$

$$f_1 \wedge f_1 = 0 \tag{A6.6}$$

$$(kf_1) \wedge f_2 = k(f_1 \wedge f_2) \tag{A6.7}$$

$$f_1 \wedge (f_2 + f_3) = f_1 \wedge f_2 + f_1 \wedge f_3 \tag{A6.8}$$

The quantity $dx \wedge dy$ is seen to have the same property as the element of algebraic area $dx\,dy$.

Starting from a euclidean space of dimension 3, in the tangent space to R^3 the following n-forms ($n = 0,1,2,3$) can be defined:

0-form: any scalar α (A6.9)
1-form: dx, dy or dz (A6.10)
2-form: $dx \wedge dy$, $dy \wedge dz$, $dz \wedge dx$ (A6.11)
3-form: $dx \wedge dy \wedge dz$ (A6.12)

Consider the 2-form

$$f = f_{ij}\, dx^i \wedge dx^j \tag{A6.13}$$

The change in variable

$$u^i = \varphi^i(x^j) \tag{A6.14}$$

means that

$$\begin{aligned} dx^i \wedge dx^j &= \frac{\partial x^i}{\partial u^\alpha}\frac{\partial x^j}{\partial u^\beta}\, du^\alpha \wedge du^\beta \\ &= \sum_{\alpha<\beta}\left(\frac{\partial x^i}{\partial u^\alpha}\frac{\partial x^j}{\partial u^\beta}\, du^\alpha \wedge du^\beta + \frac{\partial x^i}{\partial u^\beta}\frac{\partial x^j}{\partial u^\alpha}\, du^\beta \wedge du^\alpha\right) \\ &= \sum_{\alpha<\beta}\left(\frac{\partial x^i}{\partial u^\alpha}\frac{\partial x^j}{\partial u^\beta} - \frac{\partial x^i}{\partial u^\beta}\frac{\partial x^j}{\partial u^\alpha}\right) du^\alpha \wedge du^\beta = |J|\, du^\alpha \wedge du^\beta \end{aligned} \tag{A6.15}$$

Direct geometric calculation is accomplished with the help of Figure A6.1.

$$u^1 = \varphi^1(x^1,x^2) \tag{A6.16}$$

$$u^2 = \varphi^2(x^1,x^2) \tag{A6.17}$$

and

$$x^1 = \psi^1(u^1,u^2) \tag{A6.18}$$

$$x^2 = \psi^2(u^1,u^2) \tag{A6.19}$$

(φ^1 and φ^2 are supposed invertible). For the points M_i ($i = 1,2,3,4$) we see that (Figure A6.1)

$$M_1:\quad x_1^1 = \psi^1(u^1,u^2)$$
$$x_1^2 = \psi^2(u^1,u^2)$$

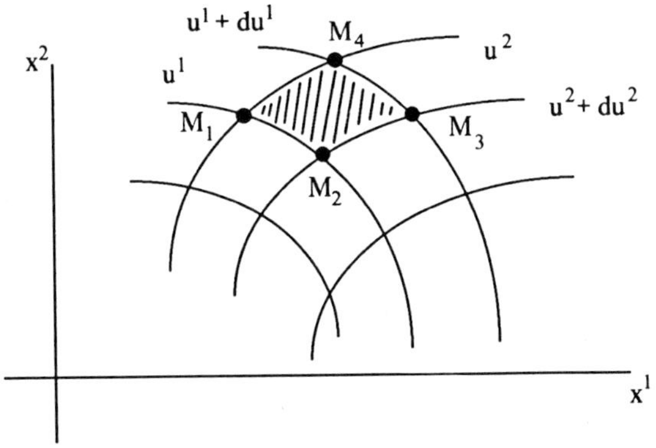

Figure A6.1 Surface element.

$$M_2:\quad x_2^1 = \psi^1(u^1, u^2 + du^2) = \psi^1(u^1,u^2) + \frac{\partial\psi^1}{\partial u^2}\,du^2 + \cdots$$

$$x_2^2 = \psi^2(u^1, u^2 + du^2) = \psi^2(u^1,u^2) + \frac{\partial\psi^2}{\partial u^2}\,du^2 + \cdots$$

$$M_3:\quad x_3^1 = \psi^1(u^1 + du^1, u^2 + du^2)$$

$$= \psi^1(u^1,u^2) + \frac{\partial\psi^1}{\partial u^1}\,du^1 + \frac{\partial\psi^1}{\partial u^2}\,du^2 + \cdots$$

$$x_3^2 = \psi^2(u^1 + du^1, u^2 + du^2)$$

$$= \psi^2(u^1,u^2) + \frac{\partial\psi^2}{\partial u^1}\,du^1 + \frac{\partial\psi^2}{\partial u^2}\,du^2 + \cdots$$

$$M_4:\quad x_4^1 = \psi^1(u^1 + du^1, u^2) = \psi^1(u^1,u^2) + \frac{\partial\psi^1}{\partial u^1}\,du^1 + \cdots$$

$$x_4^2 = \psi^2(u^1 + du^1, u^2) = \psi^2(u^1,u^2) + \frac{\partial\psi^2}{\partial u^1}\,du^1 + \cdots$$

It can be seen that the area of the parallelogram $M_1M_2M_3M_4$ is, neglecting the third-order infinitesimal terms,

$$dS = |x_1^1(x_2^2 - x_3^2) - x_1^2(x_2^1 - x_3^1) + x_2^1x_3^2 - x_3^1x_2^2|$$

and by using the preceding relations,

$$dS = \left|\frac{\partial\psi^1}{\partial u^2}\frac{\partial\psi^2}{\partial u^1} - \frac{\partial\psi^1}{\partial u^1}\frac{\partial\psi^2}{\partial u^2}\right|\,du^1\,du^2 \qquad \text{(A6.20)}$$

Comparing (A6.20) with (A6.15) we see that the surface element $du^1\, du^2$ can be replaced by the exterior product $du^1 \wedge du^2$.

A6.2 *k*-Forms and Dual

By generalizing the definition of the preceding section, the *k-form* can be expressed by

$$f = f_{i_1 \cdots i_k}\, dx^{i_1} \wedge dx^{i_2} \wedge \cdots \wedge dx^{i_k} \tag{A6.21}$$

in a space (variety) of dimension N. The form of maximal degree that can be formed is thus an N-form.

By definition, the *dual* or Hodge form is for a k-form $^*f$ such that

$$(^*f) = (^*f)_{i_1 \cdots i_{n-k}}\, dx^{i_1} \wedge \cdots \wedge dx^{i_{n-k}} \tag{A6.22}$$

with the component

$$(^*f)_{i_1 \cdots i_{n-k}} = \frac{1}{k!} \sqrt{\pm g(x)}\, \epsilon_{i_1 \cdots i_{n-k} j_1 \cdots j_{k_1}} f^{j_1 \cdots j_k} \tag{A6.23}$$

where $g(x)$ is the metric, euclidean or not.

As f is a k-form of a space of dimension N,

$$f \in \Lambda^k(M)$$

The dual form belongs to $\Lambda^{N-k}(M)$:

$$\Lambda^k(M) \xrightarrow{\ *\ } \Lambda^{N-k}(M)$$

It is used in electromagnetism and in gauge theory.

A6.3 Differential of a *k*-Form

The k-form

$$f = f_{i_1 \cdots i_k}\, dx^{i_1} \wedge dx^{i_2} \wedge \cdots \wedge dx^{i_k} \tag{A6.21}$$

differentiates into

$$df = \frac{\partial f_{i_1 \cdots i_k}}{\partial x_p}\, dx^p \wedge dx^{i_1} \wedge dx^{i_2} \wedge \cdots \wedge dx^{i_k} \tag{A6.24}$$

It is thus a $(k + 1)$-form:

$$\Lambda^k(M) \xrightarrow{\ \partial\ } \Lambda^{k+1}(M)$$

A6.4 Poincaré Theorem

According to (A6.24), we see that

$$d^2f = \frac{\partial^2 f_{i_1\cdots i_k}}{\partial x^q \, \partial x^p} \, dx^q \wedge dx^p \wedge dx^{i_1} \wedge dx^{i_2} \wedge \cdots \wedge dx^{i_k} \tag{A6.25}$$

but also (property of the derivatives)

$$d^2f = \frac{\partial^2 f_{i_1\cdots i_k}}{\partial x^p \, \partial x^q} \, dx^q \wedge dx^p \wedge dx^{i_1} \wedge dx^{i_2} \wedge \cdots \wedge dx^{i_k} \tag{A6.26}$$

By exchanging q and p in (A6.25), we find that

$$d^2f = \frac{\partial^2 f_{i_1\cdots i_k}}{\partial x^p \, \partial x^q} \, dx^p \wedge dx^q \wedge dx^{i_1} \wedge dx^{i_2} \wedge \cdots \wedge dx^{i_k} \tag{A6.27}$$

In this expression the order of dx^p and dx^q is reversed:

$$d^2f = -\frac{\partial^2 f_{i_1\cdots i_k}}{\partial x^p \, \partial x^q} \, dx^q \wedge dx^p \wedge dx^{i_1} \wedge dx^{i_2} \wedge \cdots \wedge dx^{i_k} \tag{A6.28}$$

Compare (A6.26) with (A6.28); it is seen that

$$d^2f = 0 \tag{A6.29}$$

which is Poincaré's theorem, the theorem of Liouville in physics, or the theorem of the preservation of phase space with $df = dq \wedge dp$.

A6.5 Stokes' Theorem

We begin with the linear 1-form

$$f = f_i \, dx^i \tag{A6.30}$$

We are going to operate in a space of dimension 2, that is, such that $N = 2$ and

$$f = f_x \, dx + f_y \, dy \tag{A6.31}$$

Based on our earlier discussion, we can write

$$\begin{aligned} df &= df_x \wedge dx + df_y \wedge dy \\ &= \left(\frac{\partial f_x}{\partial x} dx + \frac{\partial f_x}{\partial y} dy\right) \wedge dx + \left(\frac{\partial f_y}{\partial x} dx + \frac{\partial f_y}{\partial y} dy\right) \wedge dy \\ &= \frac{\partial f_x}{\partial y} dy \wedge dx + \frac{\partial f_y}{\partial x} dx \wedge dy = \left(\frac{\partial f_y}{\partial x} - \frac{\partial f_x}{\partial y}\right) dx \, dy \\ &= (\vec{\nabla} \wedge \vec{f}) \cdot d\vec{S} \end{aligned} \tag{A6.32}$$

as we ought to have

$$\int_{\Gamma} f = \iint_{\mathbf{D}} df \tag{A6.33}$$

Stokes' theorem can be deduced from this:

$$\int_{\Gamma} (f_x\, dx + f_y\, dy) = \iint_{\mathbf{D}} (\vec{\nabla} \wedge \vec{f})\cdot d\vec{S} \tag{A6.34}$$

To clarify the vocabulary used, according to (A6.33), Γ is simply $\partial\mathbf{D}$. If f is a k-form, df is a $(k + 1)$-form and $\mathbf{D}$ is called a $k + 1$-chain, $\partial\mathbf{D}$ a k-chain, or in symbols,

$$\mathbf{D}\ (k + 1 \text{ chain}) \longleftrightarrow f\ (\text{cochain})$$

If $\mathbf{D}$ is identical to $\partial\alpha$, we can refer to the edge, and f is an exact, total differential or co-edge $d\omega$:

$$\partial\alpha \longleftrightarrow d\omega$$

If $\partial\mathbf{D} = 0$, we are said to be dealing with a cycle and $df = 0$; we are referring to a closed form or cocycle

$$\partial\mathbf{D} = 0 \longleftrightarrow df = 0$$

If $\mathbf{D}_1 - \mathbf{D}_2 = \partial\alpha$ (homologous chain), then $f_1 - f_2 = d\omega$ (cohomologous form).

3

Covariant Formulation of Electromagnetism

3.1 CLASSICAL ELECTROMAGNETISM

If (C) is a closed curve, write (S) as any surface that leans on the curve (C) and $\vec{n}$ the normal on the surface (S) (cf. Figure 10). A positive orientation on the curve (C) is chosen; it is such that if $\vec{n}$ is taken for axis Oz in a trirectangular trihedron, the direction of circulation on (C) points the direct way in the plane xOy (counterclockwise).

If the distribution of the currents, which encompasses the curve (C), is not given by the value of the current I intensities but by the density of the current, a vector denoted $\vec{j}$, the total intensity encompassed by (C) is

$$I = \iint_S \vec{j}\cdot\overrightarrow{dS} \tag{3.1}$$

by definition of $\vec{j}$; obviously,

$$\overrightarrow{dS} = \vec{n}\, dS$$

where dS is a surface element of S. The currents cutting (S) and which go the same way as $\vec{n}$ are counted positively; those which go the opposite way are counted negatively.

The density of the current $\vec{j}$ is linked to the charge density ρ by

$$\vec{j} = \rho\vec{v} \tag{3.2}$$

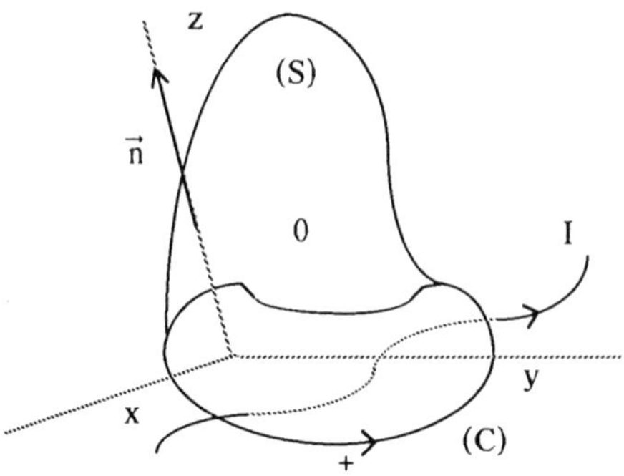

Figure 10 Conventions.

where the vector $\vec{v}$ is the drag velocity of the charge carriers and ρ is the charge density in the volume τ defined by (C) and (S):

$$Q = \iiint_{\tau} \rho \, d\tau \tag{3.3}$$

The electrical induction vector $\vec{D}$ is evaluated from the total charge Q with the help of Gauss's theorem:

$$\iint \vec{D} \cdot \vec{dS} = Q \tag{3.4}$$

Q measures the flux of vector $\vec{D}$ across the surface S. Equation (3.4) can easily be put in differential form:

$$\vec{\nabla} \cdot \vec{D} = \rho \tag{3.5}$$

The electrical induction $\vec{D}$ enables us to define the tensor of the ϵ permeabilities (rank zero tensor for the isotrope materials and a rank 2 tensor for the anisotrope materials):

$$\vec{D} = \epsilon \vec{E} \tag{3.6}$$

where $\vec{E}$ is the electrical field applied to the carriers. The experiment of the "broken magnets" shows that there is no magnetic monopole. The equivalent of (3.5) is

$$\vec{\nabla} \cdot \vec{B} = 0 \tag{3.7}$$

where $\vec{B}$ is the magnetic induction vector such that

$$\vec{B} = \mu \vec{H} \tag{3.8}$$

μ is the tensor of permeability (of rank zero for the isotrope materials and the rank 2 for the anisotrope materials), and H is the magnetic field.

Ampère's theorem links the field $\overset{\leftrightarrow}{H}$ to the currents encompassed by (C):

$$\int_{(C)} \overset{\leftrightarrow}{H} \cdot \overrightarrow{dl} = I \tag{3.9}$$

or

$$\vec{\nabla} \wedge \overset{\leftrightarrow}{H} = \vec{j} \tag{3.10}$$

Relation (3.10) must be dealt with carefully. The density of the current $\vec{j}$ is the sum of the densities of the current crossing (S).

A variable electromagnetic field in time induces a magnetic field, since according to Gauss's theorem (3.4),

$$\iint_S \frac{\partial \vec{D}}{\partial t} \cdot \overrightarrow{dS} = \frac{dQ_D}{dt} = I_D = \iint_S \vec{j}_D \cdot \overrightarrow{dS}$$

or

$$\vec{j}_D = \frac{\partial \vec{D}}{\partial t} \tag{3.11}$$

Equation (3.10) is usually expressed as

$$\vec{\nabla} \wedge \overset{\leftrightarrow}{H} = \vec{j} + \frac{\partial \vec{D}}{\partial t} \tag{3.12}$$

Lenz's law or the *cut flux law* links the temporal variation of the flux to the counterelectromotive force induced:

$$e = -\frac{\partial \Phi}{\partial t} \tag{3.13}$$

where

$$\Phi = \iint_S \overset{\leftrightarrow}{B}\, d\vec{S} \tag{3.14}$$

Φ is the flux of vector $\overset{\leftrightarrow}{B}$ across the surface S. As e is a counterelectromotive force, it can be written that

$$e = -V = -\int_{(C)} dV = -\int_{(C)} \vec{\nabla} V \cdot d\vec{r} = -\iint_{(S)} \frac{\partial \overset{\leftrightarrow}{B}}{\partial t} \cdot d\vec{S} \tag{3.15}$$

As

$$\vec{E} = -\vec{\nabla} V \tag{3.16}$$

where V is the scalar potential from which is derived the electrical field $\vec{E}$, (3.15) gives

$$\int_{(C)} \vec{E} \cdot d\vec{r} = -\iint_{(S)} \frac{\partial \vec{B}}{\partial t} \cdot d\vec{S}$$

or

$$\vec{\nabla} \wedge \vec{E} = -\frac{\partial \vec{B}}{\partial t} \tag{3.17}$$

The curl of the gradient of a scalar function f on the point is still zero:

$$\vec{\nabla} \wedge (\vec{\nabla} f) = \vec{0} \tag{3.18}$$

In view of (3.17) it is clearly advisable to write

$$\vec{B} = \vec{\nabla} \wedge \vec{A} \tag{3.19}$$

where $\vec{A}$ is a vector potential. The substitution in (3.17) leads to

$$\vec{\nabla} \wedge \left(\vec{E} + \frac{\partial \vec{A}}{\partial t} \right) = \vec{0} \tag{3.20}$$

or

$$\vec{E} + \frac{\partial \vec{A}}{\partial t} = \vec{\nabla} f \tag{3.21}$$

[cf. (3.18)], and as for $\vec{B} = \vec{0}$ we should have $\vec{A} = \vec{0}$ and that $\vec{E}$ complies with (3.16):

$$f = -V \tag{3.22}$$

and

$$\vec{E} = -\vec{\nabla} V - \frac{\partial \vec{A}}{\partial t} \tag{3.23}$$

3.2 EQUATIONS OF PROPAGATION OF ELECTROMAGNETIC FIELD

We begin with (3.12) multiplied by μ:

$$\vec{\nabla} \wedge \vec{B} = \mu \vec{j} + \mu \frac{\partial \vec{D}}{\partial t} \tag{3.24}$$

Now [cf. (3.19)]

$$\vec{\nabla} \wedge \vec{B} = \vec{\nabla} \wedge \vec{\nabla} \wedge \vec{A} = -\nabla^2\vec{A} + \vec{\nabla}(\vec{\nabla}\cdot\vec{A}) \tag{3.25}$$

and with the help of (3.23), equation (3.24) becomes

$$-\nabla^2\vec{A} + \vec{\nabla}(\vec{\nabla}\cdot\vec{A}) = \mu\vec{j} + \mu\epsilon\frac{\partial\vec{E}}{\partial t} = \mu\vec{j} - \mu\epsilon\vec{\nabla}\left(\frac{\partial V}{\partial t}\right) - \mu\epsilon\frac{\partial^2\vec{A}}{\partial t^2}$$

which we write

$$\nabla^2\vec{A} - \mu\epsilon\frac{\partial^2\vec{A}}{\partial t^2} = -\mu\vec{j} + \vec{\nabla}\left(\vec{\nabla}\cdot\vec{A} + \mu\epsilon\frac{\partial V}{\partial t}\right) \tag{3.26}$$

Now the equation

$$\vec{B} = \vec{\nabla} \wedge \vec{A} \tag{3.19}$$

does not make a unique choice of vector potential $\vec{A}$ but defines a family.

The curl of a gradient still being zero [cf. (3.18)], if $\vec{A}$ is the solution for (3.19), $\vec{A}'$ is such that

$$\vec{A}' = \vec{A} + \vec{\nabla}\chi \tag{3.27}$$

where χ is any scalar function of the point, is also the solution of (3.19).

We may thus impose an additional condition, said to be due to Maxwell:

$$\vec{\nabla}\cdot\vec{A} + \mu\epsilon\frac{\partial V}{\partial t} = 0 \tag{3.28}$$

It specifies the choice of the function χ but also and most important, it simplifies (3.26):

$$\nabla^2\vec{A} - \mu\epsilon\frac{\partial^2\vec{A}}{\partial t^2} = -\mu\vec{j} \tag{3.29}$$

This is the equation of propagation of the vector potential. That of scalar potential V is obtained from Gauss's theorem (3.5):

$$\epsilon\vec{\nabla}\cdot\vec{E} = \rho \tag{3.30}$$

and with (3.23),

$$\epsilon\nabla^2 V + \epsilon\frac{\partial}{\partial t}(\vec{\nabla}\cdot\vec{A}) = -\rho \tag{3.31}$$

The identity of (3.28) enables us to obtain the equation

$$\epsilon\left(\nabla^2 V - \mu\epsilon\frac{\partial^2 V}{\partial t^2}\right) = -\rho \tag{3.32}$$

In a vacuum

$$\epsilon \to \epsilon_0 \quad \text{(a scalar)}$$

$$\mu \to \mu_0 \quad \text{(a scalar)}$$

with

$$\mu_0\epsilon_0 = \frac{1}{c^2} \tag{3.33}$$

and the former equations of propagation become [cf. (3.29) and (3.32)]

$$\left(\frac{1}{c^2}\frac{\partial^2}{\partial t^2} - \nabla^2\right) V = \frac{\rho}{\epsilon_0} = \Box V \tag{3.34a}$$

$$\left(\frac{1}{c^2}\frac{\partial^2}{\partial t^2} - \nabla^2\right) A = \mu_0\vec{j} = \Box\vec{A} \tag{3.34b}$$

where $\Box$ is the d'Alembertian:

$$\Box = \frac{1}{c^2}\frac{\partial^2}{\partial t^2} - \nabla^2 \tag{3.35}$$

3.3 QUADRIVECTOR DENSITY OF CURRENT AND POTENTIAL

We have seen that the energy–momentum quadrivector is such that

$$P^\mu = \left(\frac{E}{c^2}, \vec{p}\right) = (m,\vec{p})$$

since

$$E = mc^2$$

Now, by definition of the momentum $\vec{p}$,

$$\vec{p} = m\vec{v}$$

a relation of the same kind links the current density to the charge density:

$$\vec{j} = \rho\vec{v}$$

A quadrivector can therefore be defined from ρ and $\vec{j}$; we take

$$J^\mu = \left(\frac{\rho}{c\epsilon}, \mu\vec{j}\right) \tag{3.36}$$

The quadrivector potential is defined by

$$A^{\mu} = \left(\frac{V}{c}, \vec{A}\right) \tag{3.37}$$

It is seen that

$$(J^2) = J_{\mu}J^{\mu} = \frac{\rho^2}{c^2\epsilon^2} - \mu^2\vec{j}^{\,2} \tag{3.38}$$

$$(A)^2 = A_{\mu}A^{\mu} = \frac{V^2}{c^2} - \vec{A}^2 \tag{3.39}$$

Remember that in our notation

$$A^{\mu} = (A_0, A_x, A_y, A_z) \tag{3.40}$$

$$A_{\mu} = (A_0, -A_x, -A_y, -A_z)$$

$$\nabla^{\mu} = \left(\frac{1}{c}\frac{\partial}{\partial t}, -\partial_i\right) = \left(\frac{1}{c}\frac{\partial}{\partial t}, -\frac{\partial}{\partial x}, -\frac{\partial}{\partial y}, -\frac{\partial}{\partial z}\right) \tag{3.41}$$

$$\nabla_{\mu} = \left(\frac{1}{c}\frac{\partial}{\partial t}, \partial_i\right) = \left(\frac{1}{c}\frac{\partial}{\partial t}, \frac{\partial}{\partial x}, \frac{\partial}{\partial y}, \frac{\partial}{\partial z}\right)$$

$$\Box = \partial_{\mu}\partial^{\mu} = \frac{1}{c^2}\frac{\partial^2}{\partial t^2} - \nabla^2 \tag{3.42}$$

3.4 FARADAY'S ELECTROMAGNETIC TENSOR

By definition, Faraday's electromagnetic tensor will be called the tensor F, of which the contravariant components $F^{\mu\nu}$ are given from the quadrivector potential by

$$F^{\mu\nu} = \partial^{\mu}A^{\nu} - \partial^{\nu}A^{\mu} \tag{3.43}$$

It is obvious that whether $\mu = 0, 1, 2,$ or 3,

$$F^{\mu\mu} = 0 \tag{3.44}$$

According to (3.37),

$$F^{01} = \partial^0 A^1 - \partial^1 A^0 = \frac{1}{c}\frac{\partial}{\partial t}A_x + \frac{\partial}{\partial x}\left(\frac{V}{c}\right) = -\frac{E_x}{c} = -\frac{E^1}{c}$$

$$F^{01} = -\frac{E^1}{c} \quad F^{02} = -\frac{E^2}{c} \quad F^{03} = -\frac{E^3}{c}$$

$$F^{10} = \frac{E^1}{c} \quad F^{12} = -B^3 \quad F^{13} = B^2$$

$$F^{20} = \frac{E^2}{c} \qquad F^{21} = B^3 \qquad F^{23} = -B^1$$

$$F^{30} = \frac{E^3}{c} \qquad F^{31} = -B^2 \quad F^{32} = B^1$$

We used

$$\vec{B} = \vec{\nabla} \wedge \vec{A} \tag{3.19}$$

and

$$\vec{E} = -\vec{\nabla}V - \frac{\partial \vec{A}}{\partial t} \tag{3.23}$$

from which the tensor

$$(F^{\mu\nu}) = \begin{bmatrix} 0 & -\frac{E^1}{c} & -\frac{E^2}{c} & -\frac{E^3}{c} \\ \frac{E^1}{c} & 0 & -B^3 & B^2 \\ \frac{E^2}{c} & B^3 & 0 & -B^1 \\ \frac{E^3}{c} & -B^2 & B^1 & 0 \end{bmatrix} \tag{3.26}$$

It is clear according to our earlier expression of $(F^{\mu\nu})$ that

$$F^{\mu\nu} = -F^{\nu\mu} \tag{3.45}$$

an antisymmetrical relation that conforms to the definition (3.43). The quantity $F^{\mu\nu}$ transforms itself, as does a quadrivector, under the Lorentz transformations.

The $F^{\mu\nu}$ define a rank 2 tensor:

$$F^{\mu\nu} = \partial^\mu A^\nu - \partial^\nu A^\mu = \Lambda^\mu_{\mu'} \partial^{\mu'} \Lambda^\nu_{\nu'} A'^{\nu'} - \Lambda^\nu_{\nu'} \partial^{\nu'} \Lambda^\mu_{\mu'} A'^{\mu'}$$

$$= \Lambda^\mu_{\mu'} \Lambda^\nu_{\nu'} (\partial^{\mu'} A'^{\nu'} - \partial^{\nu'} A'^{\mu'})$$

a transformation of the components of an antisymmetrical rank 2 contravariant tensor.

3.5 LEVI–CIVITA TENSOR

It is denoted as ϵ, of rank N, and is totally antisymmetrical. It is written in the covariant form

$$\epsilon = \frac{1}{N!}\, \epsilon_{\alpha\beta\gamma\cdots\lambda}\, \vec{e}^{\alpha}\vec{e}^{\beta}\vec{e}^{\gamma} \cdots \vec{e}^{\lambda} \tag{3.46}$$

or for the covariant components

$$\epsilon_{\alpha\beta\gamma\cdots\lambda} = \epsilon\cdot \vec{e}_{\alpha}\vec{e}_{\beta}\vec{e}_{\gamma} \cdots \vec{e}_{\lambda} \tag{3.47}$$

By definition the Levi–Civita tensor changes sign by the permutation of two consecutive indexes:

$$\epsilon_{\alpha\gamma\beta\cdots\lambda} = -\epsilon_{\alpha\beta\gamma\cdots\lambda} \tag{3.48}$$

and is such that, for example, if $N = 3$,

$$\epsilon_{\alpha\beta\gamma} = \epsilon_{\beta\gamma\alpha} = \epsilon_{\gamma\alpha\beta} = 1 \tag{3.49}$$

(the number of permutations is even),

$$\epsilon_{\beta\alpha\gamma} = \epsilon_{\gamma\beta\alpha} = \epsilon_{\alpha\gamma\beta} = -1 \tag{3.50}$$

(the number of permutations is odd), and

$$\epsilon_{\alpha\alpha\gamma} = \epsilon_{\alpha\beta\beta} = \epsilon_{\alpha\beta\alpha} = 0 \tag{3.51}$$

(two indexes are identical).

For a tensor of rank 4

$$\epsilon_{\alpha\beta\gamma\delta} = \begin{cases} +1 & \text{for an even number of permutations of 2 successive indexes} \\ -1 & \text{for an odd number of permutations of 2 successive indexes} \\ 0 & \text{if at least 2 indexes are identical} \end{cases} \tag{3.52}$$

To be exact,

$$\epsilon_{0123} = \epsilon_{1203} = \epsilon_{2130} = \epsilon_{2301} = \epsilon_{3102} = 1$$

$$\epsilon_{1023} = \epsilon_{1230} = \epsilon_{2310} = \cdots = -1$$

However, similarly,

$$\begin{aligned} \epsilon_{\alpha\beta\gamma\delta} &= \eta_{\alpha\alpha'}\epsilon^{\alpha'}{}_{\beta\gamma\delta} = \eta_{\alpha\alpha'}\eta_{\beta\beta'}\epsilon^{\alpha'\beta'}{}_{\gamma\delta} = \eta_{\alpha\alpha'}\eta_{\beta\beta'}\eta_{\gamma\gamma'}\epsilon^{\alpha'\beta'\gamma'}{}_{\delta} \\ &= \eta_{\alpha\alpha'}\eta_{\beta\beta'}\eta_{\gamma\gamma'}\eta_{\delta\delta'}\epsilon^{\alpha'\beta'\gamma'\delta} \end{aligned} \tag{3.53}$$

or

$$\begin{aligned} \epsilon_{0123} &= \eta_{0\alpha'}\eta_{1\beta'}\eta_{2\gamma'}\eta_{3\delta'}\epsilon^{\alpha'\beta'\gamma'\delta'} = \eta_{00}\eta_{11}\eta_{22}\eta_{33}\epsilon^{0123} \\ &= -\epsilon^{0123} \end{aligned}$$

More generally,

$$\epsilon_{\alpha\beta\gamma\delta} = -\epsilon^{\alpha\beta\gamma\delta} \tag{3.54}$$

a relation between covariant and contravariant components of the tensor ϵ. Indeed, ϵ is a pseudotensor; the components of the tensor ϵ are identical in all the referentials stemming from unitary transformations.

3.6 DUAL OF ELECTROMAGNETIC TENSOR

The dual $^*F$ is defined from Faraday's electromagnetic tensor F by

$$^*F^{\mu\nu} = \tfrac{1}{2}\epsilon^{\mu\nu\rho\sigma}F_{\rho\sigma} \tag{3.55}$$

To clarify the tensor $^*F$ the components $F_{\mu\nu}$ must be evaluated. It is known that

$$F^{\mu}{}_{\nu} = \eta_{\nu\alpha}F^{\mu\alpha} \tag{3.56}$$

Otherwise,

$$F^0{}_0 = F^1{}_1 = F^2{}_2 = F^3{}_3 = 0$$

$$F^0{}_i = -F^{0i} = \frac{E^i}{c}$$

$$F^i{}_0 = F^{i0} = \frac{E^i}{c}$$

$$F^1{}_2 = -F^{12} = -F^2{}_1 = F^2{}_1 = B^3$$

$$F^1{}_3 = -F^{13} = -F^3{}_1 = F^{31} = -B^2$$

$$F^2{}_3 = -F^{23} = -F^3{}_2 = F^{32} = B^1$$

or

$$(F^{\mu}{}_{\nu}) = \begin{bmatrix} 0 & \frac{E^1}{c} & \frac{E^2}{c} & \frac{E^3}{c} \\ \frac{E^1}{c} & 0 & B^3 & -B^2 \\ \frac{E^2}{c} & -B^3 & 0 & B^1 \\ \frac{E^3}{c} & B^2 & -B^1 & 0 \end{bmatrix} \tag{3.57}$$

But equally,

$$F_{\mu\nu} = \eta_{\mu\alpha}\eta_{\nu\beta}F^{\alpha\beta} \tag{3.58}$$

To be exact,

$$F_{00} = F_{11} = F_{22} = F_{33} = 0$$

$$F_{0i} = -F^{0i} = \frac{E^i}{c}$$

$$F_{i0} = -F^{i0} = -\frac{E^i}{c}$$

$$F_{12} = F^{12} = -F_{21} = -F^{21} = -B^3$$

$$F_{13} = F^{13} = -F_{31} = -F^{31} = B^2$$

$$F_{32} = F^{32} = -F_{23} = -F^{23} = B^1$$

$$(F_{\mu\nu}) = \begin{bmatrix} 0 & \frac{E^1}{c} & \frac{E^2}{c} & \frac{E^3}{c} \\ -\frac{E^1}{c} & 0 & -B^3 & B^2 \\ -\frac{E^2}{c} & B^3 & 0 & -B^1 \\ -\frac{E^3}{c} & -B^2 & B^1 & 0 \end{bmatrix} \tag{3.59}$$

The tensor is antisymmetrical: $F_{\mu\nu} = -F_{\nu\mu}$. According to (3.55) and (3.54),

$$*F^{\alpha\alpha} = \tfrac{1}{2}\epsilon^{\alpha\alpha\rho\sigma}F_{\rho\sigma} = 0 \tag{3.60}$$

whatever α, or

$$*F^{00} = *F^{11} = *F^{22} = *F^{33} = 0 \tag{3.61}$$

and similarly,

$$*F^{01} = \tfrac{1}{2}(\epsilon^{0123}F_{23} + \epsilon^{0132}F^{32}) = -F_{23} = B^1$$

$$*F^{10} = \tfrac{1}{2}(\epsilon^{1023}F_{23} + \epsilon^{1032}F_{32}) = F_{23} = -B^1$$

or, more generally,

$$*F^{0i} = -*F^{i0} = B^i$$

It can easily be shown that

$$*F^{12} = -*F^{21} = -\frac{E^3}{c}$$

$$*F^{23} = -*F^{32} = -\frac{E^1}{c}$$

$$^*F^{31} = -{}^*F^{13} = -\frac{E^2}{c}$$

The dual tensor is thus represented by

$$({}^*F^{\mu\nu}) = \begin{bmatrix} 0 & B^1 & B^2 & B^3 \\ -B^1 & 0 & -\frac{E^3}{c} & \frac{E^2}{c} \\ -B^2 & \frac{E^3}{c} & 0 & -\frac{E^1}{c} \\ -B^3 & -\frac{E^2}{c} & \frac{E^1}{c} & 0 \end{bmatrix} \tag{3.62}$$

which means that the dual tensor is antisymmetrical and is deduced from Faraday's tensor by

$$F \xrightarrow{\left| \begin{matrix} E_i \rightarrow -cB_i \\ B_i \rightarrow \frac{E_i}{c} \end{matrix} \right|} {}^*F \tag{3.63}$$

From the elements of ${}^*F^{\mu\nu}$, the ${}^*F^{\mu}{}_{\nu}$, and ${}^*F_{\mu\nu}$ can be evaluated. With the help of

$$^*F^{\mu}{}_{\nu} = \eta_{\nu\alpha}{}^*F^{\mu\alpha}$$

we obtain

$$^*F^{\alpha}{}_{\alpha} = 0$$

$$^*F^{0}{}_{i} = -{}^*F^{i}{}_{0} = -{}^*F^{0i}$$

$$^*F^{1}{}_{2} = -{}^*F^{2}{}_{1} = -{}^*F^{12}$$

$$^*F^{2}{}_{3} = -{}^*F^{3}{}_{2} = -{}^*F^{23}$$

$$^*F^{1}{}_{3} = -{}^*F^{3}{}_{1} = -{}^*F^{13}$$

and

$$({}^*F^{\mu}{}_{\nu}) = \begin{bmatrix} 0 & -B^1 & -B^2 & -B^3 \\ B^1 & 0 & \frac{E^3}{c} & -\frac{E^2}{c} \\ B^2 & -\frac{E^3}{c} & 0 & \frac{E^1}{c} \\ B^3 & \frac{E^2}{c} & -\frac{E^1}{c} & 0 \end{bmatrix} \tag{3.64}$$

In the same way,

$$^*F_{\mu\nu} = \eta_{\mu\beta}\eta_{\nu\alpha}{}^*F^{\beta\alpha}$$

and the explicit calculation shows that

$$^*F_{\alpha\alpha} = 0$$

$$^*F_{0i} = -{}^*F_{i0} = -{}^*F^{0i}$$

$$^*F_{12} = -{}^*F_{21} = {}^*F^{12}$$

$$^*F_{23} = -{}^*F_{32} = {}^*F^{23}$$

$$^*F_{13} = -{}^*F_{31} = {}^*F^{13}$$

and thus

$$({}^*F_{\mu\nu}) = \begin{bmatrix} 0 & -B^1 & -B^2 & -B^3 \\ B^1 & 0 & -\dfrac{E^3}{c} & \dfrac{E^2}{c} \\ B^2 & \dfrac{E^3}{c} & 0 & -\dfrac{E^1}{c} \\ B^3 & -\dfrac{E^2}{c} & \dfrac{E^1}{c} & 0 \end{bmatrix} \tag{3.65}$$

It can easily be shown that

$$\tfrac{1}{2}\epsilon^{\mu\nu\rho\sigma *}F_{\rho\sigma} = -F^{\mu\nu} \tag{3.66a}$$

or

$$^*({}^*F) = -F \tag{3.66b}$$

Indeed,

$$\frac{1}{2}\,\epsilon^{01\rho\sigma *}F_{\rho\sigma} = \frac{1}{2}\,(\epsilon^{0123 *}F_{23} + \epsilon^{0132 *}F_{32}) = -{}^*F_{23} = -\frac{E_1}{c} = -F^{01}$$

$$\frac{1}{2}\,\epsilon^{12\rho\sigma *}F_{\rho\sigma} = \frac{1}{2}\,(\epsilon^{1230 *}F_{30} + \epsilon^{1203 *}F_{03}) = {}^*F_{30} = B^3 = -F^{12}$$

and so on.

3.7 RELATIVISTIC INVARIANTS

From the fields $\vec{E}$ and $\vec{H}$ we can easily obtain the relativistic invariants. However, they can only be a function of the quantities $F_{\mu\nu}F^{\mu\nu}$ and $F_{\mu\nu}{}^*F^{\mu\nu}$.

3.7.1 Pseudonorm of Faraday Tensor

We have

$$F^2 = F_{\mu\nu}F^{\mu\nu} = -F_{\mu\nu}F^{\nu\mu} = -\sum_{\mu} (F^2)_{\mu\mu} = -\text{trace}(F^2) \tag{3.67}$$

Now

$$F^2 = (F_{\mu\nu})(F^{\mu\nu})$$
$$= \begin{bmatrix} 0 & \frac{E^1}{c} & \frac{E^2}{c} & \frac{E^3}{c} \\ -\frac{E^1}{c} & 0 & -B^3 & B^2 \\ -\frac{E^2}{c} & B^3 & 0 & -B^1 \\ -\frac{E^3}{c} & -B^2 & B^1 & 0 \end{bmatrix} \begin{bmatrix} 0 & -\frac{E^1}{c} & -\frac{E^2}{c} & -\frac{E^3}{c} \\ \frac{E^1}{c} & 0 & -B^3 & B^2 \\ \frac{E^2}{c} & B^3 & 0 & -B^1 \\ \frac{E^3}{c} & -B^2 & B^1 & 0 \end{bmatrix}$$

and (3.67) becomes

$$\begin{aligned} F_{\mu\nu}F^{\mu\nu} &= -\left\{\left[\left(\frac{E^1}{c}\right)^2 + \left(\frac{E^2}{c}\right)^2 + \left(\frac{E^3}{c}\right)^2\right]\right. \\ &\quad + \left[\left(\frac{E^1}{c}\right)^2 - (B^3)^2 - (B^2)^2\right] \\ &\quad + \left[\left(\frac{E^2}{c}\right)^2 - (B^3)^2 - (B^1)^2\right] \\ &\quad \left. + \left[\left(\frac{E^3}{c}\right)^2 - (B^2)^2 - (B^1)^2\right]\right\} \\ &= -2\left(\frac{E^2}{c^2} - B^2\right) \end{aligned} \tag{3.68}$$

3.7.2 Scalar Product $F \cdot {}^*F$

The scalar (dot) product between Faraday's vector and its dual is expressed by

$$F \cdot F^* = F_{\mu\nu}{}^*F^{\mu\nu} = -F_{\mu\nu}{}^*F^{\nu\mu} = -\sum_{\mu} (F^*F)_{\mu\mu} = -\text{tr}(F^*F) \tag{3.69}$$

As

$$(F_{\mu\nu})(^*F^{\mu\nu}) = \begin{bmatrix} 0 & \frac{E^1}{c} & \frac{E^2}{c} & \frac{E^3}{c} \\ -\frac{E^1}{c} & 0 & -B^3 & B^2 \\ -\frac{E^2}{c} & B^3 & 0 & -B^1 \\ -\frac{E^3}{c} & -B^2 & B^1 & 0 \end{bmatrix} \begin{bmatrix} 0 & B^1 & B^2 & B^3 \\ -B^1 & 0 & -\frac{E^3}{c} & \frac{E^2}{c} \\ -B^2 & \frac{E^3}{c} & 0 & -\frac{E^1}{c} \\ -B^3 & -\frac{E^2}{c} & \frac{E^1}{c} & 0 \end{bmatrix}$$

(3.69) gives

$$F\cdot{}^*F = F_{\mu\nu}{}^*F^{\mu\nu} = 4\,\frac{\vec{E}\cdot\vec{B}}{c} \tag{3.70}$$

3.8 COVARIANT ELECTROMAGNETISM

By the definition of Faraday's tensor,

$$F^{\mu\nu} = \partial^\mu A^\nu - \partial^\nu A^\mu \tag{3.43}$$

from which it can be deduced that

$$\begin{aligned}\partial^\alpha F^{\beta\gamma} + \partial^\beta F^{\gamma\alpha} + \partial^\gamma F^{\alpha\beta} &= \partial^{\alpha+\beta}A^\gamma - \partial^{\alpha\gamma}A^\beta + \partial^{\beta+\gamma}A^\alpha \\ &\quad - \partial^{\beta+\alpha}A^\gamma + \partial^{\gamma+\alpha}A^\beta - \partial^{\gamma+\beta}A\alpha \\ &= 0\end{aligned} \tag{3.71}$$

Riemann's curvature tensor (Section 1.11.16)

$$R^\rho{}_{\sigma\nu\mu} = \partial_\mu\Gamma^\rho_{\sigma\nu} - \partial_\nu\Gamma^\rho_{\sigma\mu} + \Gamma^\rho_{\beta\mu}\Gamma^\beta_{\sigma\nu} - \Gamma^\rho_{\beta\nu}\Gamma^\beta_{\sigma\mu}$$

complies with a condition equivalent to (3.71):

$$\partial_\alpha R^\rho{}_{\sigma\nu\mu} + \partial_\mu R^\rho{}_{\sigma\alpha\nu} + \partial_\nu R^\rho{}_{\sigma\mu\alpha} = 0$$

called *Bianchi's identity*. This means that Faraday's tensor represents the curvature tensor of the electromagnetic field.

3.8.1 Equation of Charge Preservation

According to Ampère's theorem,

$$\vec{\nabla}\wedge\vec{H} = \vec{j} + \frac{\partial\vec{D}}{\partial t} \tag{3.12}$$

Now

$$\vec{\nabla}\cdot(\vec{\nabla} \wedge \vec{H}) = 0$$

from which

$$\vec{\nabla}\cdot\vec{j} + \frac{\partial}{\partial t}\,\vec{\nabla}\cdot\vec{D} = 0$$

and with Gauss's theorem,

$$\vec{\nabla}\cdot\vec{j} + \frac{\partial \rho}{\partial t} = 0 \tag{3.72}$$

an equation of charge preservation, which is going to be written, for the moment, with the help of quadrivector density of current J^{μ}:

$$J^{\mu} = \left(\frac{\rho}{c\epsilon_0},\ \mu_0\,\vec{j}\right)$$

in a vacuum. The evaluation of $\partial_{\mu}J^{\mu}$ gives

$$\begin{aligned}\partial_{\mu}J^{\mu} &= \partial_0 J^0 + \partial_1 J^1 + \partial_2 J^2 + \partial_3 J^3 \\ &= \frac{1}{c}\frac{\partial}{\partial t}\left(\frac{\rho}{c\epsilon_0}\right) + \frac{\partial}{\partial x}(\mu_0 j_x) + \frac{\partial}{\partial y}(\mu_0 j_y) + \frac{\partial}{\partial z}(\mu_0 j_z) \\ &= \frac{1}{\epsilon_0 c^2}\frac{\partial \rho}{\partial t} + \mu_0\,\vec{\nabla}\cdot\vec{j} = \mu_0\left(\vec{\nabla}\cdot\vec{j} + \frac{\partial \rho}{\partial t}\right)\end{aligned}$$

which leads to writing, equivalent to (3.72),

$$\partial_{\mu}J^{\mu} = 0 \tag{3.73}$$

3.8.2 Theorems of Gauss and Ampère

Evaluate the quantity $\partial^{\mu}F_{\mu\nu}$. For

$$\nu = 0,$$

$$\partial^{\mu}F_{\mu 0} = \partial^0 F_{00} + \partial^1 F_{10} + \partial^2 F_{20} + \partial^3 F_{30}$$

and with the help of (3.59),

$$\partial^{\mu}F_{\mu 0} = -\frac{\partial}{\partial x}\left(-\frac{E_x}{c}\right) - \frac{\partial}{\partial y}\left(-\frac{E_y}{c}\right) - \frac{\partial}{\partial z}\left(-\frac{E_z}{c}\right) = \frac{1}{c}\,\vec{\nabla}\cdot\vec{E} \tag{3.74}$$

Now Gauss's theorem is written in a vacuum,

$$\epsilon_0\vec{\nabla}\cdot\vec{E} = \rho \tag{3.5}$$

and the equivalent is [cf. (3.74)]

$$\partial^\mu F_{\mu 0} = \frac{\rho}{\epsilon_0 c} = J_0 \tag{3.75}$$

For $\nu = 1$,

$$\partial^\mu F_{\mu 1} = \frac{1}{c^2}\frac{\partial E_x}{\partial t} + \frac{\partial B_y}{\partial z} - \frac{\partial B_z}{\partial y} = \frac{1}{c^2}\frac{\partial E_x}{\partial t} - (\vec{\nabla} \wedge \overset{\hookrightarrow}{B})_x \tag{3.76}$$

Ampère's theorem (3.12) gives by multiplying the two members by μ_0,

$$\vec{\nabla} \wedge \overset{\hookrightarrow}{B} = \mu_0 \vec{j} + \frac{1}{c^2}\frac{\partial \vec{E}}{\partial t} \tag{3.12}$$

and (3.76) becomes

$$\partial^\mu F_{\mu 1} = -\mu_0 j_x = J_1$$

More generally,

$$\partial^\mu F_{\mu i} = J_i$$

and [cf. (3.65)]

$$\partial^\mu F_{\mu\nu} = J_\nu \tag{3.77}$$

a relation that brings together Gauss's and Ampère's theorems in a covariant formulation.

3.8.3 Lorentz Laws

We clarify the quantity $\partial^{\mu *}F_{\mu\nu}$. For

$$\nu = 0,$$

$$\partial^{\mu *}F_{\mu 0} = \partial^{0*}F_{00} + \partial^{1*}F_{10} + \partial^{2*}F_{20} + \partial^{3*}F_{30}$$

and with the help of (3.65),

$$\partial^{\mu *}F_{\mu 0} = -\frac{\partial}{\partial x}(B_x) - \frac{\partial}{\partial y}(B_y) - \frac{\partial}{\partial z}(B_z) = -\vec{\nabla}\cdot\overset{\hookrightarrow}{B} \tag{3.78}$$

and with (3.7),

$$\partial^{\mu *}F_{\mu 0} = 0 \tag{3.79}$$

For, $\nu = 1$,

$$\partial^{\mu *}F_{\mu 1} = -\frac{1}{c}\frac{\partial B_x}{\partial t} + \frac{1}{c}\left(\frac{\partial E_y}{\partial z} - \frac{\partial E_z}{\partial y}\right)$$

$$= -\frac{1}{c}\frac{\partial B_x}{\partial t} - \frac{1}{c}(\vec{\nabla} \wedge \vec{E})_x \tag{3.80}$$

According to (3.17),

$$\vec{\nabla} \wedge \vec{E} = -\frac{\partial \vec{B}}{\partial t}$$

which means that [cf. (3.80)], whatever i may be,

$$\partial^{\mu *}F_{\mu i} = 0 \tag{3.81}$$

or more generally [in terms of (3.79)],

$$\partial^{\mu *}F_{\mu\nu} = 0 \tag{3.82}$$

3.8.4 Maxwell Condition

The quadrivector potential

$$A^\mu = \left(\frac{V}{c}, \vec{A}\right) \tag{3.37}$$

enables the evaluation of

$$\partial^\mu A_\mu = \partial^0 A_0 + \partial^1 A_1 + \partial^2 A_2 + \partial^3 A_3$$

$$= \frac{1}{c}\frac{\partial}{\partial t}\left(\frac{V}{c}\right) + \frac{\partial}{\partial x}A_x + \frac{\partial}{\partial y}A_y + \frac{\partial}{\partial z}A_z \tag{3.83}$$

$$= \frac{1}{c^2}\frac{\partial V}{\partial t} + \vec{\nabla}\cdot\vec{A}$$

The relation (3.28) is thus equivalent to

$$\partial^\mu A_\mu = 0 \tag{3.84}$$

3.8.5 Equations of Propagation

The d'Alembertian

$$\Box A_\mu = \partial_\nu \partial^\nu A_\mu = \frac{1}{c^2}\frac{\partial^2}{\partial t^2}A_\mu - \nabla^2 A_\mu$$

is expressed by

$\mu = 0$: $\Box A_0 = \Box(V/c) = (1/c)\Box V$ with (3.34a), $\Box V = \rho/\epsilon_0$ in a vacuum,

or

$$\Box A_0 = \frac{\rho}{c\epsilon_0} = J_0 \tag{3.85}$$

$\mu = 1$: $\Box A_1 = \Box(-A_x)$, and with (3.34b), $\Box A_x = \mu_0 j_x$ in a vacuum, or

$$\Box A_1 = -\mu_0 j_x = J_1 \tag{3.86}$$

Equations (3.85) and (3.86) become

$$\Box A_\mu = \partial_\nu \partial^\nu A_\mu = J_\mu \tag{3.87}$$

3.8.6 Poynting Vector

Consider the case of a charge q in motion in a electromagnetic field $(\vec{E}, \overset{\hookrightarrow}{H})$. The force applied is written

$$\vec{F} = q(\vec{E} + \vec{v} \wedge \overset{\hookrightarrow}{B}) \tag{3.88}$$

and during the time interval dt, the work provided by this force

$$dT = \vec{F} \cdot d\vec{r} = \vec{F} \cdot \vec{v}\, dt = q\vec{E} \cdot \vec{v}\, dt \tag{3.89}$$

The power given is

$$P = \frac{dT}{dt} = q\vec{E} \cdot \vec{v} = \rho V \vec{E} \cdot \vec{v} = V\vec{j} \cdot \vec{E} \tag{3.90}$$

which means that $\vec{j} \cdot \vec{E}$ is the power by unity of volume (power dissipated by the Joule effect).

According to Maxwell's equations,

$$\vec{\nabla} \wedge \vec{E} = -\frac{\partial \overset{\hookrightarrow}{B}}{\partial t} \tag{3.17}$$

$$\vec{\nabla} \wedge \overset{\hookrightarrow}{H} = \vec{j} + \frac{\partial \vec{D}}{\partial t} \tag{3.12}$$

and the relation, easy to confirm, is

$$-\vec{\nabla} \cdot (\vec{E} \wedge \overset{\hookrightarrow}{H}) = \vec{E} \cdot (\vec{\nabla} \wedge \overset{\hookrightarrow}{H}) - \overset{\hookrightarrow}{H} \cdot (\vec{\nabla} \wedge \vec{E}) \tag{3.91}$$

It is seen that

$$-\vec{\nabla} \cdot (\vec{E} \wedge \overset{\hookrightarrow}{H}) = \vec{j} \cdot \vec{E} + \epsilon_0 \vec{E} \cdot \frac{\partial \vec{E}}{\partial t} + \mu_0 \overset{\hookrightarrow}{H} \cdot \frac{\partial \overset{\hookrightarrow}{H}}{\partial t}$$

or, by calling $\vec{P}$ the radiant vector or Poynting's vector,

$$\vec{P} = \vec{E} \wedge \overset{\hookrightarrow}{H} \tag{3.92}$$

and w the density of energy of the electromagnetic field in a vaccum,

$$w = \tfrac{1}{2}(\epsilon_0 E^2 + \mu_0 H^2) \tag{3.93}$$

it can be seen that

$$\vec{\nabla}\cdot\vec{P} + \frac{\partial w}{\partial t} = -\vec{j}\cdot\vec{E} \tag{3.94}$$

In integral form

$$\underbrace{\iiint_V \vec{\nabla}\cdot\vec{P}\, d\tau}_{\text{radiated power}} + \underbrace{\iiint_V \frac{\partial w}{\partial t}\, d\tau}_{\text{power of the electromagnetic field}} = \underbrace{-\iiint_V \vec{j}\cdot\vec{E}\, d\tau}_{\text{power dissipated by the Joule effect}}$$

3.8.7 Maxwell Tensor of Constraints

Starting from the Lorentz force

$$\vec{F} = q(\vec{E} + \vec{v} \wedge \vec{B}) \tag{3.88}$$

a density of electromagnetic force $\vec{f}$ can be defined by

$$\vec{F} = \int_V \vec{f}\, d\tau \tag{3.95}$$

that is, such that

$$\vec{f} = \rho\vec{E} + \vec{j} \wedge \vec{B} \tag{3.96}$$

For the moment, the quantities $F_{\mu\nu}J^{\nu}$ are to be evaluated. For $\mu = 0$, [cf. (3.59)],

$$F_{0\nu}J^{\nu} = F_{01}J^1 + F_{02}J^2 + F_{03}J^3 = \frac{\mu_0}{c}\,\vec{E}\cdot\vec{j} \tag{3.97}$$

For, $\mu = 1$,

$$F_{1\nu}J^{\nu} = -\frac{\rho}{\epsilon_0 c^2}\, E_x - \mu_0 j_y B_z + \mu_0 j_z B_y$$
$$= -\mu_0(\rho\vec{E} + \vec{j} \wedge B)_x = -\mu_0 f_x$$

More generally, it can be shown that

$$F_{i\nu}J^{\nu} = -\mu_0(\vec{f})_i \tag{3.98}$$

Equations (3.97) and (3.98) can be brought together again under the form

$$F_{\mu\nu}J^{\nu} = \mu_0 f_{\mu} \tag{3.99}$$

We define f^{μ}, the quadrivector, as

$$f^{\mu} = \left(\frac{1}{c}\,\vec{E}\cdot\vec{j},\vec{f}\right) \tag{3.100}$$

It was seen earlier that

$$\partial^{\rho}F_{\rho\nu} = J_{\nu} \tag{3.77}$$

(the Gauss and Ampère theorems), which is written with the help of a metric tensor

$$\eta^{\rho\alpha}\eta_{\rho\beta}\eta_{\nu\gamma}\partial_{\alpha}F^{\beta\gamma} = \eta_{\nu\delta}J^{\delta}$$

or

$$\partial_{\alpha}F^{\alpha\gamma} = J^{\gamma} \tag{3.101}$$

when (3.77) goes to (3.101), which corresponds to a general rule of equivalent formulas. For the moment we are going to simplify (3.99) in terms of (3.101):

$$F_{\mu\nu}\partial_{\alpha}F^{\alpha\nu} = \mu_0 f_{\mu} \tag{3.102}$$

Now

$$F_{\mu\nu}\partial_{\alpha}F^{\alpha\nu} = \partial_{\alpha}(F_{\mu\nu}F^{\alpha\nu}) - F^{\alpha\nu}\partial_{\alpha}F_{\mu\nu} \tag{3.103}$$

$$= \partial_{\alpha}(F_{\mu\nu}F^{\alpha\nu}) - \tfrac{1}{2}(F^{\alpha\nu}\partial_{\alpha}F_{\mu\nu} + \underbrace{F^{\nu\alpha}\partial_{\nu}F_{\mu\alpha}}_{\text{exchange } \nu \to \alpha}) \tag{3.104}$$

Bianchi's identity,

$$\partial_{\alpha}F_{\beta\gamma} + \partial_{\beta}F_{\gamma\alpha} + \partial_{\gamma}F_{\alpha\beta} = 0 \tag{3.71}$$

enables the transformation of (3.104) into

$$\begin{aligned} F_{\mu\nu}\partial_{\alpha}F^{\alpha\nu} &= \partial_{\alpha}(F_{\mu\nu}F^{\alpha\nu}) - \tfrac{1}{2}F^{\alpha\nu}(\partial_{\alpha}F_{\mu\nu} + \partial_{\nu}F_{\mu\alpha}) \\ &= \partial_{\alpha}(F_{\mu\nu}F^{\alpha\nu}) + \tfrac{1}{2}F^{\alpha\nu}\partial_{\mu}F_{\nu\alpha} \\ &= \partial_{\alpha}(F_{\mu\nu}F^{\alpha\nu}) + \tfrac{1}{4}(F^{\alpha\nu}\partial_{\mu}F_{\nu\alpha} + \underbrace{F^{\nu\alpha}\partial_{\mu}F_{\alpha\nu}}_{\text{exchange } \alpha \to \nu}) \\ &= \partial_{\alpha}(F_{\mu\nu}F^{\alpha\nu}) + \tfrac{1}{4}(F^{\alpha\nu}\partial_{\mu}F_{\nu\alpha} + F_{\nu\alpha}\partial_{\mu}F^{\alpha\nu}) \\ &= \partial_{\alpha}(F_{\mu\nu}F^{\alpha\nu}) + \tfrac{1}{4}\partial_{\mu}(F^{\alpha\nu}F_{\nu\alpha}) \end{aligned} \tag{3.104a}$$

We then rewrite (3.102) as

$$\partial_{\alpha}(F_{\mu\nu}F^{\alpha\nu}) + \tfrac{1}{4}\partial_{\mu}(F^{\alpha\nu}F_{\nu\alpha}) = \mu_0 f_{\mu} \tag{3.105}$$

or

$$\eta_{\alpha\rho}\partial^{\rho}(F_{\mu\nu}F^{\alpha\nu}) + \tfrac{1}{4}\eta_{\mu\rho}\partial^{\rho}(F^{\alpha\nu}F_{\nu\alpha}) = \mu_0 f_{\mu}$$

Otherwise,

$$\partial^{\rho}(\eta_{\alpha\rho}F_{\mu\nu}F^{\alpha\nu} + \tfrac{1}{4}\eta_{\mu\rho}F^{\alpha\nu}F_{\nu\alpha}) = \mu_0 f_{\mu}$$

which is written

$$\partial^{\rho}T_{\rho\mu} = \mu_0 f_{\mu} \tag{3.106}$$

with

$$T_{\rho\mu} = \eta_{\alpha\rho}F_{\mu\nu}F^{\alpha\nu} + \tfrac{1}{4}\eta_{\mu\rho}F^{\alpha\nu}F_{\nu\alpha} \tag{3.107}$$

a component of Maxwell's tensor of constraints T. Note that

$$\begin{aligned} T_{00} &= \eta_{\alpha 0}F_{0\nu}F^{\alpha\nu} + \tfrac{1}{4}\eta_{00}F^{\alpha\nu}F_{\nu\alpha} = F_{0\nu}F^{0\nu} + \tfrac{1}{4}F^{\alpha\nu}F_{\nu\alpha} \\ &= -\frac{E_x^2}{c^2} - \frac{E_y^2}{c^2} - \frac{E_z^2}{c^2} + \frac{1}{4}(-1)(-2)\left(\frac{E^2}{c^2} - B^2\right) \\ &= -\frac{1}{2}\left(\frac{E^2}{c^2} + B^2\right) \\ &= -\tfrac{1}{2}\mu_0(\epsilon_0 E^2 + \mu_0 H^2) = -\mu_0 w \end{aligned} \tag{3.108}$$

Similarly,

$$T_{0i} = \eta_{\alpha 0}F_{i\nu}F^{\alpha\nu} + \tfrac{1}{4}\eta_{i0}F^{\alpha\nu}F_{\nu\alpha}$$

with $i = 1, 2, 3$ and thus $\eta_{\alpha 0} = 0$ except for $\alpha = 0$:

$$\eta_{i0} = 0 \qquad \text{whatever } i$$

and

$$T_{0i} = F_{i\nu}F^{0\nu} \tag{3.109}$$

In particular,

$$T_{01} = \frac{B_z E_y}{c} - \frac{B_y E_z}{c} = \frac{1}{c}(\vec{E} \wedge \vec{B})_x = \frac{\mu_0}{c}P_x$$

which becomes

$$T_{0i} = \frac{\mu_0}{c}(\vec{P})_i \tag{3.110}$$

but equally,

$$\begin{aligned} T_{i0} &= \eta_{\alpha i}F_{0\nu}F^{\alpha\nu} + \tfrac{1}{4}\eta_{0i}F^{\alpha\nu}F_{\nu\alpha} = F_{0\nu}F^{i\nu} = -F^{0\nu}(-F_{i\nu}) \\ &= F^{0\nu}F_{i\nu} = T_{0i} \end{aligned} \tag{3.111}$$

We have

$$T_{11} = \eta_{\alpha 1} F_{1\nu} F^{\alpha\nu} + \tfrac{1}{4}\eta_{11} F^{\alpha\nu} F_{\nu\alpha} = -F_{1\nu} F^{1\nu} - \tfrac{1}{4} F^{\alpha\nu} F_{\nu\alpha}$$

$$= \left(\frac{E_x}{c}\right)^2 + (B_x)^2 - \frac{1}{2}\left(\frac{E^2}{c^2} + B^2\right)$$

and more generally,

$$T_{ii} = \left[\left(\frac{\vec{E}}{c}\right)_i\right]^2 + [(\vec{B})_i]^2 - \frac{1}{2}\left(\frac{E^2}{c^2} + B^2\right)$$

Now evaluate

$$T_{12} = \eta_{\alpha 1} F_{2\nu} F^{\alpha\nu} + \frac{1}{4}\eta_{21} F^{\alpha\nu} F_{\nu\alpha} = -F_{2\nu} F^{1\nu} = \frac{E_x}{c}\frac{E_y}{c} + B_x B_y$$

that is,

$$T_{ij} = \left(\frac{\vec{E}}{c}\right)_i \left(\frac{\vec{E}}{c}\right)_j + (\vec{B})_i(\vec{B})_j - \frac{1}{2}\delta_{ij}\left(\frac{E^2}{c^2} + B^2\right)$$

with

$$T_{ij} = T_{ji}$$

The Maxwell tensor of constraints thus takes the form

$$(T_{\mu\nu}) = \begin{bmatrix} -\mu_0 w & \frac{\mu_0}{c} P_x & \frac{\mu_0}{c} P_y & \frac{\mu_0}{c} P_z \\ \frac{\mu_0}{c} P_x & — & — & — \\ \frac{\mu_0}{c} P_y & — & T_{ij} & — \\ \frac{\mu_0}{c} P_z & — & — & — \end{bmatrix}$$

According to (3.106),

$$\partial^\rho T_{\rho\mu} = \mu_0 f_\mu$$

and for $\mu = 0$,

$$\partial^\rho T_{\rho 0} = \partial^0 T_{00} + \partial^1 T_{10} + \partial^2 T_{20} + \partial^3 T_{30} = \mu_0 f_0$$

with

$$f_0 = \frac{1}{c}\vec{E}\cdot\vec{j} \qquad \text{[cf. (3.100)]}$$

if

$$\frac{1}{c}\frac{\partial}{\partial t}(-\mu_0 w) - \frac{\partial}{\partial x}\left(\frac{\mu_0}{c}P_x\right) - \frac{\partial}{\partial y}\left(\frac{\mu_0}{c}P_y\right) - \frac{\partial}{\partial z}\left(\frac{\mu_0}{c}P_z\right) = \frac{\mu_0}{c}\vec{E}\cdot\vec{j}$$

or

$$\frac{\partial w}{\partial t} + \vec{\nabla}\cdot\vec{P} = -\vec{j}\cdot\vec{E}$$

We come back to (3.94).

To sum up, we have devised the following table of the equations of electromagnetism.

Classical formulation	Covariant formulation
$\vec{\nabla}\wedge H = \vec{j} + \frac{\partial \vec{D}}{\partial t}$	
$\vec{\nabla}\cdot\vec{D} = \rho$	$\partial^{\mu}F_{\mu\nu} = J_{\nu}$
$\vec{\nabla}\wedge\vec{E} = -\frac{\partial B}{\partial t}$	
$\vec{\nabla}\cdot B = 0$	$\partial^{\mu}{}^{*}F_{\mu\nu} = 0$
$\frac{1}{c^2}\frac{\partial V}{\partial t} + \vec{\nabla}\cdot\vec{A} = 0$	$\partial^{\mu}A_{\mu} = 0$
$\Box V = \frac{\rho}{\epsilon_0}$	
$\Box\vec{A} = \mu_0\vec{j}$	$\partial_{\nu}\partial^{\nu}A_{\mu} = \Box A_{\mu} = J_{\mu}$
$\frac{\partial w}{\partial t} + \vec{\nabla}\cdot\vec{P} = -\vec{j}\cdot\vec{E}$	
$(\vec{f})_i = (\rho\vec{E} + \vec{j}\wedge B)_i$	$\partial^{\rho}T_{\rho\mu} = \mu_0 f_{\mu}$

3.8.8 Gauge Transformation

Maxwell's condition enables us to define the *Lorentz gauge transformation*. According to Section 3.1, the potentials V and $\vec{A}$ comply with

$$\vec{E} = -\vec{\nabla}V - \frac{\partial}{\partial t}\vec{A}$$

$$\vec{\nabla}\cdot\vec{A} + \mu\epsilon\frac{\partial}{\partial t}V = 0$$

It is obvious that if we take

$$V'(\vec{r},t) = V(\vec{r},t) - \frac{\partial}{\partial t}F(\vec{r},t) \tag{3.112}$$

and

$$\vec{A}'(\vec{r},t) = \vec{A}(\vec{r},t) + \vec{\nabla}F(\vec{r},t) \tag{3.113}$$

where $F(\vec{r},t)$ is an arbitrary function, we can only obtain the same fields $\vec{E}$ and $\vec{B}$ and V and $\vec{A}$, since

$$\vec{E}' = -\vec{\nabla}V + \vec{\nabla}\frac{\partial}{\partial t}F - \frac{\partial}{\partial t}\vec{A} - \frac{\partial}{\partial t}\vec{\nabla}F = \vec{E}$$

$$\vec{B}' = \vec{\nabla} \wedge \vec{A}' = \vec{\nabla} \wedge \vec{A} + \vec{\nabla} \wedge (\vec{\nabla}F) = \vec{B}$$

By choosing the function F the relation can be set between V and $\vec{A}$. We are said to be making a choice of gauge. There are basically two of them.

Lorentz Gauge

We chose Maxwell's condition

$$\vec{\nabla}\cdot\vec{A} + \mu\epsilon\frac{\partial}{\partial t}V = 0 \tag{3.114}$$

which means that the function $F(\vec{r},t)$ complies with

$$\vec{\nabla}\cdot\vec{A} + \vec{\nabla}\cdot(\vec{\nabla}F) + \mu\epsilon\frac{\partial}{\partial t}V - \mu\epsilon\frac{\partial^2}{\partial t^2}F = 0$$

or

$$\nabla^2 F = \mu\epsilon\frac{\partial^2}{\partial t^2}F \tag{3.115}$$

and equation of propagation in a vacuum. The covariance of the electromagnetic equations is preserved.

Coulomb Gauge

Take the condition

$$\vec{\nabla}\cdot\vec{A}(\vec{r},t) = 0 \tag{3.116}$$

In this case (cf. Section 3.1)

$$\nabla^2\vec{A} - \mu\epsilon\frac{\partial^2}{\partial t^2}\vec{A} = -\mu\vec{j} + \vec{\nabla}\left(\mu\epsilon\frac{\partial}{\partial t}V\right) \tag{3.117}$$

$$\nabla^2 V = -\frac{\rho}{\epsilon} \tag{3.118}$$

Equation (3.118) is *Poisson's equation*. Equation (3.117) is also written

$$\Box\vec{A} = \mu\vec{j} - \mu\epsilon \frac{\partial}{\partial t}(\vec{\nabla}V)$$

The covariance is lost, but the interest of this gauge is obvious due to the quantification of the electromagnetic fields. If V and $\vec{A}$ are known in Coulomb's gauge, we can go to Lorentz's gauge, in which they are denoted V' and $\vec{A}'$. Which condition must be imposed on the function F?

$$\vec{\nabla}\cdot\vec{A}' + \mu\epsilon \frac{\partial}{\partial t} V' = 0$$

with

$$\vec{A}' = \vec{A} + \vec{\nabla}F$$

$$V' = V - \frac{\partial}{\partial t} F$$

from which

$$\vec{\nabla}\cdot\vec{A} + \nabla^2 F + \mu\epsilon \frac{\partial}{\partial t} V - \mu\epsilon \frac{\partial^2}{\partial t^2} F = 0$$

Now in Coulomb's gauge,

$$\vec{\nabla}\cdot\vec{A} = 0$$

or

$$\Box F = \mu\epsilon \frac{\partial}{\partial t} V$$

Gauge Invariance

When to describe an electromagnetic field, a particular set of potentials is chosen, we are said to be making a choice of gauge. In quantum mechanics especially, the equations of motion of a system cause intervention of the potentials V and $\vec{A}$, not intervention of the fields $\vec{E}$ and $\overleftrightarrow{B}$:

$$i\hbar \frac{\partial}{\partial t}|\Psi\rangle = \left[\frac{1}{2m}(\vec{p} - q\vec{A})^2 + V\right]|\Psi\rangle$$

The following problem arises: Do the physics results predicted by the theory depend on the values of the fields E and $\overleftrightarrow{B}$ in every point of the space, or do they depend on the gauge—in other words, on the choice of the potentials that is used to write the equations?

In the latter case it would clearly be necessary each time to specify the gauge in which equations are valid. We prove that the physics results, in quantum mechanics as in classical mechanics, are not altered by a change in gauge. The scalar and vector potentials thus appear as the intermediaries of calculation; only the values of the fields $\vec{E}$ and $\vec{B}$ in any point in space count.

The mechanics have the property of gauge invariance.

3.9 APPLICATIONS

3.9.1 The Photon

According to the general formula,

$$E^2 = p^2c^2 + m^2c^4 \tag{1}$$

and for a photon of velocity c,

$$E = h\nu = \hbar kc \tag{2}$$

where

$$k = \frac{2\pi}{\lambda}$$

is the modulus of the wave vector.

As

$$p = \hbar k$$

is the momentum of the photon

$$E = pc \tag{3}$$

and by comparing (3) with (1), it is seen that

$$m = 0$$

The photon is a zero mass particle. Note that the pseudonorm of the energy–momentum quadrivector

$$P^\mu = \left(\frac{E}{c}, \vec{p}\right)$$

gives

$$P^2 = \frac{E^2}{c^2} - p^2 \tag{4}$$

As P^2 must have the same value in the "proper frame," we also have

$$P^2 = m_0c^2 \tag{5}$$

The identification of the relations (4) and (5) enables us to go back to formula (1).

3.9.2 Absorption and Emission of a Photon Through a Nucleus

A nucleus of energy E, with $E = \gamma E_0$ is the energy at rest, has a speed $\vec{v}$ in a referential linked to an observer O. A photon of energy E_{ph} of which the trajectory forms an angle α with vector $\vec{v}$ (cf. Figure 11) is absorbed. Evaluate the energy at rest, E_0', of the nucleus after the absorbtion of the photon as well as its speed $\vec{v}\,'$. The preservation of the quadrivector energy–momentum

$$P^{\mu} = \left(\frac{E}{c}, \vec{p}\right)$$

gives the two identities

$$E' = E + E_{ph} \tag{1}$$

$$m'\gamma'\vec{v}\,' = m\gamma\vec{v} + \hbar\vec{k} \tag{2}$$

or

$$E_0'\gamma' = E_0\gamma + E_{ph} \tag{3}$$

bearing in mind the fact that

$$\vec{p} = m\gamma\vec{v} = \frac{E}{c^2}\vec{v} = \frac{E\vec{\beta}}{c} \tag{4}$$

with

$$\vec{\beta} = \frac{\vec{v}}{c}$$

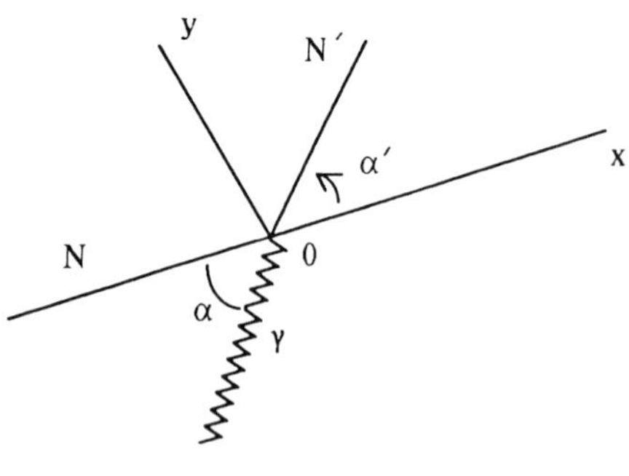

Figure 11 Photon.

The identity (2) projected onto Ox is written

$$\frac{E'\beta'}{c}\cos\alpha' = \frac{E\beta}{c} + \frac{E_{ph}}{c}\cos\alpha \tag{5}$$

and onto Oy,

$$\frac{E'\beta'}{c}\sin\alpha' = \frac{E_{ph}}{c}\sin\alpha \tag{6}$$

(5) is squared and (6) is used:

$$E'^2\beta'^2(1 - \sin^2\alpha') = E'\beta'^2\left(1 - \frac{E_{ph}^2\sin^2\alpha}{E'^2\beta'^2}\right) = (E\beta + E_{ph}\cos\alpha)^2$$

by developing

$$E'^2\beta'^2 = E_{ph}^2 + E^2\beta^2 + 2E\beta E_{ph}\cos\alpha$$

or with (1),

$$\beta' = \sqrt{\frac{E_{ph}^2 + E^2\beta^2 + 2\beta E E_{ph}\cos\alpha}{(E + E_{ph})^2}} \tag{7}$$

and if $\alpha = 0$,

$$\beta_0' = \frac{E_{ph} + \beta E}{E_{ph} + E}$$

As, by definition,

$$\gamma^2 = \frac{1}{1 - \dfrac{v^2}{c^2}} = \frac{1}{1 - \beta^2}$$

we see that

$$\gamma' = \frac{E + E_{ph}}{\sqrt{E^2(1 - \beta^2) + 2EE_{ph}(1 - \beta\cos\alpha)}} \tag{8}$$

And, according to (3),

$$E_0' = \frac{E_0\gamma + E_{ph}}{\gamma'} = \sqrt{E_0[E_0 + 2\gamma E_{ph}(1 - \beta\cos\alpha)]} \tag{9}$$

When the energy of the nucleus is strong in relation to that of the photon, we calculate the variation Δm of the nucleus mass. As

$$E_0' = m'c^2 = E_0 \sqrt{1 + 2\gamma \frac{E_{ph}}{E_0} (1 - \beta \cos \alpha)}$$

$$= E_0 \left[1 + \gamma \frac{E_{ph}}{E_0} (1 - \beta \cos \alpha)\right] \tag{10}$$

and

$$\Delta m' = \frac{E_0' - E_0}{c^2} = \gamma \frac{E_{ph}}{c^2} (1 - \beta \cos \alpha) \tag{11}$$

In the case of a nucleus supposed at rest ($\vec{v} = \vec{0}$, $\beta = 0$, $\gamma = 1$),

$$\Delta m' = \frac{E_{ph}}{c^2} \tag{12}$$

all of the photon energy is transformed into mass.

After awhile the nucleus will send out a photon with energy E_{ph}. The speed of the nucleus is $\vec{v}\,'$ and its energy is E' before its emission. These two quantities are given by equations (7) and (10) in the approximation $E' \gg E_{ph}$. In this case the photon and nucleus after emission have approximately the same direction of motion (identical to that of the nucleus before emission). The preservation of the quadrivector energy–momentum gives [cf. (1) and (5)]

$$E_0'\gamma' = E_0''\gamma'' + E_{ph} \tag{13}$$

(E'' is the energy of the nucleus after the emission of the photon) and

$$E_0'\gamma'\beta' = E_0''\gamma''\beta'' + E_{ph} \tag{14}$$

From this it can be deduced that

$$\beta'' = \frac{E_0'\gamma'\beta' - E_{ph}}{E_0'\gamma' - E_{ph}}$$

If we want to evaluate the energy E_0'', we must first establish γ'':

$$\gamma'' = \frac{1}{\sqrt{1 - \beta''^2}} = \frac{E_0'\gamma' - E_{ph}}{\sqrt{[E_0'\gamma'(1 + \beta') - 2E_{ph}][E_0'(1 - \beta'\gamma')]}}$$

from which [cf. (13)]

$$E_0'' = \frac{E_0'\gamma' - E_{\text{ph}}}{\gamma''} = E_0' \sqrt{1 - 2\frac{E_{\text{ph}}}{E_0'}\gamma'(1-\beta')}$$

$$\simeq E_0' - E_{\text{ph}}\gamma'(1-\beta')$$

and the variation $\Delta m''$ of the mass is obtained by

$$\Delta m'' = \frac{E_0'' - E_0'}{c^2} = -E_{\text{ph}}\frac{\gamma'(1-\beta')}{c^2}$$

which we compare with (11):

The sign is opposing.
We have supposed that $\alpha = 0$.

If the nucleus is at rest, it will lose a mass equal to E_{ph}/c^2.

3.9.3 Invariants

Write the complex vector

$$\vec{X} = \frac{\vec{E}}{c} + i\overleftrightarrow{B} \tag{1}$$

It is seen that

$$X^2 = \frac{E^2}{c^2} - B^2 + 2\frac{i}{c}\vec{E}\cdot\overleftrightarrow{B} \tag{2}$$

According to formula (3.68),

$$F^2 = -2\left(\frac{E^2}{c^2} - B^2\right) \tag{3.68}$$

and according to (3.70),

$$F\cdot{}^*F = 4\frac{\vec{E}\cdot\overleftrightarrow{B}}{c} \tag{3.70}$$

We can thus write X^2 in the form

$$X^2 = -\tfrac{1}{2}(F^2 - iF\cdot{}^*F)$$

As X^2 is an invariant, we see the two electromagnetic invariants of Section 3.7 again. Note that, by definition, invariants are called the coefficients of the *characteristic polynomial*:

$$P(\lambda) = \det(F_{\mu\upsilon} - \lambda g_{\mu\upsilon}) = -\lambda^4 + \left(\frac{E^2}{c^2} - B^2\right)\lambda^2 + \frac{\vec{E}\cdot\overleftrightarrow{B}}{c}$$

3.9.4 Antisymmetrical Tensors

Angular Velocity

For a circular movement characterized by the angular speed $\overleftrightarrow{\omega}$, we know that

$$\vec{v} = \overleftrightarrow{\omega} \wedge \vec{r} \tag{1}$$

where $\vec{v}$ is the linear velocity of the mobile M such that $\vec{r} = \overrightarrow{OM}$. With the help of the Levi–Civita tensor, the components v_i are written

$$v_i = \epsilon_{ijk}\omega^j x^k \tag{2}$$

v_i can equally be written in the form

$$v_i = \Omega_{ik} x^k \tag{3}$$

from which it is deduced that

$$\Omega_{ik} = \epsilon_{ijk}\omega^j \tag{4}$$

but also that

$$\Omega_{ki} = \epsilon_{kji}\omega^j \tag{5}$$

According to the properties of Levi–Civita tensor,

$$\epsilon_{kji} = \epsilon_{ikj} = -\epsilon_{ijk} \tag{6}$$

which means that

$$\Omega_{ik} = -\Omega_{ki} \tag{7}$$

The tensor (Ω_{ij}) is thus antisymmetrical:

$$(\Omega_{ij}) = \begin{bmatrix} 0 & -\Omega_{21} & \Omega_{13} \\ \Omega_{21} & 0 & -\Omega_{32} \\ -\Omega_{13} & \Omega_{32} & 0 \end{bmatrix} \tag{8}$$

Starting from (4), which is multiplied by ϵ^{ilk},

$$\Omega_{ik}\epsilon^{ilk} = \epsilon_{ijk}\epsilon^{ilk}\omega^j$$

As

$$\epsilon_{ijk}\epsilon^{ilk} = 2\delta_j^l \tag{9}$$

$$\Omega_{ik}\epsilon^{ilk} = 2\omega^l$$

or

$$\omega^l = \tfrac{1}{2}\epsilon^{ilk}\Omega_{ik} \tag{10}$$

We can extend this procedure to four dimensions. We are going to give several examples taken from mechanics and electromagnetism. Remember that Faraday's tensor is antisymmetrical, due to the way it is made.

Dipolar Tensor

By definition of the polarization vector $\vec{P}$,

$$\vec{D} = \epsilon_0 \vec{E} + \vec{P} \tag{1}$$

Similarly for the magnetization $\overset{\hookrightarrow}{M}$:

$$\overset{\hookrightarrow}{B} = \mu_0 \overset{\hookrightarrow}{H} + \overset{\hookrightarrow}{M} \tag{2}$$

Equations (1) and (2) can be brought together in a polarization–magnetization tensor. According to (1) and (2),

$$\epsilon_0 E_x = D_x - P_x$$

$$B_x = \mu_0 H_x + M_x$$

and according to (3.44),

$$E_x = cF^{10}$$

$$B_x = F^{32}$$

or

$$F^{10} = \frac{D_x}{\epsilon_0 c} - \frac{P_x}{\epsilon_0 c} = f^{10} + M^{10}$$

$$F^{32} = \mu_0 H_x + M_x = f^{32} + M^{32}$$

More generally, we have

$$(f^{\mu\upsilon}) = \mu_0 c \begin{bmatrix} 0 & -D_x & -D_y & -D_z \\ D_x & 0 & -\dfrac{H_z}{c} & \dfrac{H_y}{c} \\ D_y & \dfrac{H_z}{c} & 0 & -\dfrac{H_x}{x} \\ D_z & -\dfrac{H_y}{c} & \dfrac{H_x}{c} & 0 \end{bmatrix}$$

$$(M^{\mu\upsilon}) = \mu_0 c \begin{bmatrix} 0 & P_x & P_y & P_z \\ -P_x & 0 & -\dfrac{M_z}{c\mu_0} & \dfrac{M_y}{c\mu_0} \\ -P_y & \dfrac{M_z}{c\mu_0} & 0 & -\dfrac{M_x}{c\mu_0} \\ -P_z & -\dfrac{M_y}{c\mu_0} & \dfrac{M_x}{c\mu_0} & 0 \end{bmatrix}$$

Otherwise,

$$F^{\mu\upsilon} = f^{\mu\upsilon} + M^{\mu\upsilon}$$

The reading of tensor $(M^{\mu\upsilon})$ shows that the polarization $\vec{P}$ is a vector and that the magnetization is a pseudovector. The reason this representation is of interest is that it simplifies the final calculation of the polarization and magnetization vectors during a Lorentz transformation.

Angular Momentum Tensor

We know that the angular (kinetic) momentum, calculated at a point O associated to a particle of momentum $\vec{p}$, is

$$\overset{\hookrightarrow}{L} = \vec{r} \wedge \vec{p}$$

or

$$L_i = \epsilon_{ijk} x^j p^k$$

More simply,

$$L_k = x_i p_j - x_j p_i$$

Otherwise, by writing L_{ij} as the component L_k,

$$L_{ij} = x_i p_j - x_j p_i$$

and by going to four dimensions,

$$L_{\mu\nu} = X^{\mu} P^{\nu} - X^{\nu} P^{\mu}$$

Recall that

$$X^{\mu} = (ct, \vec{r})$$

$$P^{\mu} = \left(\frac{E}{c}, \vec{p}\right)$$

To be exact,

$$L^{01} = ctp_x - x\frac{E}{c}$$

$$L^{12} = xp_y - yp_x$$

from which the antisymmetrical tensor is

$$(L^{\mu\nu}) = \begin{bmatrix} 0 & c(tp_x - Ex/c^2 & c(tp_y - Ey/c^2) & c(tp_z - Ez/c^2) \\ -c(tp_x - Ex/c^2) & 0 & L_z & -L_y \\ -c(tp_y - Ey/c^2) & -L_z & 0 & L_x \\ -c(tp_z - Ez/c^2) & L_y & -L_x & 0 \end{bmatrix}$$

Remember that, for example,

$$L_z = xp_y - yp_x$$

with

$$p_x = \gamma m_0 v_x$$

so

$$L_z = \gamma m_0 (xv_y - yv_x)$$

Torque Tensor

In the same way as that used in Section 3.9.3, we obtain from the definition of the torque at a point O

$$\overleftrightarrow{M} = \vec{r} \wedge \vec{F}$$

The tensor

$$(M^{\mu\upsilon}) =$$

$$\begin{bmatrix} 0 & c\left(tF_x - \dfrac{\vec{F}\cdot\vec{v}}{c^2}x\right) & c\left(tF_y - \dfrac{\vec{F}\cdot\vec{v}}{c^2}y\right) & c\left(tF_z - \dfrac{\vec{F}\cdot\vec{v}}{c^2}z\right) \\ -c\left(tF_x - \dfrac{\vec{F}\cdot\vec{v}}{c^2}x\right) & 0 & M_z & -M_y \\ -c\left(tF_y - \dfrac{\vec{F}\cdot\vec{v}}{c^2}y\right) & -M_z & 0 & M_x \\ -c\left(tF_z - \dfrac{\vec{F}\cdot\vec{v}}{c^2}z\right) & M_y & -M_x & 0 \end{bmatrix}$$

3.9.5 Lorentz Transformation

For a special Lorentz transformation

$$X'^{\mu'} = l_\nu^{\mu'} X^\nu \tag{1}$$

with

$$(l_\nu^{\mu'}) = \begin{bmatrix} \Gamma & 0 & 0 & -\Gamma\beta \\ 0 & 1 & 0 & 0 \\ 0 & 0 & 1 & 0 \\ -\Gamma\beta & 0 & 0 & \Gamma \end{bmatrix}$$

we clearly have

$$X^2 = X^\mu \vec{e}_\mu \cdot X^\nu \vec{e}_\nu = g_{\mu\nu} X^\mu X^\nu$$

The invariance of the norm of the quadrivector position imposes

$$X^2 = X'^2$$

either to be exact

$$g_{\mu\nu}X^{\mu}X^{\nu} = l_{\mu}^{\mu'}X^{\mu}\vec{e}_{\mu'}\cdot l_{\nu}^{\nu'}X^{\nu}\vec{e}_{\nu'} = l_{\mu}^{\mu'}g_{\mu'\nu'}l_{\nu}^{\nu'}X^{\mu}X^{\nu}$$

or

$$g_{\mu\nu} = l_{\mu}^{\mu'}g_{\mu'\nu'}l_{\nu}^{\nu'} \tag{2}$$

in a euclidean geometry and in matrix notation (2) is written by lowering all the indices, that is, by taking

$$l_{\mu}^{\mu'} \longrightarrow l_{\mu\mu'}$$

$$l_{\mu}^{\mu'}g_{\mu'\nu'}l_{\nu}^{\nu'} \longrightarrow l_{\mu\mu'}g_{\mu'\nu'}l_{\nu\nu'}$$

An incorrect relation as the matrix product must give

$$g_{\mu\nu} = l_{\mu\mu'}g_{\mu'\nu'}l_{\nu'\nu}$$

Thus in matrix notation (2) is written

$$(g) = (l)(g)^{t}(l) \tag{3}$$

In the same way, we can use the relation

$$l_{\mu}^{\mu'} \longrightarrow l_{\mu'\mu}$$

$$l_{\mu}^{\mu'}g_{\mu'\nu'}l_{\nu}^{\nu'} \longrightarrow l_{\mu'\mu}g_{\mu'\nu'}l_{\nu'\nu}$$

which means, for the same reasons as before (the first index of $l_{\mu'\mu}$ should be μ):

$$(g) = {}^{t}(l)(g)(l) \tag{4}$$

Let's go back to (3):

$$(l)^{-1} = (g)^{t}(l)(g) \tag{5}$$

having borne in mind that

$$(g)^{-1} = (g)$$

In a euclidean space

$$g_{ij} = \delta_{ij}$$

and

$$(l)^{-1} = {}^{t}(l)$$

a definition of the orthogonal group of dimension 4 denoted O(4). In restricted relativity the group is *pseudo-orthogonal* and denoted O(1,3); the metric comes into the calculation of the inverse matrix of the Lorentz transformation.

3.9.6 Transformation of the Electromagnetic Field

The vectors $\vec{E}$, $\vec{D}$, $\overleftrightarrow{H}$, and $\overleftrightarrow{B}$ are defined as characteristics of an electromagnetic field in a vacuum. Denote by (R) the laboratory frame. A change in referential (Lorentz transformation) transforms the previous vectors into $\vec{E}'$, $\vec{D}'$, $\overleftrightarrow{H}'$, and $\overleftrightarrow{B}'$ in (R') the new frame. We now want to identify the relations linking $\vec{E}'$, $\overleftrightarrow{B}'$ to $\vec{E}$ and $\overleftrightarrow{B}$.

Under the action of the Lorentz transformation l,

$$l = \begin{bmatrix} \Gamma & 0 & 0 & -\Gamma\beta \\ 0 & 1 & 0 & 0 \\ 0 & 0 & 1 & 0 \\ -\Gamma\beta & 0 & 0 & \Gamma \end{bmatrix}$$

Faraday's tensor $F^{\mu\nu}$ complies with the law

$$(F') = (l)(F)^{t}(l)$$

(l) and ${}^{t}(l)$ are symmetrical matrices:

$$(l) = {}^{t}(l)$$

and $(F'_{\mu'\nu'})$ will be an antisymmetrical tensor. We need simply to evaluate six terms. We obtain

$$(F'_{\mu'\nu'}) =$$

$$\begin{bmatrix} 0 & \Gamma\left(-\frac{E_x}{c} + \beta B_x\right) & \Gamma\left(-\frac{E_y}{c} - \beta B_x\right) & -\frac{E_z}{c} \\ \Gamma\left(\frac{E_x}{c} - \beta B_x\right) & 0 & -B_z & \Gamma\left(B_y - \beta\frac{E_x}{c}\right) \\ \Gamma\left(\beta\frac{E_y}{c} + B_y\right) & \Gamma\left(-B_y + \beta\frac{E_x}{c}\right) & 0 & -\Gamma\left(\beta\frac{E_y}{c} + B_x\right) \\ \frac{E_z}{c} & \Gamma\left(\beta\frac{E_x}{c} - B_y\right) & \Gamma\left(\beta\frac{E_y}{c} + B_x\right) & 0 \end{bmatrix}$$

By identification it may be written that

$$\frac{E'_x}{c} = \Gamma\left(\frac{E_x}{c} - \beta B_y\right)$$

$$\frac{E'_y}{c} = \Gamma\left(\frac{E_y}{c} + \beta B_x\right)$$

$$\frac{E'_z}{c} = \frac{E_z}{c}$$

and

$$B'_x = \Gamma\left(\beta\frac{E_y}{c} + B_x\right)$$

$$B'_y = -\Gamma\left(\beta\frac{E_x}{c} - B_y\right)$$

$$B'_z = B_z$$

3.9.7 Parallel Electric and Magnetic Fields

Consider the fields $\vec{E}$ and $\vec{B}$ orthogonal to Oz (cf. Figure 12). According to Section 3.9.6, by the Lorentz transformation following the axis Ox,

$$\frac{E'_x}{c} = \Gamma\left(\frac{E_x}{c} - \beta B_y\right)$$

$$\frac{E'_y}{c} = \Gamma\left(\frac{E_y}{c} + \beta B_x\right)$$

$$\frac{E'_z}{c} = \frac{E_z}{c}$$

and

$$B'_x = \Gamma\left(\beta\frac{E_y}{c} + B_x\right)$$

$$B'_y = -\Gamma\left(\beta\frac{E_x}{c} - B_y\right)$$

$$B'_z = B_z$$

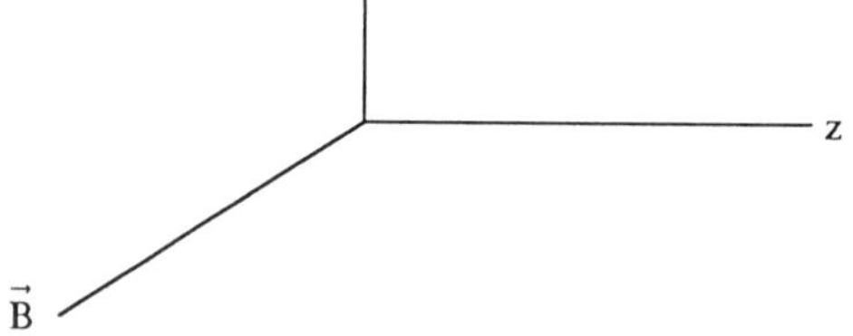

Figure 12 Fields.

Only after the transformation do we insist that the fields $\vec{E}'$ and $\vec{B}'$ be parallel, that is,

$$\vec{E}' = \lambda\vec{B}'$$

or

$$\lambda = \frac{E'_x}{B'_x} = \frac{E'_y}{B'_y}$$

To be precise,

$$\frac{E_x - \beta c B_y}{\beta(E_y/c) + B_x} = \frac{E_y + \beta c B_x}{-\beta(E_x/c) + B_y}$$

Otherwise,

$$(1 + \beta^2)(\vec{E} \wedge \vec{B})_z = \frac{\beta}{c}(E^2 + c^2B^2)$$

More generally,

$$\frac{\vec{\beta}}{c(1 + \beta^2)} = \frac{\vec{E} \wedge \vec{B}}{E^2 + c^2B^2}$$

a relation that gives the translation velocity of the referential so as to make electric and magnetic fields parallel.

3.9.8 Green's Functions and Retarded Potential

To find the solutions of the differential equation in partial derivatives,

$$D_{\vec{r}} f(\vec{r}) = h(\vec{r}) \tag{1}$$

D_x being the operator

$$D_x = \sum_{p=1}^{m} a_p \left(\frac{\partial}{\partial x}\right)^p \tag{2}$$

Remember that the convolution product of $g(x)$ and $h(x)$ is, by definition,

$$(h*g)(x) = \int_{-\infty}^{+\infty} h(x')g(x - x')\, dx' \tag{3}$$

and that Dirac's function, δ, is the neutral element of this algebra:

$$(g*\delta)(x) = \int_{-\infty}^{+\infty} g(x')\delta(x - x')\, dx' = g(x) \tag{4}$$

The property (4) enables replacement of (1) by

$$D_x(\delta^* f)(x) = h(x) \tag{5}$$

By generalizing the notion of the dot product, if F is a distribution and φ an infinitely differentiable function of rapid decline (called a *test function*), we write

$$F(\varphi) = \langle F,\varphi \rangle = \int_{-\infty}^{+\infty} F(x)\varphi(x)\, dx \tag{6}$$

and in particular,

$$\langle \delta,\varphi \rangle = \int_{-\infty}^{+\infty} \delta(x)\varphi(x)\, dx = \varphi(0) \tag{7}$$

Note that

$$\langle \delta, D_x f \rangle = -\langle D_x\delta, f \rangle \tag{8}$$

Equation (8) is shown by simplifying ($D_x \to \partial/\partial x$) to one dimension:

$$\int_{-\infty}^{+\infty} \delta(x) \frac{\partial}{\partial x} f(x)\, dx = -\int_{-\infty}^{+\infty} \left[\frac{\partial}{\partial x} \delta(x)\right] f(x)\, dx \tag{9}$$

an identity obtained by integrating by parts the integral of the left-hand member, which confirms the relation (8).

Evaluate the quantity $(D\delta)*f$

$$\begin{aligned}(D\delta)^* f &= \int_{-\infty}^{+\infty} \left[\frac{\partial}{\partial x'} \delta(x')\right] f(x - x')\, dx' \\ &= -\int_{-\infty}^{+\infty} \delta(x') \frac{\partial}{\partial x'} f(x - x')\, dx' = \int_{-\infty}^{+\infty} \delta(x - u) \frac{\partial}{\partial u} f(u)\, du \\ &= Df^*\delta \end{aligned} \tag{10}$$

from which

$$(D\delta)^* f(x) = Df(x)^*\delta = h(x) \tag{11}$$

and by multiplying on the left by $(D\delta)^{-1}$, we have

$$f(x) = (D\delta)^{-1*} h(x) \tag{12}$$

Green's function G is written by definition:

$$G = (D\delta)^{-1} \tag{13}$$

and (12) becomes

$$f(x) = G^*h(x) = \int_{-\infty}^{+\infty} G(x')h(x - x')\,dx'$$
$$= \int_{-\infty}^{+\infty} G(x - u)h(u)\,du \tag{14}$$

We write $u = x - x'$. The solution of the equation (1) is given by (14). For each specific differential operator of (1) we must specify a Green's function or *propagator*.

Equation (13) multiplied by $(D\delta)$ on the left gives

$$D\delta^*g = (D\delta)^*(D\delta)^{-1} = \delta \tag{15}$$

and with (10),

$$Dg^*\delta = \delta \tag{16}$$

Now δ is the neutral element; (16) is also written

$$Dg = \delta \tag{17}$$

the usual definition of Green's function. All the previous results can be generalized to

$$D_{\vec{r}} f(\vec{r}) = h(\vec{r}) \tag{1}$$

where $\vec{r}$ is a vector of N dimensions.

The vector potential $\vec{A}(\vec{r},t)$ as well as the scalar potential $V(\vec{r},t)$ comply with the differential equation

$$\Box A_\mu(x_\nu) = J_\mu(x_\nu) \tag{18}$$

A_μ being the quadripotential $A_\mu = (V/c,A)$, J_μ the quadrivector density of current $J_\mu = (\rho/c\epsilon_0,\mu_0\vec{j})$, and $\Box = (1/c^2)(\partial^2/\partial t^2) - \nabla^2$ the d'Alembertian.

The solution of (18) is, then [cf. (14)],

$$A_\mu(x_\nu) = A_\mu^0(x_\nu) + \int_{\Omega_{x'_\nu}} G(x_\nu - x'_\nu)J_\mu(x'_\nu)\,d^4x' \tag{19}$$

$A_\mu^0(x_\mu)$ is the solution to the limits of $\Omega_{x'_\nu}$; $\Omega_{x'_\nu}$ is the volume in Minkowski space, and G is given by

$$\Box G(x_\nu) = \delta^4(x_\nu) \tag{20}$$

By Fourier's transform,

$$\Box\left(\frac{1}{2\pi}\right)^2 \int_{\Omega_{k_\nu}} G(\vec{k},\omega)e^{i(\vec{k}\cdot\vec{r}-\omega t)}\,d^3k\,d\omega = \left(\frac{1}{2\pi}\right)^2 \int_{\Omega_{k_\nu}} e^{i(\vec{k}\cdot\vec{r}-\omega t)}\,d^3k\,d\omega$$

or

$$\left(k^2 - \frac{\omega^2}{c^2}\right) G(\vec{k},\omega) = 1$$

Green's function is thus

$$G(\vec{k},\omega) = \frac{1}{k^2 - \omega^2/c^2} \tag{21}$$

$k \neq \omega/c$ and Fourier's integral can be expressed by

$$\begin{aligned} G(x_\nu) &= \left(\frac{1}{2\pi}\right)^2 \int_\Omega G(\vec{k},\omega) e^{i(\vec{k}\cdot\vec{r}-\omega t)}\, d^3k\, d\omega \\ &= \left(\frac{1}{2\pi}\right)^2 \int_\Omega \frac{e^{i(\vec{k}\cdot\vec{r}-\omega t)}}{k^2 - \omega^2/c^2}\, d^3k\, d\omega \end{aligned}$$

and with the calculation of the integral using the method of residues, we come to the Green's function associated with the problem set. A more direct method is to characterize (18) as vector potential $\vec{\mathrm{A}}(\vec{r},t)$. We obviously have

$$\vec{A}(\vec{r},t) = \left(\frac{1}{2\pi}\right)^2 \int_\Omega \vec{A}(\vec{k},\omega) e^{i(\vec{k}\cdot\vec{r}-\omega t)}\, d^3k\, d\omega \tag{22}$$

by definition of Fourier's integral; the equivalent of (21) is

$$\vec{A}(\vec{r},t) = \frac{\mu_0 \vec{j}(\vec{k},\omega)}{k^2 - \omega^2/c^2} \tag{23}$$

or [cf. (22)]

$$\vec{A}(\vec{r},t) = \mu_0 \left(\frac{1}{2\pi}\right)^2 \int_\Omega \frac{\vec{j}(\vec{k},\omega)}{k^2 - \omega^2/c^2} e^{i(\vec{k}\cdot\vec{r}-\omega t)}\, d^3k\, d\omega$$

and by writing Fourier's transform of $\vec{j}(\vec{k},\omega)$,

$$\vec{A}(\vec{r},t) = \mu_0 \left(\frac{1}{2\pi}\right)^4 \int_\Omega \frac{\vec{j}(\vec{r}\,',t')}{k^2 - \omega^2/c^2} e^{i\vec{k}\cdot(\vec{r}-\vec{r}\,')-i\omega(t-t')}\, d^3k\, d\omega\, d^3r'\, dt'$$

We put

$$\begin{aligned} \vec{r} - \vec{r}\,' &= \vec{R} \\ t - t' &= \tau \end{aligned}$$

Now

$$d^3k = 2\pi \sin\theta\, d\theta\, k^2\, dk$$

θ being the angle between the vectors $\vec{R}$ and $\vec{k}$. We now make the assumption

$$\tau > \frac{R}{c}$$

This is true for retarded potentials:

$$\vec{A}(\vec{r},t) = \mu_0 \left(\frac{1}{2\pi}\right)^3 \int_\Omega \frac{\vec{j}(\vec{r} - \vec{R}, t - \tau)}{k^2 - \omega^2/c^2} \times e^{ikR\cos\theta - i\omega\tau} \sin\theta \, d\theta \, k^2 \, dk \, d^3R \, d\tau \, d\omega$$

$$= \mu_0 \left(\frac{1}{2\pi}\right)^3 \int_\Omega \frac{\vec{j}(\vec{r} - \vec{R}, t - \tau)}{ikR(k^2 - \omega^2/c^2)} \times (e^{ikR} - e^{-ikR})k^2 \, dk \, d^3R \, d\tau \, d\omega$$

We integrate on k by the method of residues:

$$I = \int_0^\infty \frac{e^{ikR} - e^{-ikR}}{i(k^2 - \omega^2/c^2)} k \, dk = \int_{-\infty}^{+\infty} \frac{e^{ikR}}{i(k^2 - \omega^2/c^2)} k \, dk = \pi e^{i(\omega R/c)}$$

or

$$\vec{A}(\vec{r},t) = \frac{\mu_0}{8\pi^2} \int_\Omega \frac{\vec{j}(\vec{r} - \vec{R}, t - \tau)}{R} e^{i\omega(R/c - \tau)} \, d^3R \, d\tau \, d\omega$$

By integrating on ω and using the representation of the Dirac function δ with Fourier integrals,

$$\vec{A}(\vec{r},t) = \frac{\mu_0}{4\pi} \int \frac{\vec{j}(\vec{r} - \vec{R}, t - \tau)\delta(R/c - \tau)}{R} \, d^3R \, d\tau$$

the integration on τ gives

$$\vec{A}(\vec{r},t) = \frac{\mu_0}{4\pi} \int \frac{\vec{j}(\vec{r} - \vec{R}, t - R/c)}{R} \, d^3R$$

or, remembering that $\vec{R} = \vec{r} - \vec{r}\,'$,

$$\vec{A}(\vec{r},t) = \frac{\mu_0}{4\pi} \int_V \frac{\vec{j}\left(\vec{r}\,', t - \dfrac{|\vec{r} - \vec{r}\,'|}{c}\right)}{|\vec{r} - \vec{r}\,'|} \, d^3r'$$

a formula well known for retarded potential.

Note. Dirac's function δ is introduced into physics so as to permit the association of a physical magnitude such as that of mass or charge to a point (according to Euclid).

One of the representations of Dirac's function δ is given by

$$\delta(x) = \lim_{\epsilon \to 0} f_\epsilon(x)$$

with

$$f_\epsilon(x) = \begin{cases} 0 & \text{for } |x| > \epsilon \\ \dfrac{1}{\epsilon} & \text{for } |x| \leq \epsilon \end{cases}$$

(cf. Figure 13) According to the figure, we have not only

$$\int_{-\infty}^{+\infty} \delta(x)\, dx = 1 \tag{24a}$$

but also

$$F(a) = \int_{-\infty}^{+\infty} F(x)\delta(x - a)\, dx \tag{25}$$

a relation which necessitates only that the functon $F(x)$ defined and continuous in the neighborhood of $x = a$. Dirac's function δ, by definition, is such that

$$\delta(x) = 0 \qquad \text{for } x \neq 0 \tag{26a}$$

$$\delta(x) \longrightarrow \infty \qquad \text{for } x = 0 \tag{27a}$$

However, (24a) must still be respected.

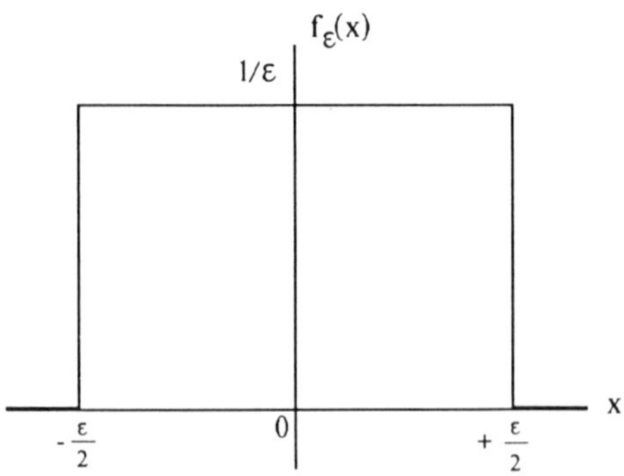

Figure 13 Function δ.

If we envisage a function $F(x)$ such that

$$F(x) = k\delta(x)$$

where k can be any scalar other than zero, the definition domain of $F(x)$ is identical to that of the function δ:

$$F(x) = 0 \qquad \text{for } x \neq 0 \tag{26b}$$

$$F(x) \longrightarrow \infty \qquad \text{for } x = 0 \tag{27b}$$

One should therefore satisfy the equivalent of (24a) or

$$\int_{-\infty}^{+\infty} F(x)\, dx = 1 \tag{24b}$$

which means that $k = 1$, which is contrary to the hypothesis.

Functions such as δ and $k\delta(x)$ are not true functions and are called *distributions*. As our aim is not to use this last notion, we indicate only that the use of distributions in physics requires a new definition of the notion of integration (Lebesgue integrals).

3.9.9 Retarded Potential

Take a conductor of section S and of infinite length (cf. Figure 14). We evaluate the fields $\vec{E}$ and $\overleftrightarrow{B}$ created at point M by the conductor, supposed to carry a current density $\vec{j}$. The system is put into a vacuum. Remember that Maxwell's equations give $\vec{E}$ and $\overleftrightarrow{B}$:

$$\vec{E} = -\vec{\nabla} V - \frac{\partial \vec{A}}{\partial t} \tag{3.23}$$

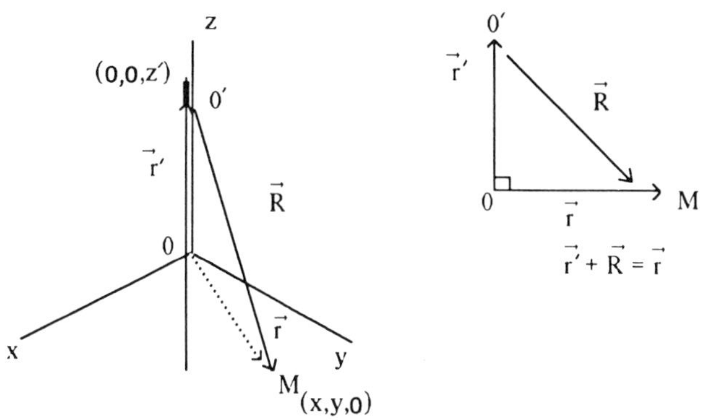

Figure 14 Thread.

and

$$\vec{B} = \vec{\nabla} \wedge \vec{A} \tag{3.19}$$

Now

$$\vec{A}(\vec{r},t) = \frac{\mu_0}{4\pi} \int_\tau \frac{\vec{j}(\vec{r}\,', t - R/c)}{R}\, d^3r'$$

where

$$R = |\vec{r}\,' - \vec{r}\,'|$$

and

$$\vec{r}\,' = (0,0,z')$$

At the moment t, the point undergoes the action of the two points obtained by the condition

$$R = ct$$

the intersections of the circle of radius R and of the thread. Note a and $-a$ as their respective altitudes. Supposing that at the moment $t = 0$ the thread is crossed by the current I such that

$$\vec{j}\; d^3r' = I\; dz'\; \vec{k}$$

it can be seen that

$$\vec{A} = \frac{\mu_0 I}{4\pi} \int_{-a}^{a} \frac{\vec{k}\; dz'}{\sqrt{r^2 + z'^2}}$$

According to Figure 14,

$$a = \sqrt{R^2 - r^2}$$

and thus

$$\vec{A} = \frac{\mu_0 I \vec{k}}{2\pi} \log \frac{R + \sqrt{R^2 - r^2}}{r}$$

from which we can deduce the electric field $\vec{E}$ by supposing the scalar potential V to be zero:

$$\vec{E} = -\frac{\partial \vec{A}}{\partial t} = -\frac{\mu_0 I \vec{k}}{2\pi} \frac{c}{\sqrt{c^2t^2 - r^2}} = -\frac{\mu_0 I c}{2\pi a} \vec{k}$$

Similarly, for the magnetic induction $\vec{B}$,

$$\vec{B} = \vec{i}\, \frac{\partial \mathrm{A}}{\partial y} - \vec{j}\, \frac{\partial \mathrm{A}}{\partial x}$$

Now, for example,

$$\frac{\partial A}{\partial x} = \frac{\partial A}{\partial r}\frac{\partial r}{\partial x} = \frac{x}{r}\frac{\partial A}{\partial x}$$

and therefore (cf. Figure 15)

$$\vec{B} = (\vec{i}\sin\theta - \vec{j}\cos\theta)\frac{\partial A}{\partial r}$$

but also

$$\vec{u}_r = \vec{i}(\vec{u}_r\cdot\vec{i}) + \vec{j}(\vec{u}_r\cdot\vec{j}) = \vec{i}\cos\theta + \vec{j}\sin\theta$$
$$\vec{u}_\theta = \vec{i}(\vec{u}_\theta\cdot\vec{i}) + \vec{j}(\vec{u}_\theta\cdot\vec{j}) = -\vec{i}\sin\theta + \vec{j}\cos\theta$$

which enables us to write

$$\vec{B} = -\vec{u}_\theta\frac{\partial A}{\partial r} = \frac{\mu_0 IR}{2\pi ra}\vec{u}_\theta$$

We assume the time t to be toward infinity, which is equivalent to

$$a \longrightarrow \infty$$
$$R \longrightarrow a$$

Then

$$\vec{E} \longrightarrow \vec{0}$$
$$\vec{B} \longrightarrow \frac{\mu_0 I}{2\pi r}\vec{u}_\theta$$

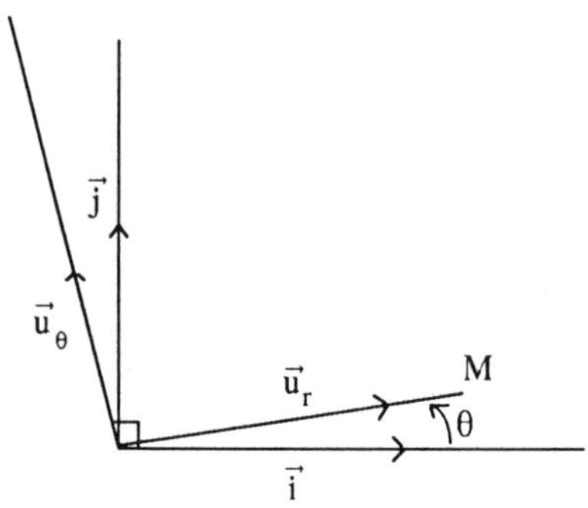

Figure 15 Polar coordinates.

We come back to the static case, that is, the formula traditionally deduced with the help of Ampère's theorem. We now evaluate Poynting's vector:

$$\vec{P} = \vec{E} \wedge \vec{B} = -\frac{\mu_0 I^2 cR}{4\pi^2 a^2 r}\vec{k} \wedge \vec{u}_\theta = \frac{\mu_0 I^2 c^2 t}{4\pi^2 a^2 r}\vec{u}_r$$

The flux of Poynting's vector across a cylinder of height H and radius r (Figure 16) or radiant energy is

$$\Phi = \oiint_\Sigma \vec{P}\cdot\vec{dS}$$

as

$$d\vec{S} = \vec{u}_r hr\, d\theta \qquad 0 \leqslant \theta \leqslant 2\pi$$

The flux Φ is written

$$\Phi = \frac{\mu_0 h I^2 c^2 t}{2\pi a^2} = \frac{\mu_0 h I^2 c^2 t}{2\pi(c^2t^2 - r^2)} \qquad ct > r$$

If we trace the curve $\Phi = \Phi(t)$ (Figure 17), we observe a lateness in the radiation, which is what justifies the title of retarded potentials.

3.9.10 Lienard–Wiechert Potential

Evaluate the retarded potential created by a point charge in motion. The charge density is

$$\rho(\vec{r},t) = q\delta(\vec{X}(t) - \vec{r}) \tag{1}$$

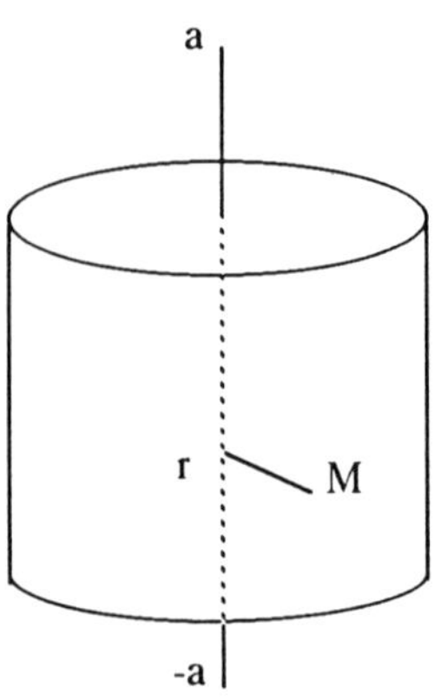

Figure 16 Cylinder.

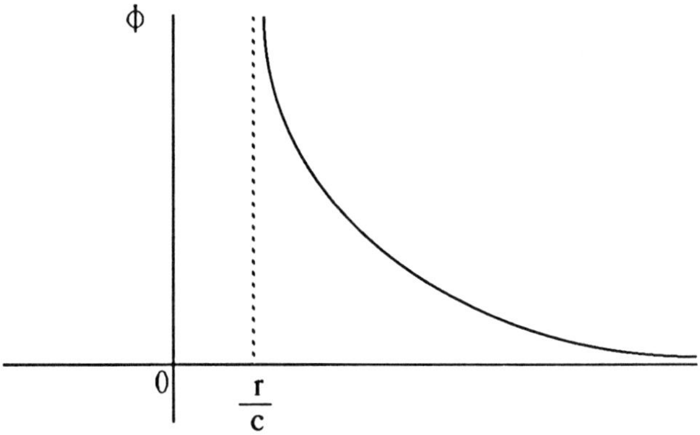

Figure 17 Flux variation.

where $\vec{X}(t)$ defines the trajectory of the particle by the intermediary of the parametric equations of motion.

For the scalar potential (cf. Section 3.9.9), by analogy with the equation of retarded vector potential and by definition of the potential and current density quadrivectors,

$$\vec{A} \longrightarrow \frac{V}{c}$$

$$\mu_0 \vec{j} \longrightarrow \frac{\rho}{c\epsilon_0}$$

We instantly have

$$V(\vec{r},t) = \frac{1}{4\pi\epsilon_0} \int_\tau \frac{\rho\left(\vec{r}\,', t - \dfrac{|\vec{r} - \vec{r}\,'|}{c}\right)}{|\vec{r} - \vec{r}\,'|} d^3r' \qquad (2)$$

By writing $\vec{R} = \vec{r} - \vec{r}\,'$ and $\tau = R/c$, we obtain

$$V(\vec{r},t) = \frac{1}{4\pi\epsilon_0} \int \frac{\rho(\vec{r} - \vec{R}, t - \tau)\delta(\tau - R/c)}{R} d^3R\, d\tau$$

and with (1),

$$V(\vec{r},t) = \frac{q}{4\pi\epsilon_0} \int \frac{\delta(\vec{X}(t - \tau) - \vec{r} + \vec{R})\delta(\tau - R/c)}{R} d^3R\, d\tau \qquad (3)$$

Then integration on $\vec{R}$ leads to

$$V(\vec{r},t) = \frac{q}{4\pi\epsilon_0} \int \frac{\delta(\tau - |\vec{r} - \vec{X}(t-\tau)|/c)}{|\vec{r} - \vec{X}(t-\tau)|} d\tau$$

It is important to write $t' = t - \tau$:

$$V(\vec{r},t) = \frac{q}{4\pi\epsilon_0} \int \frac{\tau(t - t' - |\vec{r} - \vec{X}(t')|/c)}{|\vec{r} - \vec{X}(t')|} dt' \tag{4}$$

but also

$$\vec{a} = \vec{r} - \vec{X}(t') \tag{5}$$

$$\lambda = t' + \frac{\|\vec{a}\|}{c} \tag{6}$$

and thus

$$\frac{d\lambda}{dt'} = 1 + \frac{1}{c}\frac{d}{dt'}\|\vec{a}\|$$

Now

$$\frac{d}{dt'}\|\vec{a}\| = -\frac{\vec{a}\cdot\vec{v}(t')}{\|\vec{a}\|}$$

from which, by writing $\vec{\beta} = \vec{v}/c$,

$$\frac{d\lambda}{dt'} = 1 - \frac{\vec{a}\cdot\vec{\beta}}{\|\vec{a}\|}$$

and therefore,

$$V(\vec{r},t) = \frac{q}{4\pi\epsilon_0} \int \frac{\delta(t - \lambda(t'))}{\|\vec{a}(t')\| - \vec{a}(t')\cdot\vec{\beta}(t')} d\lambda(t') \tag{7}$$

The function δ sets the time t'; it is denoted t_+ and is given by the equation (6)

$$t' = t_+ = t - \frac{\|\vec{a}(t_+)\|}{c}$$

Equation (7) can then be integrated and gives

$$V(\vec{r},t) = \frac{q}{4\pi\epsilon_0} \frac{1}{(\mathrm{a}(t_+) - \vec{a}(t_+)\cdot\vec{\beta}(t_+)}$$

and we can deduce from it, with the help of Maxwell's equations,

$$\vec{E} = -\vec{\nabla} V - \frac{\partial \vec{A}}{\partial t}$$

and

$$\vec{B} = \vec{\nabla} \wedge \vec{A}$$

the electric and magnetic fields.

3.9.11 Dipolar Radiation and Bohr Model

We take a dipole of length $\|\vec{d}\|$ and of momentum $\vec{\mu}(t)$ which moves according to Oz (Figure 18). The vectors $\vec{\mu}$, $\dot{\vec{\mu}}$, $\ddot{\vec{\mu}}$ are collinear. Evaluate the potential created by the dipole at any point M.

It was seen earlier that

$$\vec{A}(\vec{r},t) = \frac{\mu_0}{4\pi} \int \frac{\vec{j}\left(\vec{r}\,', t - \dfrac{\|\vec{r} - \vec{r}\,'\|}{c}\right)}{\|\vec{r} - \vec{r}\,'\|} d^3r' \tag{1}$$

For the dipole we suppose that $\|\vec{r}\| \gg \|\vec{r}\,'\|$ and that

$$\vec{j}\left(\vec{r}\,', t - \frac{\|\vec{r} - \vec{r}\,'\|}{c}\right) = \vec{j}\left(\vec{r}\,', t - \frac{\|\vec{r}\|}{c}\right)$$

with

$$\vec{j}\left(\vec{r}\,', t - \frac{\|\vec{r}\|}{c}\right) = \vec{v}\left(t - \frac{\|\vec{r}\|}{c}\right) \rho \left(\vec{r}, t - \frac{\|\vec{r}\|}{c}\right) \tag{2}$$

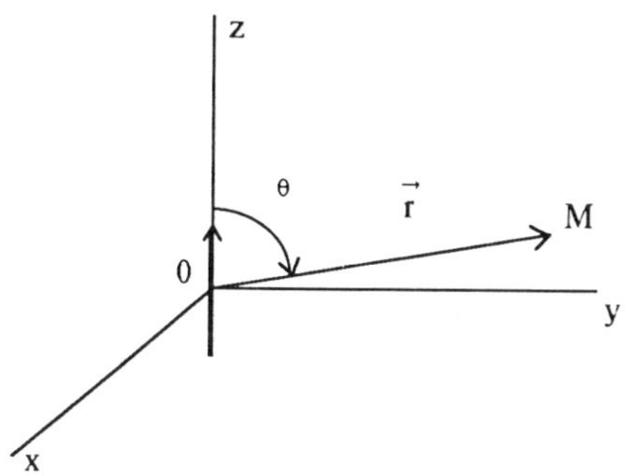

Figure 18 Dipole.

where $\vec{v}$ is the velocity of the charges in motion. We evaluate the vector potential with the help of (1):

$$\vec{A}(\vec{r},t) = \frac{\mu_0}{4\pi}\vec{v}\left(t - \frac{\|\vec{r}\|}{c}\right)\int \frac{\rho(\vec{r}\,', t - \|\vec{r}\|/c)}{\|\vec{r}\|} d^3r' + \cdots$$

$$= \frac{\mu_0}{4\pi}\frac{\vec{v}}{\|\vec{r}\|} q\left(t - \frac{\|\vec{r}\|}{c}\right)$$

By definition the dipolar momentum is the vector

$$\vec{\mu} = q\vec{d}\left(t - \frac{\|\vec{r}\|}{c}\right) \tag{3}$$

and thus

$$\dot{\vec{\mu}} = q\vec{v}\left(t - \frac{\|\vec{r}\|}{c}\right) \tag{4}$$

The dipolar momentum $\vec{\mu}$ can vary, either due to the distance $\vec{d}$ separating the two charges $-q$ and $+q$, or due to the charge q itself. The first case gives, for example, Thomson's atomic model. The second case corresponds to the study of the radiation of an antena.

In the Lorentz gauge,

$$\vec{\nabla}\cdot\vec{A} + \frac{1}{c^2}\frac{\partial V}{\partial t} = 0 \tag{5}$$

or, after calculation,

$$V = \frac{1}{4\pi\epsilon_0}\frac{\vec{r}\cdot\vec{\mu}^*}{r^3} \tag{6}$$

with

$$\vec{\mu}^* = \vec{\mu} + \frac{r}{c}\dot{\vec{\mu}} \tag{7}$$

we come back to the classic calculation of the dipole for $\dot{\vec{\mu}} = \vec{0}$.

The magnetic field $\vec{B}$ is easily obtained by

$$\vec{B} = \vec{\nabla}\wedge\vec{A} = \frac{\mu_0}{4\pi r^3}\dot{\vec{\mu}}^*\wedge\vec{r} \tag{8}$$

It can equally be seen that

$$\vec{E} = \frac{1}{4\pi\epsilon_0}\left[-\frac{\vec{\mu}^*}{r^3} + \frac{3\vec{r}(\vec{r}\cdot\vec{\mu}^*)}{r^5} + \frac{(\ddot{\vec{\mu}}\wedge\vec{r})\wedge\vec{r}}{c^2r^3}\right] \tag{9}$$

Note T, the dipole period of oscillation, and suppose that $r/c \ll T$: the oscillations are very quick; the movement of the dipole then enables its comparison to a current:

$$
\begin{aligned}
\vec{\mu}(t) &= \vec{\mu}\left(t - \frac{r}{c} + \frac{r}{c}\right) = \vec{\mu}\left(t - \frac{r}{c}\right) + \frac{r}{c}\,\dot{\vec{\mu}}\left(t - \frac{r}{c}\right) \\
&\quad + \frac{r^2}{c^2}\,\ddot{\vec{\mu}}\left(t - \frac{r}{c}\right) + \cdots \\
&\simeq \vec{\mu}^*\left(t - \frac{r}{c}\right) \qquad (10)
\end{aligned}
$$

This is called the approximation of the field of velocities, the accelerations having been neglected. We also find that

$$\overset{\hookrightarrow}{B} = \frac{\mu_0}{4\pi}\,\dot{\vec{\mu}} \wedge \frac{\vec{r}}{r^3} \qquad (11)$$

the Biot–Savart law and the field of an electrostatic dipole

$$\vec{E} = \frac{1}{4\pi\epsilon_0}\left[-\frac{\vec{\mu}}{r^3} + \frac{3\vec{r}(\vec{r}\cdot\vec{\mu})}{r^5}\right] \qquad (12)$$

By taking $r/c \gg T$, we have the case of the field of accelerations,

$$\dot{\vec{\mu}}^* = \dot{\vec{\mu}} + \frac{r}{c}\,\ddot{\vec{\mu}} \simeq \frac{r}{c}\,\ddot{\vec{\mu}}$$

and

$$\overset{\hookrightarrow}{B} = \frac{1}{4\pi\epsilon_0 r^2 c^3}\,\ddot{\vec{\mu}} \wedge \vec{r} \qquad (13)$$

$$\vec{E} = c(\overset{\hookrightarrow}{B} \wedge \vec{u}_r) \qquad (14)$$

with

$$\vec{u}_r = \frac{\vec{r}}{r}$$

the unit vector in the vector direction $\overrightarrow{OM}$. We note [cf. (13) and (14)] that the propagation of the electromagnetic field is transverse (of plane-wave type).

Poynting's vector, giving the expression of radiated power, is, in the case of long distances,

$$\vec{P} = \vec{E} \wedge \vec{H} = \mu_0 E^2 \vec{u}_r = \frac{c}{\mu_0} B^2 \vec{u}_r = \frac{\sin^2\theta}{16\pi^2\epsilon_0 c^3 r^2} \ddot{\mu}^2 \vec{u}_r \tag{15}$$

θ being the angle between the vectors $\vec{\mu}$ and $\vec{r}$.

If the law of evolution during the time of dipolar vector momentum $\vec{\mu}$ is of the form

$$\vec{\mu} = \vec{\mu}_0 e^{i\omega t} \tag{16}$$

the average value, over a period of time, of Poynting's vector is

$$\overline{\vec{P}} = \frac{\omega^4 \mu_0^2 \sin^2\theta}{32\pi^2\epsilon_0 c^3} \vec{u}_r \tag{17}$$

The average radiation electromagnetic energy is

$$-\frac{\partial \overline{W}}{\partial t} = \iint_S \overline{\vec{P}} \cdot d\vec{S} = \frac{\omega^4 \mu_0^2}{12\pi\epsilon_0 c^3} \tag{18}$$

having used the fact that at long distances the dipole can be assumed to be a point and we thus take for surface S a sphere centered on the origin constituted by the dipole.

Let's return to the calculations that enabled Bohr to improve Rutherford's model. For an electron in motion on an extra nuclear orbit, the previous results give, by their similarity to a dipole,

$$-\frac{\partial E}{\partial t} = \iint_S \vec{P} \cdot \overrightarrow{dS} \tag{19}$$

E being the total mechanical energy of the electron, that is, the kinetic energy plus potential energy. According to (15),

$$-\frac{\partial E}{\partial t} = \int_0^{2\pi} \int_0^{\pi} \frac{\sin^2\theta \ddot{\mu}^2}{16\pi^2\epsilon_0 c^3} \sin\theta \, d\theta \, d\varphi$$

As $\mu = er$, that is $\ddot{p} = e\ddot{r}$,

$$-\frac{\partial E}{\partial t} = \frac{1}{4\pi\epsilon_0} \frac{2e^2}{3c^3} \ddot{r}^2 \tag{20}$$

According to the fundamental principle of dynamics, $\|\vec{F}\| = e\|\vec{E}\|$ or

$$\ddot{r} = \frac{e^2 Z}{4\pi\epsilon_0 m r^2}$$

We can easily check that

$$E = -\frac{e^2 Z}{8\pi\epsilon_0 r}$$

Equation (20) becomes

$$-\frac{\partial E}{\partial t} = 4\pi\epsilon_0 \frac{32E^4}{3c^3(mZe)^2}$$

which is integrated, giving

$$\frac{1}{E^3} = \frac{1}{(E_0)^3} + 4\pi\epsilon_0 \frac{32t}{c^3(mZe)^2}$$

with

$$E_0 = -\frac{Ze^2}{8\pi\epsilon_0 a}$$

the energy value when the electron is on Bohr's orbit of radius a. If we extend the distance r toward zero, the electron should fall onto the nucleus through a loss of radiated energy. The energy E tends to infinity, and from it can be deduced that the time of the electron on the nucleus is

$$t_{\text{fall}} = \frac{(4\pi\epsilon_0)^2 m^2 c^3 a^3}{4Ze^4}$$

This calculation gives a result that is contrary to experiment and caused Bohr to postulate the quantification rules.

3.9.12 Demonstration of Maxwell Equation According to Feynman

Recall several of the notions on Poisson's brackets (cf. A. Charlier, A. Bérard, and M. F. Charlier, *Analytic Mechanics*, Editions Ellipses, Paris, 1989). In the phase space (q_k, p_k), if f is a function such that

$$f = f(q_k, p_k; t) \tag{1}$$

the q_k being the generalized coordinates of the system studied and the p_k their canonically combined momentums, then

$$\frac{df}{dt} = \frac{\partial f}{\partial t} + [H, f] \tag{2}$$

where H is the hamiltonian of the system and

$$[H,f] = \frac{\partial H}{\partial p_l}\frac{\partial f}{\partial q_l} - \frac{\partial H}{\partial q_l}\frac{\partial f}{\partial p_l} \tag{3}$$

by the definition of Poisson's bracket formed with the functions H and f, $l = 1, 2, \ldots, g$, where g is the number of degrees of freedom of the system.

These brackets are antisymmetrical:

$$[f,g] = -[g,f] \tag{4}$$

and such that

$$[p_i,f] = \frac{\partial f}{\partial q_i} \tag{5}$$

$$[q_i,f] = -\frac{\partial f}{\partial p_i} \tag{6}$$

with, similarly,

$$[q_i,q_j] = 0 \tag{7}$$

$$[p_i,p_j] = 0 \tag{8}$$

$$[p_i,q_j] = \delta_{ij} \tag{9}$$

Poisson's brackets comply with the Jacobi identity:

$$[f,[g,h]] + [g,[h,f]] + [h,[f,g]] = 0 \tag{10}$$

A particle that undergoes a force $\vec{F} = \vec{F}(\vec{r},\vec{v},t)$ complies, in the nonrelativist hypothesis, with the equations of motion

$$F_i(x_j,\dot{x}_j,t) = m\ddot{x}_i \tag{11}$$

The commutators

$$[x_i,x_j] = 0 \tag{12}$$

$$[\dot{x}_i,x_j] = \delta_{ij} \tag{13}$$

were written axiomatically by Feynman but are only the transcription of equations (7) and (9). The quantities x_i and $\dot{x}_i$ are not the operators of quantum mechanics, at this point of the demonstration, but measures of the coordinates and of the components of the particle speed.

By differentiating (13) with respect to time, then by using (11) we obtain

$$[F_i,x_j] + m[\dot{x}_i,\dot{x}_j] = 0 \tag{14}$$

The preceding equation is put into the form

$$[x_l,[F_i,x_j]] + m[x_l,[\dot{x}_i,\dot{x}_j]] = 0 \tag{15}$$

with Jacobi's identity,

$$[x_l,[\dot{x}_j,\dot{x}_k]] + [\dot{x}_j,[\dot{x}_k,x_l]] + [\dot{x}_k,[\dot{x}_l,\dot{x}_j]] = 0 \tag{16}$$

and with the help of (13), we obtain

$$[x_l,[\dot{x}_j,\dot{x}_k]] = 0 \tag{17}$$

and thus (15) is equivalent to

$$[x_l,[F_i,x_j]] = 0 \tag{18}$$

Equation (14) enables us to prove the antisymmetry of the commutator $[x_j,F_k]$ by exchanging the indices j and k. Indeed, according to (14),

$$[F_i,x_j] = -m[\dot{x}_i,\dot{x}_j] \tag{19}$$

and as

$$[\dot{x}_i,\dot{x}_j] = -[\dot{x}_j,\dot{x}_i]$$

and (19) gives

$$[F_i,x_j] = -[F_j,x_i] \tag{20}$$

The tensor $[F_i,x_j]$ is therefore antisymmetrical and can be expressed in its dual form by a relation such as

$$[F_i,x_j] = \epsilon_{ij}{}^k H_k \tag{21}$$

H_l is the component of a certain pseudovector $\overleftrightarrow{H}$ which from now on will depend on the vectors $\vec{r}$ and $\vec{v}$ and the time t.

According to (18),

$$[x_l,H_k] = 0 \tag{22}$$

which means that the vector $\overleftrightarrow{H}$ depends only on the particle position and the time. The components of vector $\vec{F}$ can still be put into the form (by taking $\mu = 1$)

$$F_j = E_j + \epsilon_j{}^{kl}\dot{x}_k H_l \tag{23}$$

The values of $\vec{E}$ and of $\overleftrightarrow{H}$ clearly being important, these vectors are not independent. We clarify the relations between vectors $\vec{E}$ and $\overleftrightarrow{H}$ defined by (23). According to (21), by using (22) and (23),

$$[E_i,x_j] = 0 \tag{24}$$

a result which means that the electric field, like the magnetic field, depends on the distance and time.

The definition of the field $\overleftrightarrow{H}$ [cf. (21)] gives, with the help of (14),

$$\epsilon_{ij}{}^{k} H_k = -m[\dot{x}_i, \dot{x}_j] \tag{25}$$

We have seen in Section 3.9.4 that

$$\epsilon_{ijk}\epsilon^{ilk} = 2\delta_j^l \tag{26}$$

or [cf. (25)]

$$H_i = -\frac{m}{2}\,\epsilon_i{}^{jk}[\dot{x}_j, \dot{x}_k] \tag{27}$$

It can easily be proved, by using Jacobi's identity and explaining the summations, that

$$\epsilon^{jkl}[\dot{x}_l,[\dot{x}_j, \dot{x}_k]] = 0 \tag{28}$$

When (27) is introduced into (26), we see that

$$[\dot{x}_l, H_l] = 0 \tag{29}$$

or, according to (5),

$$\frac{\partial H_l}{\partial x_l} = 0 \tag{30}$$

which is equivalent to Maxwell's equation

$$\text{div}\, \overleftrightarrow{H} = 0 \tag{31}$$

by differentiating (27) with respect to time, it can be seen that

$$\frac{\partial H_i}{\partial t} + \frac{\partial H_i}{\partial x_j}\dot{x}_j = -m\epsilon_i{}^{jk}[\ddot{x}_j, \dot{x}_k] \tag{32}$$

and by using (11) and (23),

$$\begin{aligned}
\frac{\partial H_i}{\partial t} + \frac{\partial H_i}{\partial x_j}\dot{x}_j &= -m\epsilon_i{}^{jk}[E_j, \dot{x}_k] - m\epsilon_i{}^{jk}[\epsilon_j{}^{mn}\dot{x}_m H_n, \dot{x}_k] \\
&= \epsilon_i{}^{jk}\frac{\partial E_j}{\partial x_k} - m(\delta_m^k\delta_i^n - \delta_n^k\delta_i^m)[\dot{x}_m H_n, \dot{x}_k] \\
&= \epsilon_i{}^{jk}\frac{\partial E_j}{\partial x_k} - m([\dot{x}_k H_i, \dot{x}_k] + [\dot{x}_i H_k, \dot{x}_k]) \\
&= \epsilon_i{}^{jk}\frac{\partial E_j}{\partial x_k} - m\dot{x}_k[H_i, \dot{x}_k] + m\dot{x}_i[H_k, \dot{x}_k]
\end{aligned} \tag{33}$$

The last term on the right-hand side is zero, as the Poisson's bracket which intervenes is the divergence of vector $\overleftrightarrow{H}$, a zero-quantity according to (30):

$$\frac{\partial H_i}{\partial t} + \frac{\partial H_i}{\partial x_j} \dot{x}_j = \epsilon_i^{\ jk} \frac{\partial E_j}{\partial x_k} - m\dot{x}_k[H_i, \dot{x}_k] \tag{34}$$

According to (5),

$$m[H_i, \dot{x}_k] = -\frac{\partial H_i}{\partial x_k} \tag{35}$$

and thus

$$\frac{\partial H_i}{\partial t} = \epsilon_i^{\ jk} \frac{\partial E_j}{\partial x_k} \tag{36}$$

Formula (36) is clearly equivalent to Maxwell's equation

$$\vec{\nabla} \wedge \vec{E} = -\frac{\partial \vec{B}}{\partial t}$$

for $\mu = 1$.

Feynman was trying to find a new quantum mechanics. The previous demonstration uses classical mechanics (invariant under the transformations of Galileo's group) and brings us to the relativistic equations of Maxwell, invariant under a Lorentz transformation. Of great interest in this regard is F. J. Dyson's article in *Am. J. Phys. 58*(3), 209 (1990).

APPENDIX 7 TENSORS AND PSEUDOTENSORS

A7.1 Pseudotensors

In a vector space of dimension 3, we examine the components of an antisymmetrical tensor of order 3:

$$T = T^{ijk}\vec{e}_i\vec{e}_j\vec{e}_k \tag{A7.1}$$

The diagonal elements are zero by the definition of tensor T; the $C_3^3 = 1$ component must now be given. It is written $T^{123} = t$. If the tensor were nonspecific and nonantisymmetrical, it would be necessary to write $3^3 = 27$ components. During a linear transformation the tensor T becomes T' and the element T'^{123} will be such that

$$T'^{1'2'3'} = t' = \Lambda_l^{1'}\Lambda_m^{2'}\Lambda_n^{3'}T^{lmn} \tag{A7.2}$$

The tensor T' is clearly antisymmetrical [cf. (A7.2)]. According to (A7.2), we should also have

$$T^{123} = t = \Lambda_{l'}^{1}\Lambda_{m'}^{2}\Lambda_{n'}^{3}T'^{l'm'n'} = \Lambda_{l'}^{1}\Lambda_{m'}^{2}\Lambda_{n'}^{3}\epsilon'^{l'm'n'}t' \tag{A7.3}$$

or

$$t = (\det \Lambda)t' \tag{A7.4}$$

by the definition of the determinant of a matrix. The quantity t is not a true scalar; $\det(\Lambda)$ can differ from unity.

To obtain a true scalar, t must be multiplied by another pseudoscalar s (formed in the same way as t, starting from an antisymmetrical tensor of the third order), in such a way that

$$ts = t's' \tag{A7.5}$$

with s such that

$$s = \frac{1}{\det \Lambda} s' \tag{A7.6}$$

The elements of a tensor complying with the transformation law (A7.4) [respectively, (A7.5)] are called *capacity* (respectively, *density*). We can give as examples:

The volume $v = v' \det \Lambda$.

The divergence of a vector $\operatorname{div} \vec{A} = \dfrac{1}{\det \Lambda} (\operatorname{div} \vec{A})'$.

In a vector space of dimension n, an antisymmetrical tensor of rank m will have C_m^n different components (any tensor will, a priori, have n^m components). The previous notions are generalized to tensor capacities and tensor densities.

A7.2 Levi–Civita Tensor

We saw earlier that a pseudoscalar can be associated with an antisymmetrical tensor of rank 3. We are going to take the case of $t = \pm 1$.

The Levi–Civita tensor has been defined as [cf. (3.46)]

$$\epsilon = \frac{1}{N!} \epsilon^{ijk\cdots} \vec{e}_i \vec{e}_j \vec{e}_k \cdots \tag{A7.7}$$

This is a differential form

$$\epsilon = dx^1 \wedge dx^2 \wedge dx^3 \cdots \wedge dx^n \tag{A7.8}$$

a specific case, valid only in a euclidean space. When generalizing to non-cartesian systems of (A7.8), we have

$$\epsilon = \sqrt{g}\, dx^1 \wedge dx^2 \wedge dx^3 \cdots \wedge dx^n \tag{A7.9}$$

where g is the determinant of the metric tensor. For Minkowski space we have

$$\epsilon = \sqrt{-g}\, dx^1 \wedge dx^2 \wedge dx^3 \cdots \wedge dx^n \tag{A7.10}$$

One of the basic properties of the components of tensor ϵ is

$$\epsilon^{ijk\cdots}\epsilon_{lmn\cdots} = \begin{vmatrix} \delta^i_l & \delta^i_m & \delta^i_n & \cdots \\ \delta^j_l & \delta^j_m & \delta^j_n & \cdots \\ \delta^k_l & \delta^k_m & \delta^k_n & \cdots \\ \cdot & \cdot & \cdot & \cdots \\ \cdot & \cdot & \cdot & \cdots \\ \cdot & \cdot & \cdot & \cdots \end{vmatrix} \tag{A7.11}$$

We have, for example,

$$\epsilon^{ij}\epsilon_{ik} = \begin{vmatrix} \delta^i_i & \delta^i_k \\ \delta^j_i & \delta^j_k \end{vmatrix} = \delta^i_i\delta^j_k - \delta^i_k\delta^j_i = 2\delta^j_k - \delta^j_k = \delta^j_k \tag{A7.12}$$

a relation that gives in particular

$$\epsilon^{ij}\epsilon_{ij} = \delta^i_j = 2 = 2! \tag{A7.13}$$

Equally,

$$\epsilon^{ijk}\epsilon_{ijl} = 3(3\delta^k_l - \delta^k_l) - 3\delta^l_k + \delta^l_k + \delta^l_k - 3\delta^l_k = 2\delta^l_k \tag{A7.14}$$

and thus

$$\epsilon^{ijk}\epsilon_{ijk} = 2\delta^k_k = 6 = 3! \tag{A7.15}$$

We can easily show the relations

$$\epsilon_{ijk}\epsilon^{ijk} = 6 \qquad \text{A7.16)}$$

and

$$\epsilon_{ijk}\epsilon^{ilm} = \delta^l_j\delta^m_k - \delta^m_j\delta^l_k \tag{A7.17}$$

In Minkowski space,

$$\epsilon_{\alpha\beta\gamma\delta} = g_{\alpha\alpha'}g_{\beta\beta'}g_{\gamma\gamma'}g_{\delta\delta'}\epsilon^{\alpha'\beta'\gamma'\delta'}$$

or

$$\epsilon_{0123} = g_{0\alpha'}g_{1\beta'}g_{2\gamma'}g_{3\gamma'}\epsilon^{\alpha'\beta'\gamma'\delta'} = g_{00}g_{11}g_{22}g_{33}\epsilon^{0123} = -\epsilon^{0123}$$

or even

$$\epsilon_{\alpha\beta\gamma\delta} = -\epsilon^{\alpha\beta\gamma\delta} \tag{A7.18}$$

The relation (A7.18) is written, generally, in a space of dimension 4:

$$\epsilon_{\alpha\beta\gamma\delta} = -g\epsilon^{\alpha\beta\gamma\delta} \tag{A7.19}$$

where g is such that

$$g = -\det(g_{\mu\nu}) \tag{A7.20}$$

Note that if A is a 3×3 matrix,

$$\epsilon_{pqr} \det(A) = \epsilon_{ijk} A^i_p A^j_q A^k_r \tag{A7.21}$$

a relation equivalent to

$$\det(A) = \epsilon_{ijk} A^i_1 A^j_2 A^k_3 \tag{A7.22}$$

the development according to a row or a column [(A7.21) with $p = 1$, $q = 2$, $r = 3$].

We can easily check in the euclidean space of dimension 3 the classic formulas

$$(\vec{A} \wedge \vec{B})_i = \epsilon_{ijk} A^j B^k \tag{A7.23}$$

$$\vec{A}\cdot(\vec{B} \wedge \vec{C}) = A^i \epsilon_{ijk} B^j C^k \tag{A7.24}$$

$$\vec{\nabla}\cdot(\vec{\nabla} \wedge \vec{A}) = \partial_i \epsilon^{ij}{}_k \partial_j A^k \tag{A7.25}$$

$$[\vec{\nabla} \wedge (\vec{\nabla}\varphi)]_i = \epsilon_i{}^{jk} \partial_j \partial_k \varphi \tag{A7.26}$$

$$\begin{aligned}[\vec{A} \wedge (\vec{B} \wedge \vec{C})]_i &= \epsilon_{kij} \epsilon^{klm} A^j B_l C_m = (\delta^l_i \delta^m_j - \delta^m_i \delta^l_j) A^j B_l C_m \\ &= B_i(\vec{A}\cdot\vec{C}) - C_i(\vec{A}\cdot\vec{B})\end{aligned} \tag{A7.27}$$

but

$$\begin{aligned}[(\vec{A} \wedge \vec{B}) \wedge \vec{C})]_i &= -\epsilon_{kij} \epsilon^{klm} C^j A_l B_m \\ &= B_i(\vec{A}\cdot\vec{C}) - A_i(\vec{C}\cdot\vec{B})\end{aligned} \tag{A7.28}$$

$$[\vec{\nabla}\cdot(\vec{A}\cdot\vec{B})]_i = A_j \partial_i B^j + B_j \partial_i A^j \tag{A7.29}$$

$$[\vec{A} \wedge (\vec{\nabla} \wedge \vec{B})]_i = A_j \partial_i B^j - (\vec{A}\cdot\vec{\nabla}) B_i \tag{A7.30}$$

so

$$\begin{aligned}[\vec{\nabla}(\vec{A}\cdot\vec{B})]_i &= [\vec{A} \wedge (\vec{\nabla} \wedge \vec{B})]_i + [\vec{A}\cdot\vec{\nabla}] B_i \\ &+ [\vec{B} \wedge (\vec{\nabla} \wedge \vec{A})]_i + [\vec{B}\cdot\vec{\nabla}] A_i\end{aligned} \tag{A7.31}$$

A7.3 Surface Element

According to Appendix A6 (Sections A6.1 and A6.2), the element of area of the parallelogram formed from two vectors $d\vec{x}$ and $d\vec{y}$ is written with the help of vector $d\vec{S}$ such that

$$d\vec{S} = d\vec{x} \wedge d\vec{y} \tag{A7.32}$$

or by introducing the Levi–Civita tensor

$$(d\vec{S})_i = \epsilon_{ijk}\, dx^j\, dx^k \tag{A7.33}$$

a pseudotensor of rank 2. Now $dx^j\, dx^k$ becomes $dx^j \wedge dx^k$ by replacing the surface element with an exterior product:

$$(d\vec{S})_i = \tfrac{1}{2}\epsilon_{ijk}\, dx^j \wedge dx^k \tag{A7.34}$$

(the factor $\frac{1}{2}$ comes from the double counting). Note that the relation (A7.34) corresponds to the definition of the dual (A6.23):

$$({}^*S)_{i_1} = \tfrac{1}{2}\epsilon_{i_1 j_1 j_2} S^{j_1 j_2} \tag{A7.35}$$

where S is a 1-form and $S^{j_1 j_2}$ are the tensor components of a 2-form, a relation equivalent to the formulation

$$^*S = ({}^*S)_{i_1}\, dx^{i_1} \tag{A7.36}$$

Formula (A7.36) is a generalization of a classic equation of the components of a cross product. For example, for kinetic momentum or any other physical quantity being expressed, in its definition, by the intermediary of a cross product

$$L_i = (\vec{L})_i = (\vec{r} \wedge \vec{p})_i = \epsilon_{ijk} x^j p^k = \epsilon_{ijk} L^{jk} \tag{A7.37}$$

with

$$L^{jk} = x^j p^k \tag{A7.38}$$

Observations

1. The relation (A7.10) causes the intervention of the *volume form*, of which the dual is the usual element of volume:

$$d\tau = \frac{1}{3!}\,\epsilon_{ijk}\, dx^i \wedge dx^j \wedge dx^k \tag{A7.39}$$

   by analogy with (A7.10). It can be seen that the right-hand side of (A7.39) is quite simply the form that defines the Levi–Civita tensor.
2. The definition of the volume form

$$V = \epsilon = \frac{1}{n!}\,\epsilon_{i_1 \cdots i_n}\, dx^{i_1} \wedge \cdots \wedge dx^{i_n}$$
$$= dx^1 \wedge dx^2 \wedge \cdots \wedge dx^n$$

   V is the euclidean volume n-form.

By moving into dual space, $\epsilon \to {}^*\epsilon$, which is a scalar equal to 1, the usual geometric volume τ:

$$^*\epsilon = {}^*V = 1 = \tau$$

As V is an n-form, we can apply it to a basis vector of the tangent space

$$V(\vec{e}_i) = S_i = \frac{1}{n!}\,\epsilon_{i_1\cdots i_n}\,dx^{i_1}\wedge dx^{i_2}\wedge\cdots\wedge dx^{i_n}(\vec{e}_i)$$

For example, in three dimensions,

$$V = \frac{1}{3!}\,\epsilon_{ijk}\,dx^i\wedge dx^j\wedge dx^k$$

$$S_2 = V(\vec{e}_2) = \frac{1}{3!}(-\epsilon_{123}\,dx^1\wedge dx^3\wedge dx^2 + \epsilon_{132}\,dx^1\wedge dx^3\wedge dx^2$$
$$+ \epsilon_{213}\,dx^1\wedge dx^3\wedge dx^2 + \epsilon_{231}\,dx^3\wedge dx^1\wedge dx^2$$
$$+ \epsilon_{312}\,dx^3\wedge dx^1\wedge dx^2 - \epsilon_{321}\,dx^3\wedge dx^1\wedge dx^2)(\vec{e}_2)$$

while bearing in mind that

$$dx^i(\vec{e}_j) = \delta^i_j$$

We have

$$S_2 = \frac{1}{2!}\,dx^1\wedge dx^2$$

By generalizing to three dimensions,

$$S_i = \frac{1}{2!}\,\epsilon_{ijk}\,dx^j\wedge dx^k$$

and in n dimensions,

$$S_i = \frac{1}{(n-1)!}\,\epsilon_{j_1\cdots j_{n-1}i}\,dx^{j_1}\wedge\cdots\wedge dx^{j_{n-1}}$$

at once enabling us to write the form of the surface (hypersurface of dimension $n - 1$)

$$S_j = \frac{1}{(n-1)!}\,\epsilon_{i_1\cdots i_{n-1}i_j}\,dx^{i_1}\wedge\cdots\wedge dx^{i_{n-1}}$$

a hypersurface of dimension $n - 1$ associated with a volume of dimension n.

In three dimensions, by projecting S_2 onto $(\vec{e}_1)$, for example:

$$S_2(\vec{e}_1) = L_{21} = \frac{1}{2!}\epsilon_{2jk}\, dx^j \wedge dx^k(\vec{e}_1)$$
$$= \tfrac{1}{2}(\epsilon_{213}\, dx^1 \wedge dx^3 + \epsilon_{231}\, dx^3 \wedge dx^1)(\vec{e}_1) = dx^3 \wedge dx^1(\vec{e}_1)$$
$$= dx^3$$

We once again find the element of length. It is clear that

$$S_1(\vec{e}_{21}) = -dx^3$$

(the sense of the permutation of the three axes is respected).

APPENDIX 8 DIFFERENTIAL FORMS AND MAXWELL'S EQUATIONS

f being a k-form, it was seen in Appendix 6 that the differential of a k-form is a $(k + 1)$-form. This passage is equivalent to the one obtained by applying the curl to a vector, an operation that causes a vector to correspond to a tensor of second order. The change of a vector into a scalar, by divergence, has its equivalent for the forms. We will clarify this notation.

The *codifferential* of a form by the operation is defined such that

$$\delta\colon \Lambda^k(M) \longrightarrow \Lambda^{k-1}(M)$$

that is,

$$\delta f \in \Lambda^{k-1}(M)$$

Remember that

$$d\colon \Lambda^k(M) \longrightarrow \Lambda^{k+1}(M)$$
$$*\colon \Lambda^k(M) \longrightarrow \Lambda^{n-k}(M)$$

δ is expressed with the help of the dual operator and the differential d. From what came before it follows that

$$d\colon \Lambda^k(M) \longrightarrow \Lambda^{n-k+1}(M)$$
$$*d*\colon \Lambda^k(M) \longrightarrow \Lambda^{n-(n-k+1)}(M) = \Lambda^{k-1}(M)$$

We will thus write by definition

$$\delta = (-1)^{k(n-k+1)} * d * \qquad \text{(A8.1)}$$

The phase is introduced for convenience (permutations).

We can show that the laplacian is written

$$\Delta = \delta d + d\delta \qquad \text{(A8.2)}$$

and that

$$\delta^2 = 0 \tag{A8.3}$$

this last relation simply being Poincaré's theorem.

We now give several useful formulas (T is a k-form):

$$**T = (-1)^{k(n-k)}T \tag{A8.4}$$

$$*\delta T = (-1)^{n-k+1}\, d*T \tag{A8.5}$$

$$\delta * T = (-1)^{n-k}\, *dT \tag{A8.6}$$

Given are several indications for demonstration of the previous results. By definition for a k-form T,

$$(*T)^{i_1\cdots i_{n-k}} = \frac{1}{k!}\, \epsilon^{i_1\cdots i_{n-k}j_1\cdots j_k}T_{j_1\cdots j_k}$$

and

$$\begin{aligned}(**T)_{l_1\cdots l_k} &= \frac{1}{(n-k)!}\, \epsilon_{l_1\cdots l_k i_1\cdots i_{n-k}}(*T)^{i_1\cdots i_{n-k}} \\ &= \frac{1}{k!(n-k)!}\, \epsilon^{i_1\cdots i_{n-k}j_1\cdots j_k}\epsilon_{l_1\cdots l_k i_1\cdots i_{n-k}}(T)^{j_1\cdots j_k} \\ &= \frac{(-1)^{k(n-k)}}{k!(n-k)!}\, \epsilon_{i_1\cdots i_{n-k}l_1\cdots l_k}\epsilon^{i_1\cdots i_{n-k}j_1\cdots j_k}T_{j_1\cdots j_k} \\ &= \frac{(-1)^{k(n-k)}}{k!(n-k)!}\,(n-k)!\ \mathrm{sign}\begin{vmatrix} j_1 & \cdots & j_k \\ l_1 & \cdots & l_k \end{vmatrix}\ T_{j_1\cdots j_k}\end{aligned}$$

with

$$\mathrm{sign}\begin{vmatrix} j_1 & \cdots & j_k \\ l_1 & \cdots & l_k \end{vmatrix} = k!\delta^{j_1}_{l_1}\cdots\delta^{j_k}_{l_k}$$

and thus

$$(**T)_{l_1\cdots l_k} = (-1)^{k(n-k)}T_{l_1\cdots l_k}$$

In conclusion,
If $k(n-k)$ is even, $(**T)_{l_1\cdots l_k}$ is self-dual.
If $k(n-k)$ is odd, $(**T)_{l_1\cdots l_k}$ is an anti-self-dual relation.

Supposing that the coderivative takes into account the introduction of sign$(-1)^{k(n-k)}$:

$$\delta T = (-1)^{k(n-k)+1}\, *d*T$$

By defining the *Maxwell form* associated with the antisymmetrical tensor

$$F^{\mu\nu} = \partial^\mu A^\nu - \partial^\nu A^\mu \tag{3.43}$$

by the 2-form

$$F = -\tfrac{1}{2}F_{\mu\nu}\, dx^\mu \wedge dx^\nu \tag{A8.7}$$

with

$$(F_{\mu\nu}) = \begin{bmatrix} 0 & \dfrac{E^1}{c} & \dfrac{E^2}{c} & \dfrac{E^3}{c} \\ -\dfrac{E^1}{c} & 0 & -B^3 & B^2 \\ -\dfrac{E^2}{c} & B^3 & 0 & -B^1 \\ -\dfrac{E^3}{c} & -B^2 & B^1 & 0 \end{bmatrix} \tag{A8.8}$$

relation (A8.7) is expressed by

$$F = (E_i\, dx^i) \wedge dt + \epsilon_{ijk} B_i\, dx^j \wedge dx^k \tag{A8.9}$$

The dual form $^*F$ is such that

$$\begin{aligned} ^*F &= -\tfrac{1}{2}{}^*F_{\mu\nu}(dx^\mu \wedge dx^\nu) \\ &= (B_i\, dx^i) \wedge dt - \epsilon_{ijk} E_i\, dx^j \wedge dx^k \end{aligned} \tag{A8.10}$$

Equation (A8.7) with (3.43) gives

$$F = -\tfrac{1}{2}(\partial_\mu A_\nu - \partial_\nu A_\mu)\, dx^\mu \wedge dx^\nu \tag{A8.11}$$

If we write the 1-form

$$A = A_\mu\, dx^\mu \tag{A8.12}$$

we see that

$$\begin{aligned} dA &= \partial_\nu A_\mu\, dx^\nu\, dx^\mu = \tfrac{1}{2}\partial_\nu A_\mu\, dx^\nu \wedge dx^\mu + \tfrac{1}{2}\partial_\mu A_\nu\, dx^\mu \wedge dx^\nu \\ &= \tfrac{1}{2}(\partial_\nu A_\mu - \partial_\mu A_\nu)\, dx^\nu \wedge dx^\mu = -F \end{aligned} \tag{A8.13}$$

F is a coboundary. We evaluate the differential of the dual of F by using the relation (A8.10)

$$\begin{aligned} d^*F = {} & \partial_j B_i\, dx^j \wedge dx^i \wedge c\, dt - \frac{1}{c}\,\epsilon^i_{jk}\partial_l E_i\, dx^l \wedge dx^j \wedge dx^k \\ & - \frac{1}{c}\,\epsilon^i{}_{jk}\partial_0 E_i c\, dt \wedge dx^j \wedge dx^k \end{aligned} \tag{A8.14}$$

It is seen that

$$\frac{1}{c}\,\epsilon^{i}{}_{jk}\partial_l E_i\, dx^l \wedge dx^j \wedge dx^k$$

$$= \frac{1}{c}\,\epsilon^{i}{}_{jk}\partial_l E_i(\tfrac{1}{2}\epsilon^{ljk}\, dx \wedge dy \wedge dz)$$

$$= \frac{1}{c}\,\delta^{il}\partial_l E_i\, dx \wedge dy \wedge dz = \frac{1}{c}\,\vec{\nabla}\cdot\vec{E}\, dx \wedge dy \wedge dz$$

$$= \frac{\rho}{\epsilon_0 c}\, dx \wedge dy \wedge dz = J_0\, dx \wedge dy \wedge dz \qquad \text{(A8.15)}$$

remembering that

$$J^{\mu} = \left(\frac{\rho}{\epsilon_0 c},\ \mu_0 \vec{j}\right)$$

Thus

$$d^*F = \partial_j B_i\, dx^j \wedge dx^i \wedge c\, dt - J_0\, dx \wedge dx \wedge dx$$
$$- \frac{1}{c^2}\,\epsilon^{i}{}_{jk}\,\frac{\partial E_i}{\partial t}\, c\, dt \wedge dx^j \wedge dx^k \qquad \text{(A8.16)}$$

Ampère's theorem,

$$\vec{\nabla} \wedge \overleftrightarrow{H} = \vec{j} + \frac{\partial \vec{D}}{\partial t}$$

is written with the help of the Levi–Civita tensor:

$$\epsilon_l{}^{mn}\partial_m H_n = j_l + \epsilon_0\,\frac{\partial E_l}{\partial t}$$

By multiplying the two members by $\epsilon^{l}{}_{pq}$ it can be seen that

$$(\delta^m_p\delta^n_q - \delta^m_q\delta^n_p)\partial_m H_n = \epsilon^{l}{}_{pq} j_l + \epsilon_0\epsilon^{l}{}_{pq}\,\frac{\partial E_l}{\partial t}$$

a relation that gives

$$\partial_p B_q - \partial_q B_p - \epsilon_0\mu_0\epsilon^{l}{}_{pq}\,\frac{\partial E_l}{\partial t} = \epsilon^{l}{}_{pq} J_l \qquad \text{(A8.17)}$$

and so, as

$$\epsilon_0\mu_0 = \frac{1}{c^2}$$

and

$$\partial_j B_i \, dx^j \wedge dx^i \wedge c\,dt = \tfrac{1}{2}(\partial_j B_i - \partial_i B_j)\, dx^j \wedge dx^i \wedge c\,dt \qquad \text{(A8.18)}$$

(A8.16), (A8.17), and (A8.18) mean that

$$d^*F = \epsilon^l{}_{ji} J_l \, dx^j \wedge dx^i \wedge c\,dt - J_0 \, dx \wedge dy \wedge dz \qquad \text{(A8.19)}$$

a relation that is written

$$d^*F = J \qquad \text{(A8.20)}$$

where by definition

$$J = \epsilon^l{}_{ji} J_l \, dx^j \wedge dx^i \wedge c\,dt - J_0 \, dx \wedge dx \wedge dx \qquad \text{(A8.21)}$$

is the 3-form density of current.

We also evaluate the quantity dF starting from (A8.9):

$$dF = (\partial_j E_i + \epsilon^p{}_{ji}\partial_0 B_p)\, dx^j \wedge dx^i \wedge dt + \epsilon^p{}_{ji}\partial_l B_p \, dx^l \wedge dx^j \wedge dx^i$$

The second term on the right-hand side is, as we have just seen [cf. (A8.15)], the quantity $\vec{\nabla}\cdot\overleftrightarrow{B}\, dx \wedge dy \wedge dz$, thus zero, according to the experiment of the broken magnets and

$$dF = (\partial_j E_i + \epsilon^p{}_{ji}\partial_0 B_p)\, dx^j \wedge dx^i \wedge dt \qquad \text{(A8.22)}$$

According to Maxwell's equation,

$$\vec{\nabla} \wedge \vec{E} = -\frac{\partial B}{\partial t}$$

that is,

$$\epsilon^{ij}{}_k \partial_j E^k = -\frac{\partial B^i}{\partial t}$$

By multiplying the two members by $\epsilon_{il}{}^m$,

$$(\delta^j_l \delta^m_k - \delta^j_m \delta^k_l)\partial_j E^k = -\epsilon_{il}{}^m \frac{\partial B_i}{\partial t}$$

or

$$\partial_l E^m - \partial_m E^l + \epsilon_{il}{}^m \frac{\partial B^i}{\partial t} = 0$$

and (A8.22) is written

$$dF = 0 \qquad \text{(A8.23)}$$

To sum up the different ways of writing Maxwell's equations:

Classic	Covariant	Form
$\vec{\nabla} \wedge \overset{\hookrightarrow}{H} = \vec{j} + \dfrac{\partial \vec{D}}{\partial t}$		$d^*F = J$
	$\partial^\mu F_{\mu\nu} = J_\nu$	
$\vec{\nabla} \cdot \vec{D} = \rho$		$-\delta E = \dfrac{\rho}{\epsilon_0}$
$\vec{\nabla} \wedge \vec{E} = -\dfrac{\partial \overset{\hookrightarrow}{B}}{\partial t}$		$dF = 0$
	$\partial^{\mu *} F_{\mu\nu} = 0$	
$\vec{\nabla} \cdot \overset{\hookrightarrow}{B} = 0$		$dB = 0$

In the table, $E = E_i \, dx^i$, an 1-form, and B is a 2-form with $B = B_{ij} \, dx^i \wedge dx^j$ and $B_{ij} = F_{ij}$.

Observations

1. We find directly from (A8.13) condition (A8.23):

$$dF = -d(dA) = 0$$

2. The 1-form associated with the quadrivector density of current is written

$$J = J_\mu \, dx^\mu$$

with, of course,

$$J^\mu = \left(\frac{\rho}{\epsilon_0 c}, \mu_0 \vec{j} \right)$$

The dual of $^*J$ can be obtained based on (A6.22). It is an $(n - k)$-form with $n = 4$ (dimension of Minkowski space) and $k = 1$ (degree of the form) or $n - k = 3$, and the dual of J is a 3-form. For the dual of Maxwell's tensor $F^{\mu\nu}$, the fact that $n = 4$ and $k = 2$ means that for the dual $^*F^{\mu\nu}$ the associated form is equally a 2-form.

The condition $n - k = k$, that is, $n = 2k$, requires that the form and its dual have the same degree; the space must have an even dimension. We check that for the angular momentum (Section A7.3)

$$L^{jk} = x^j p^k \qquad \text{(a 2-form)}$$

The dual,

$$L_i = \epsilon_{ijk} L^{jk}$$

is a 1-form ($n = 3$, $k = 2$).

3. The forms such that $^*F = F$ (respectively, $^*F = -F$) are said to be *self dual* (respectively, *antidual*). These conditions can be checked only in space of dimension $4n$, $n = 1, 2, 3, \ldots$.

4. The 3-form dual of J (cf. note 2)

$$^*J = {}^*J_{\mu\nu\rho}\, dx^\mu \wedge dx^\nu \wedge dx^\rho$$

has for components the $(^*J_{\rho\nu\sigma})$ obtained by (A6.23):

$$(^*J_{\mu\nu\rho}) = \frac{1}{1!}\, \epsilon_{\mu\nu\rho}{}^{\sigma} J_\sigma$$

a relation that can be reversed:

$$J_\mu = \frac{1}{3!}\, \epsilon_\mu{}^{\rho\nu\sigma}(^*J_{\rho\nu\sigma})$$

or

$$^*J_{0xy} = \epsilon_{0xy}{}^{z} J_z$$

$$^*J_{xyz} = \epsilon_{xyz}{}^{0} J_0 = \frac{\rho}{\epsilon_0}\, \epsilon_{xyz}{}^{0}$$

$$^*J_{0ij} = \epsilon_{0ij}{}^{k} J_k = \epsilon_{ij}{}^{k} J_k = \mu_0 j_k \epsilon_{ij}{}^{k}$$

and thus

$$^*J = \epsilon_{xyz}{}^{0} J_0\, dx \wedge dy \wedge dz + \epsilon_{ij}{}^{k} J_k\, dx^i \wedge dx^j \wedge c\, dt$$

a relation that completes the definition (A8.21).

5. *The field of Yang and Mills*. The theories of gauge used up until now in electromagnetism can easily be extended to theories that cause the intervention of nonlinear terms. It is the case of gravitation and of strong interactions. The field of Yang and Mills is defined by the 2-form

$$F = dA + A^2$$

where, by convention,

$$A^2 = A \wedge A$$

F is the curvature of Yang and Mills. To clarify dA,

$$dA = d(A_\nu\, dx^\nu) = \partial_\mu A_\nu\, dx^\mu \wedge dx^\nu = \tfrac{1}{2}(\partial_\mu A_\nu - \partial_\nu A_\mu)\, dx^\mu \wedge dx^\nu$$

and A^2,

$$A^2 = A_\mu \, dx^\mu \wedge A_\nu \, dx^\nu = \tfrac{1}{2}[A_\mu, A_\nu] \, dx^\mu \wedge dx^\nu$$

Indeed, A_μ and A_ν must be considered as if they were matrices; the theory is nonabelian. We see that

$$F = \tfrac{1}{2}\{(\partial_\mu A_\nu - \partial_\nu A_\mu) + [A_\mu, A_\nu]\} \, dx^\mu \wedge dx^\nu$$

We write

$$F = \tfrac{1}{2} F_{\mu\nu} \, dx^\mu \wedge dx^\nu$$

with

$$F = (\partial_\mu A_\nu - \partial_\nu A_\mu) + [A_\mu, A_\nu]$$

We return to the definition of Yang and Mills and evaluate dF:

$$dF = d(dA + A^2) = d^2A + d(A \wedge A)$$

Now

$$d^2A = 0$$

and by using the general relation

$$d(G \wedge H) = dG \wedge H + (-1)^k G \wedge dH$$

where G is a k-form and H a 1-form; indeed,

$$\begin{aligned}
G \wedge H &= G_{i_1 \cdots i_k} H_{j_1 \cdots j_l} \, dx^{i_1} \wedge \cdots \wedge dx^{i_k} \wedge dx^{j_1} \wedge \cdots \wedge dx^{j_l} \\
d(G \wedge H) &= \partial_p G_{i_1 \cdots i_k} H_{j_1 \cdots j_l} \, dx^p \wedge dx^{i_1} \cdots \wedge dx^{i_k} \wedge dx^{j_1} \wedge \cdots \wedge dx^{j_l} \\
&\quad + G_{i_1 \cdots i_k} \partial_p H_{j_1 \cdots j_l} \, dx^p \wedge dx^{i_1} \cdots \wedge dx^{i_k} \\
&\quad \wedge dx^{j_1} \wedge \cdots \wedge dx^{j_l} \\
d(G \wedge H) &= dG \wedge H + G_{i_1 \cdots i_k} \partial_p H_{j_1 \cdots j_l} \, dx^{i_1} \wedge \cdots \wedge dx^{i_k} \\
&\quad \wedge dx^p \wedge dx^{j_1} \wedge \cdots \wedge dx^{j_l} \\
&= dG \wedge H + (-1)^k G \wedge dH
\end{aligned}$$

In particular, as A is a 1-form,

$$d^2A = dA \wedge A - A \wedge dA$$

By adding and subtracting the same term A^3 such that

$$A^3 = A \wedge A^2 = A^2 \wedge A$$

the quantity dF is written

$$dF = dA \wedge A + A^2 \wedge A - (A \wedge dA + A \wedge A^2)$$
$$= (dA + A^2) \wedge A - A \wedge (dA + A^2)$$
$$= (dA + A^2) \wedge A - A \wedge (dA + A^2) = F \wedge A - A \wedge F$$

which is denoted

$$dF + [A,F] = 0$$

or even

$$DF = 0$$

a relation equivalent to Maxwell's equation where the differential D is a generalization of the covariant derivative. Note that

$$d(\text{tr}(F^2)) = \text{tr}(d(F^2)) = \text{tr}[dF \wedge F + (-1)^2 F \wedge dF]$$

We take $k = 2$ since F is a 2-form and

$$d(\text{tr}(F^2)) = 2\ \text{tr}(dF \wedge F) = -2\ \text{tr}([A,F] \wedge F) = 0$$

the nullity stemming from the fact that trace $[f,g] = 0$ regardless of f and g. This implies that trace $F \wedge F$ is an invariant scalar: a closed 4-form.

As the trace of $F \wedge F$ is preserved, locally a 3-form ω_3^0 exists, such that

$$\text{tr}(F \wedge F) = d\omega_3^0$$

Moreover, we show that

$$\omega_3^0 = \text{tr}(AF - \tfrac{1}{3}A^3)$$

We refer to chiral anomaly, which is important in the field theory calculations. We can choose a Lagrange density l such as

$$l = \text{tr}(F \wedge F)$$

4

The Cayley–Klein Parameters

4.1 GROUP O_3^+ OR SO(3)

In the geometric space of dimension 3, a referential of orthogonal axes, point M is located in relation to the origin O by its coordinates x_1, x_2, and x_3. If we change referential, $R \rightarrow R'$, we have

$$(x_1,x_2,x_3) \longrightarrow (x_1',x_2',x_3')$$

By preserving the origin we should have

$$\sum_{i=1}^{3} x_i^2 = \sum_{i=1}^{3} x_i'^2 \tag{4.1}$$

which expresses the preservation of the length of the vector $\overrightarrow{OM}$. The transformation being linear, nonspecific, we can write

$$x_1' = a_{11}x_1 + a_{12}x_2 + a_{13}x_3$$

and more generally,

$$x_i' = a_{ij}x_j \qquad i, j = 1, 2, 3 \tag{4.2}$$

The condition (4.1) requires that

$$\sum_{i=1}^{3} a_{ij}a_{il} = \delta_{jl} \qquad j, l = 1, 2, 3 \tag{4.3}$$

a relation that defines an orthogonal transformation. It is written

$$\vec{r}\,' = A\vec{r} \tag{4.4}$$

with

$$A = \begin{bmatrix} a_{11} & a_{12} & a_{13} \\ a_{21} & a_{22} & a_{23} \\ a_{31} & a_{32} & a_{33} \end{bmatrix}$$

The matrix A is said to be orthogonal. The condition (4.3) means that

$$\tilde{A} = A^{-1}$$

and therefore that

$$\tilde{A}A = I \tag{4.5}$$

The tilde symbolizes the transposition, that is, the exchange of rows and columns. A matrix is said to be unitary if

$$A^{+}A = I \tag{4.6}$$

(if the elements of the matrix are real: $A^{+} = \tilde{A}^{*} = \tilde{A}$).

If we indicate by $|A|$ the determinant of the matrix A, for the orthogonal matrix,

$$|\tilde{A}A| = 1 \tag{4.7}$$

and as

$$|\tilde{A}| = |A| \tag{4.8}$$

since a determinant is not affected by the exchange of the rows and columns

$$|A|^2 = 1 \qquad \text{and} \qquad |A| = \pm 1 \tag{4.9}$$

Matrices such that $|A| = -1$ do not form a group; O_3^+ is also defined as the subgroup of the 3×3 orthogonal matrices with determinant equal to $+1$. It is, for example, the group of rotations in real space. It is also written SO(3) (the special three-dimensional orthogonal group), a "special" group since it is made up of matrices of determinant equal to $+1$.

4.2 GROUP SU(2)

This is the group of special, unitary, 2×2 matrices. Let's consider a space of dimension 2 but one for which the axes are complex and denoted by u and v. A linear transformation of these axes can be written in the form

$$\begin{aligned} u' &= \alpha u + \beta v \\ v' &= \gamma u + \delta v \end{aligned} \tag{4.10}$$

The coefficients α, β, γ, and δ belong to $\mathbb{C}_2$, the field of complex numbers of dimension 2. The matrix of the transformation is

$$Q = \begin{bmatrix} \alpha & \beta \\ \gamma & \delta \end{bmatrix} \tag{4.11}$$

Limit the transformations Q to the unitary transformations of determinant +1. This requires two particular conditions:

Unitarity of Q: $Q^+Q = I$ or $Q^+ = Q^{-1}$, which means that

$$\det Q = |Q| = \pm 1 \tag{4.12}$$

Determinant +1 leads to

$$\alpha\delta - \gamma\beta = 1 \tag{4.13}$$

As we are in a space $\mathbb{C}_2$, eight quantities must be specified in order to pinpoint the four elements α, β, γ, and δ. The condition $QQ^+ = I$ gives

$$\alpha\alpha^* + \beta\beta^* = 1 \tag{4.14}$$

$$\gamma\gamma^* + \delta\delta^* = 1 \tag{4.15}$$

$$\alpha\gamma^* + \beta\delta^* = 0 \tag{4.16}$$

$$\gamma\alpha^* + \delta\beta^* = 0 \tag{4.17}$$

Equations (4.16) and (4.17) are combined; (4.14) and (4.15) are real. Equations (4.14) to (4.17) therefore represent four equations.

Equation (4.8) gives an additional constraint, or a total of five equations of constraints with eight coefficients: the matrix Q has only three independent quantities. From (4.12) we can see that

$$\frac{\delta}{\gamma} = -\frac{\alpha^*}{\beta^*} \tag{4.18}$$

which is substituted in (4.13)

$$-(\alpha\alpha^* + \beta\beta^*)\,\frac{\gamma}{\beta^*} = 1$$

and which with (4.14) leads to

$$\gamma = -\beta^* \qquad \text{and [cf. (4.18)]} \qquad \delta = \alpha^*$$

from which

$$Q = \begin{bmatrix} \alpha & \beta \\ -\beta^* & \alpha^* \end{bmatrix} \tag{4.19}$$

with

$$\alpha\alpha^* + \beta\beta^* = 1 \tag{4.20}$$

which is the equation of a four-dimensional sphere.

4.3 PAULI MATRICES

Let us now consider in this space the u, v matrix P such that its determinant is equal to the norm squared, omitting the sign, of the vector $\overrightarrow{OM}$ in (R). Let's choose

$$P = \begin{bmatrix} z & x - iy \\ x + iy & -z \end{bmatrix} \tag{4.21}$$

x, y, and z all real. We confirm that

$$|P| = -z^2 - x^2 - y^2 = -r^2 \tag{4.22}$$

Obviously,

$$\mathrm{tr}(P) = 0 \tag{4.23}$$

and

$$P^+ = P \tag{4.24}$$

The matrix is hermitian.

If the matrix P works on vector V, it gives a vector W such that

$$W = PV \qquad \text{where} \quad V \xrightarrow{P} W \tag{4.25}$$

We transform the axes u, v into u', v', by Q a unitary transformation:

$$(u, v) \xrightarrow{Q} (u', v') \tag{4.26}$$

so

$$QW = QPV = QP(Q^+Q)V \tag{4.27}$$

as $Q^+Q = I$, and if we write

$$W' = P'V' \tag{4.28}$$

in the new axes

$$W' = QW \qquad V' = QV \tag{4.29}$$

it can be seen that

$$P' = QPQ^{+} = QPQ^{-1} \tag{4.30}$$

P therefore undergoes an equivalent transformation.

Obviously,

$$P'^{+} = P' \tag{4.31}$$

and

$$\det(P') = |P'| = |P| \tag{4.32}$$

but also

$$\mathrm{tr}(P') = \mathrm{tr}(QPQ^{-1}) = \mathrm{tr}(P) = 0 \tag{4.33}$$

If we write

$$P' = \begin{bmatrix} a_1 + ia_2 & b_1 + ib_2 \\ c_1 + ic_2 & d_1 + id_2 \end{bmatrix} \tag{4.34}$$

The condition $\mathrm{tr}(P') = 0$ leads to

$$a_2 = d_2 = 0$$

$$c_1 = b_1 \qquad \text{and} \qquad c_2 = -b_2$$

$$P' = \begin{bmatrix} a_1 & b_1 + ib_2 \\ b_1 - ib_2 & -a_1 \end{bmatrix} \tag{4.35}$$

To use the preservation of the value of the determinant, we write

$$a_1 = z' \qquad b_1 = x' \qquad b_2 = -y' \tag{4.36}$$

x', y', z' all real. $|P| = |P'|$ is then written

$$x^2 + y^2 + z^2 = x'^2 + y'^2 + z'^2 \tag{4.37}$$

which is the condition of orthogonality: preservation of the norm of the vector $\overrightarrow{OM}$. It is known that any rotation in the geometric space preserves the vector norm (point group).

Note that every matrix M,

$$M = \begin{bmatrix} m_{11} & m_{12} \\ m_{21} & m_{22} \end{bmatrix} \tag{4.38}$$

can be put into the form

$$M = a_0 I_2 + a_k\sigma_k = a_0 I_2 + a_1\sigma_1 + a_2\sigma_2 + a_3\sigma_3 \tag{4.39}$$

where I_2 is the 2×2 unity matrix and the σ_k are the Pauli matrices:

$$I_2 = \begin{bmatrix} 1 & 0 \\ 0 & 1 \end{bmatrix} \qquad \sigma_1 = \begin{bmatrix} 0 & 1 \\ 1 & 0 \end{bmatrix}$$
$$\sigma_2 = \begin{bmatrix} 0 & -i \\ i & 0 \end{bmatrix} \qquad \sigma_3 = \begin{bmatrix} 1 & 0 \\ 0 & -1 \end{bmatrix} \tag{4.40}$$

Equation (4.39) requires that

$$a_0 = \frac{m_{11} + m_{22}}{2} \qquad a_2 = \frac{i(m_{12} - m_{21})}{2}$$
$$a_1 = \frac{m_{12} + m_{21}}{2} \qquad a_3 = \frac{m_{11} - m_{22}}{2} \tag{4.41}$$

For the matrix P we get

$$P = x\sigma_1 + y\sigma_2 + z\sigma_3 \tag{4.42}$$

with

$$\det(P) = -(x^2 + y^2 + z^2) = -r^2 \tag{4.43}$$

P can be considered as the writing of vector position $\overrightarrow{OM}$ in the base of the vectors I_2, σ_1, σ_2, and σ_3. By inverting (4.42) it can be seen that

$$x_i = \tfrac{1}{2}\,\mathrm{tr}(\sigma_i P)$$

For example,

$$z = \frac{1}{2}\,\mathrm{tr}(\sigma_z P) = \frac{1}{2}\,\mathrm{tr}\left[\begin{pmatrix} 1 & 0 \\ 0 & -1 \end{pmatrix}\begin{pmatrix} z & x - iy \\ x + iy & -z \end{pmatrix}\right]$$
$$= \frac{1}{2}\,\mathrm{tr}\begin{pmatrix} z & x - iy \\ -x - iy & z \end{pmatrix} = z$$

Equation (4.42) is a generalization of the relation $\vec{r} = x^i\vec{e}_i$ at a *matrix base*.

4.4 HOMOMORPHISM

By B_1 we denote the real matrix of orthogonal transformation such that

$$\vec{r}\,' = B_1\vec{r} \tag{4.44}$$

and by Q_1 the associated unitary matrix in the complex space

$$P' = Q_1 P Q_1^+ \tag{4.45}$$

Similarly, if B_2 is such that

$$\vec{r}'' = B_2\vec{r}' \tag{4.46}$$

with an associated matrix Q_2,

$$P'' = Q_2P'Q_2^+ \tag{4.47}$$

then

$$\vec{r}'' = B_2B_1\vec{r} = B\vec{r} \tag{4.48}$$

if we write

$$B = B_2B_1 \tag{4.49}$$

so

$$P'' = Q_2Q_1PQ_1^+Q_2^+ = Q_2Q_1P(Q_2Q_1)^+ = QPQ^+ \tag{4.50}$$

if we suppose that

$$Q = Q_2Q_1 \tag{4.51}$$

The correspondence between the complex 2×2 matrix and the 3×3 matrix is homomorphic since the product Q_2Q_1 corresponds to the product B_2B_1.

4.5 ORTHOGONAL MATRIX AND CAYLEY–KLEIN PARAMETERS

We can therefore write the elements of an orthogonal matrix in terms of the elements of a homomorphic matrix Q:

$$P' = QPQ^+ \tag{4.52}$$

We have

$$Q = \begin{bmatrix} \alpha & \beta \\ \gamma & \delta \end{bmatrix} \qquad Q^+ = \begin{bmatrix} \delta & -\beta \\ -\gamma & \alpha \end{bmatrix} \tag{4.53}$$

since $\delta = \alpha^*$ and $\gamma = -\beta^*$ and $P' = \begin{bmatrix} p_{11} & p_{12} \\ p_{21} & p_{22} \end{bmatrix}$

$$\begin{aligned} \text{with} \quad (\alpha\delta + \beta\gamma)z - \alpha\gamma(x - iy) + \beta\delta(x + iy) &= p_{11} \\ 2\gamma\delta z - \gamma^2(x - iy) + \delta^2(x + iy) &= p_{21} \\ -2\alpha\beta z + \alpha^2(x - iy) + \beta^2(x + iy) &= p_{12} \\ -(\alpha\delta + \beta\gamma)z + \alpha\gamma(x - iy) - \beta\gamma(x + iy) &= p_{22} \end{aligned} \tag{4.54}$$

By making the matrix elements equal we obtain identical

$$x' = \frac{1}{2}(\alpha^2 - \gamma^2 + \delta^2 - \beta^2)x + \frac{i}{2}(\gamma^2 - \alpha^2 + \delta^2 - \beta^2)y + (\gamma\delta - \alpha\beta)z \tag{4.55}$$

$$y' = \frac{i}{2}(\alpha^2 + \gamma^2 - \beta^2 - \delta^2)x + \frac{1}{2}(\alpha^2 + \gamma^2 + \beta^2 + \delta^2)y - i(\alpha\beta + \gamma\delta)z \tag{4.56}$$

$$z' = (\delta\beta - \alpha\gamma)x + i(\alpha\gamma + \delta\beta)y + (\alpha\delta + \beta\gamma)z \tag{4.57}$$

This enables us to clarify the matrix of the transformation B associated with the matrix $Q = \left[\begin{smallmatrix}\alpha & \beta\\ \gamma & \delta\end{smallmatrix}\right]$:

$$B = \begin{bmatrix} \frac{1}{2}(\alpha^2 - \gamma^2 + \delta^2 - \beta^2) & \frac{i}{2}(\gamma^2 - \alpha^2 + \delta^2 - \beta^2) & \gamma\delta - \alpha\beta \\ \frac{i}{2}(\alpha^2 + \gamma^2 - \beta^2 - \delta^2) & \frac{1}{2}(\alpha^2 + \gamma^2 + \beta^2 + \delta^2) & -i(\alpha\beta + \gamma\delta) \\ \beta\delta - \alpha\gamma & i(\alpha\gamma + \beta\delta) & \alpha\delta + \beta\gamma \end{bmatrix} \tag{4.58}$$

4.6 RELATION WITH EULER ANGLES

The matrix B gives the precise orientation of a solid body in terms of α, β, γ, and δ. These coefficients play a role equivalent to those of Euler angles. They are called Cayley–Klein parameters.

All the coefficients of the matrix B are real [cf. (4.58) and (4.14) to (4.17)]:

$$\begin{aligned} b_{11} &= \alpha^2 - \beta^2 & b_{12} &= 0 & b_{13} &= -2\mathbf{R}(\alpha\beta) \\ b_{21} &= 0 & b_{22} &= \alpha^2 + \beta^2 & b_{23} &= -2\mathbf{R}(\alpha\beta) \\ b_{31} &= 2\mathbf{R}(\beta\delta) & b_{32} &= 2i\mathbf{I}(\beta\delta) & b_{33} &= \alpha\alpha^* - \beta\beta^* \end{aligned} \tag{4.59}$$

R indicates "real part" and **I** "imaginary part."

We can for the moment join the Cayley–Klein parameters to Euler angles. If φ is counterclockwise rotation around axis $0z$,

$$\begin{aligned} x' &= x\cos\varphi + y\sin\varphi \\ y' &= -x\sin\varphi + y\cos\varphi \\ z' &= z \end{aligned} \tag{4.60}$$

and

$$\begin{aligned} x' + iy' &= (x + iy)e^{-i\varphi} \\ x' - iy' &= (x - iy)e^{i\varphi} \\ z' &= z \end{aligned} \tag{4.61}$$

Let's make a comparison with those obtained from relations (5.55) to (5.57):

$$\begin{aligned} x' + iy' &= (\delta^2 - \gamma^2)x + i(\delta^2 + \gamma^2)y + 2\gamma\delta z \\ x' - iy' &= (\alpha^2 - \beta^2)x - i(\alpha^2 + \beta^2)y - 2\alpha\beta z \\ z' &= (\beta\delta - \alpha\gamma)x + i(\beta\delta + \alpha\gamma)y + (\alpha\delta + \beta\gamma)z \end{aligned} \tag{4.62}$$

It can be seen that

$$\gamma\delta = 0 \qquad \gamma^2 = 0 \qquad \delta = e^{-i\varphi/2} \qquad \beta^2 = 0$$

$$\alpha\beta = 0 \qquad \alpha = e^{i\varphi/2} \qquad \alpha\gamma = 0 \qquad \beta\delta = 0$$

Otherwise,

$$\alpha = e^{i\varphi/2} \qquad \beta = 0 \qquad \gamma = 0 \qquad \delta = e^{-i\varphi/2}$$

We confirm that

$$\alpha\delta = 1$$

and

$$Q_\varphi = \begin{bmatrix} e^{i\varphi/2} & 0 \\ 0 & e^{-i\varphi/2} \end{bmatrix} \tag{4.63}$$

Under the action of θ, rotation around Ox, the same reasoning leads to

$$Q_\theta = \begin{bmatrix} \cos\dfrac{\theta}{2} & -i\sin\dfrac{\theta}{2} \\ i\sin\dfrac{\theta}{2} & \cos\dfrac{\theta}{2} \end{bmatrix} \tag{4.64}$$

The third rotation, of angle ψ around the new axis Oz, brings us to

$$Q_\psi = \begin{bmatrix} e^{i\psi/2} & 0 \\ 0 & e^{-i\psi/2} \end{bmatrix} \tag{4.63}$$

The complete orthogonal matrix is given by

$$B = B(\psi)B(\theta)B(\varphi)$$

It follows from the homorphism that the associated matrix is given by

$$Q = Q_\psi Q_\theta Q_\varphi = \begin{bmatrix} e^{i(\psi+\varphi)/2} \cos \dfrac{\theta}{2} & ie^{i(\psi-\varphi)/2} \sin \dfrac{\theta}{2} \\ ie^{-i(\psi-\varphi)/2} \sin \dfrac{\theta}{2} & e^{-i(\psi+\varphi)/2} \cos \dfrac{\theta}{2} \end{bmatrix} \tag{4.64}$$

from which we obtain the Cayley–Klein parameters in relation to Euler angles.

$$\alpha = \cos \frac{\theta}{2} e^{i(\psi+\varphi)/2}$$

$$\beta = \sin \frac{\theta}{2} e^{i(\psi-\varphi)/2}$$

$$\gamma = i \sin \frac{\theta}{2} e^{-i(\psi-\varphi)/2}$$

$$\delta = \cos \frac{\theta}{2} e^{-i(\psi+\varphi)/2}$$

It is clear that Q_θ, Q_φ, and Q_ψ are expressed as linear combinations of I_2 and the σ_i, the Pauli matrices:

$$Q_\theta = I_2 \cos \frac{\theta}{2} + i\sigma_1 \sin \frac{\theta}{2} \tag{4.65}$$

$$Q_\varphi = I_2 \cos \frac{\theta}{2} + i\sigma_3 \sin \frac{\varphi}{2} \tag{4.66}$$

σ_1 is associated with a rotation around the x axis.
σ_2 is associated with a rotation around the y axis.
σ_3 is associated with a rotation around the z axis.

Note that

$$e^{i\sigma_1(\theta/2)} = I_2 + i\sigma_1 \frac{\theta}{2} + \frac{1}{2!}\left(i\sigma_1 \frac{\theta}{2}\right)^2 + \cdots$$

Now

$$\sigma_1^2 = I_2$$

from which

$$e^{i\sigma_1(\theta/2)} = I_2 + i\sigma_1 \frac{\theta}{2} - \frac{1}{2!} I_2 \left(\frac{\theta}{2}\right)^2 + \cdots$$
$$= I_2 \cos\frac{\theta}{2} + i\sigma_1 \sin\frac{\theta}{2} \tag{4.67}$$

but also [cf. (4.65)]

$$Q_\theta = e^{i\sigma_1(\theta/2)} \tag{4.68}$$

σ_1 should be considered as the axis around which the rotation of angle θ takes place in the space $I_2, \sigma_1, \sigma_2, \sigma_3$.

When we carry out a rotation of $2k\pi$ around the axis Oz in the real space, three-dimensional the matrix associated with the rotation is the identity matrix.

$$I_3 = \begin{bmatrix} 1 & 0 & 0 \\ 0 & 1 & 0 \\ 0 & 0 & 1 \end{bmatrix}$$

The system evidently remains unchanged.

In complex space, two matrices correspond to a rotation $2k\pi$, either matrix I_2 or $-I_2$, depending on whether k is odd or even.

$$Q_0 = \begin{bmatrix} 1 & 0 \\ 0 & 1 \end{bmatrix} = I_2$$
$$Q_{2\pi} = \begin{bmatrix} e^{i\pi} & 0 \\ 0 & e^{-i\pi} \end{bmatrix} = \begin{bmatrix} -1 & 0 \\ 0 & -1 \end{bmatrix} = -I_2$$

This means that two matrices of SU(2) correspond to the matrix I_3 of O^3_+. In general, if a matrix Q corresponds to a real orthogonal matrix, $-Q$ corresponds to the same matrix. Moreover, the homorphism between the two groups implies a biunivocal correspondence between the 3×3 matrix and the pair of matrices $(Q, -Q)$ and not between individual matrices. We can say that the matrix Q is a function with two values or a bivalued matrix of the corresponding two-dimensional matrix:

$$O_3^+ \longrightarrow SU(2) \times SU(2)$$
$$A \longrightarrow (Q, -Q)$$

SU(2) is thus linked to the spin of the electrons, since two values of spin $\pm\frac{1}{2}$ correspond to an internal rotation. The representation SU(2) is said to be "unfaithful."

APPENDIX 9 LIE GROUPS

A9.1 Lie Group

A group G is made up of the set of elements $(A,B,\ldots)$ for which an internal composition law (e.g., multiplication) is given and are such that if

$$A \in G$$
$$B = G$$

then

$$AB = G$$

The following properties must be verified:

$$A(BC) = (AB)C \text{ (associativity)} \qquad \forall\, A,\, B,\, C$$

The neutral element E is such that

$$EA = AE = A \qquad \forall\, A$$

The inverse element A^{-1} is such that

$$A^{-1}A = AA^{-1} = E \qquad \forall\, A$$

We distinguish:

Continuous groups. The elements of the group depend on parameters that vary in a continuous way: for example, the Euler angles θ, φ, and ψ.
Lie groups. These are continuous groups such that

$$(A,B) \xrightarrow{f} AB$$
$$A \xrightarrow{f'} A^{-1}$$

where f and f' are *analytic.*

A9.2 Lie Algebra

We associate with every Lie group G an algebra such that if A, B, and C are elements of G,

$$[A,[B,C]] + [B,[C,A]] + [C,[A,B]] = 0 \tag{A9.1}$$

a Jacobi relation with, for example,

$$[A,B] = AB - BA \tag{A9.2}$$

An alternative definition of Lie algebra that is often used is that if $A \in G$, α is such that $A = e^{\alpha}$, where α is a matrix. Now

$$e^{\alpha} = 1 + \alpha + \frac{1}{2!}\alpha^2 + \cdots \frac{1}{n!}\alpha^n + \cdots \tag{A9.3}$$

If the group G is abelian,

$$AB = BA$$

If the group G is not abelian,

$$AB \neq BA$$

and we introduce an element C of group G such that

$$C = AB(BA)^{-1}$$

C is equal to 1, the identity element, if G is abelian, and C is not unity if G is not abelian. Note that

$$B = e^{\beta} \qquad \text{and} \qquad C = e^{\gamma}$$

so

$$e^{\gamma} = e^{\alpha}e^{\beta}(e^{\beta}e^{\alpha})^{-1} = e^{\alpha}e^{\beta}e^{-\alpha}e^{-\beta}$$

by developing in the first order

$$\begin{aligned} e^{\alpha} &= (1 + \alpha)(1 + \beta)(1 - \alpha)(1 - \beta) \\ &= (1 - \beta\alpha + \alpha\beta) \end{aligned}$$

Otherwise,

$$\gamma = \alpha\beta - \beta\alpha = [\alpha,\beta] \tag{A9.4}$$

The matrices α, β, and γ define in (A9.4) an algebra said to be a Lie algebra. The preceding matrices can be expressed in terms of a basis of the algebra; for example,

$$\gamma = \gamma^i X_i \tag{A9.5}$$

where the γ^i are the parameters of the group and the X_i are the generators of the group. The relation (A9.4) is then expressed in the form

$$\gamma^i X_i = (\alpha\beta)^i X_i - (\beta\alpha)^i X_i$$

Now

$$\gamma = [\alpha,\beta]$$

or

$$\gamma = [\alpha^j X_j, \beta^k X_k] = \alpha^j \beta^k [X_j, X_k]$$

and thus

$$\gamma^i X_i = \alpha^j \beta^k [X_j, X_k] \tag{A9.6}$$

which enables us to write, for example,

$$[X_j, X_k] = f_{jk}{}^i X_i \tag{A9.7}$$

where the $f_{jk}{}^i$ are the *structure constants of the group*. The Jacobi relation

$$[X_i,[X_j,X_k]] + [X_j,[X_k,X_i]] + [X_k,[X_i,X_j]] = 0 \tag{A9.8}$$

is expressed by

$$f_{jk}{}^l f_{il}{}^p X_p + f_{ki}{}^m f_{jm}{}^q X_q + f_{ij}{}^n f_{kn}{}^r X_r = 0$$

Otherwise,

$$f_{jk}{}^l f_{il}{}^p + f_{ki}{}^m f_{jm}{}^p + f_{ij}{}^n f_{kn}{}^p = 0 \tag{A9.9}$$

as the X_p form a basis; it is a relation on the structure constants.

Example 1. The group of translations of three dimensions, T_3, is a Lie group with three parameters, t_1, t_2, and t_3. Each element A of the group will depend on these parameters:

$$A = A(\vec{X}(t_i)) \qquad i = 1,2,3$$

Let's develop A in the neighborhood of $\vec{X}_0$:

$$\begin{aligned} A(\vec{X}) &= A(\vec{X} - \vec{X}_0 + \vec{X}_0) = A(\vec{X}_0) + (\vec{X} - \vec{X}_0)\left(\frac{\partial A}{\partial \vec{X}}\right)_{\vec{X}=\vec{X}_0} + \cdots \\ &= A(\vec{X}_0) + \left[\vec{t}\cdot\left(\frac{\partial A}{\partial \vec{X}}\right)\right]_{\vec{X}=\vec{X}_0} + \cdots \\ &= \left[\left(1 + \vec{t}\cdot\frac{\partial}{\partial \vec{X}}\right)A\right]_{\vec{X}=\vec{X}_0} + \cdots = \left[\exp\left(\vec{t}\cdot\frac{\partial}{\partial \vec{X}}\right)A\right]_{\vec{X}=\vec{X}_0} \end{aligned}$$

$\vec{p}$ is written as a generator such that

$$A(\vec{X}) = [(e^{i\vec{t}\cdot\vec{p}})A]_{\vec{X}=\vec{X}_0}$$

In quantum mechanics,

$$\vec{p} = -i\hbar\vec{\nabla}$$

The vectors $\vec{t}$ are the parameters of the group. With the group of translations being abelian, the structure constants $f_{\mu\nu}{}^{\sigma}$ are all zero.

Example 2. Consider the group of rotations of three dimensions, SO(3). Under an infinitesimal rotation the coordinates of a vector are transformed according to the law,

$$X' = AX$$

with

$$A = 1 + \frac{\partial A}{\partial \alpha^k} d\alpha_k$$

or for the coordinates

$$X'_k = a_{km} X^m$$

where

$$a_{km} = \delta_{km} + \omega_{km}$$

and

$$\omega_{km} = \frac{\partial a_{km}}{\partial \alpha_k} d\alpha_k$$

We write that

$$A^t A = 1$$

an orthogonal matrix

$$(\delta_{km} + \omega_{km})(\delta_{mn} + \omega_{mn}) = \delta_{kn}$$

In the first order we arrive at

$$\omega_{kn} = -\omega_{nk}$$

the antisymmetry of the tensor of the second order, which enables us to move to the dual

$$\omega_{kl} = \epsilon_{kml} \alpha^m$$

and thus

$$X'_k = (\delta_{km} + \omega_{km}) X^m = (\delta_{km} + \epsilon_{knm} \alpha^n) X^m$$
$$= X_k + (\vec{\alpha} \wedge \vec{r})_k$$

By returning to the reasoning used for the group of translations

$$A(\vec{X}) = A(\vec{X}_0) + \epsilon_{knm} \alpha^n X^m = A(\vec{X}_0 + \epsilon_{knm} \alpha^n X^m \vec{e}^{\,k})$$

and a Taylor's development gives

$$A(\vec{X}) = A(\vec{X}_0) + \epsilon_{knm}\alpha^n X^m \left(\frac{\partial A}{\partial x_k}\right)_{\vec{X}=\vec{X}_0} + \cdots$$

$$\simeq \left[\left(1 + \epsilon_{knm}\alpha^n X^m \frac{\partial}{\partial X^k}\right)\right]_{\vec{X}=\vec{X}_0} = \left[\left(e^{\epsilon_{knm}\alpha^n x^m} \frac{\partial}{\partial X_k}\right)A\right]_{\vec{X}=\vec{X}_0}$$

a relation that is written in the form

$$A(\vec{X}) = e^{i\alpha^n \mathcal{L}^n}$$

where $\mathcal{L}^n$ is a component of angular momentum, the $\mathcal{L}^n$ are generators of group SO(3), and the $\mathcal{L}^n$ are the three parameters (Euler angles) of the group.

An operator that commutes with all the generators is called the *Casimir operator*. Here $\mathcal{L}^2$ meets this requirement. The Casimir operator is important in quantum mechanics for the construction of a complete set of observables that commute, as well as being important in the research on grand unified theories. The number of Casimir operators, for a given group, defines the rank of the group.

A9.3 Construction of Lie Algebras

A9.3.1 SL(n,R)

Take matrix A of which all the elements are real, SL(n,r), is, by definition, the special linear group of dimension n:

$$\det A = 1$$

and

$$A = 1 + X\beta + \cdots$$

where X belongs to the Lie algebra of sl(n,r). We use the general formula

$$\lim_{\beta=0} \det(1 + X\beta) = 1 + \mathrm{tr}(X\beta) + \det(X\beta) + \cdots$$

By ignoring the term det$(X\beta)$ in front of tr$(X\beta)$,

$$\mathrm{tr}(\beta X) = 0 \qquad \text{or} \qquad \mathrm{tr}\, X = 0$$

The Lie algebra is made up of the set of zero trace matrices.

A9.3.2 O(n,R)

The orthogonal group of dimension n requires that

$$A^{-1} = {}^tA$$

Otherwise, in a Lie algebra,

$$1 - X\beta = 1 + \beta{}^{t}X$$

that is,

$${}^{t}X = -X$$

The algebra is formed by the antisymmetrical matrices.

A9.3.3 SO(n,R)

The special orthogonal group is such that

$$\det A = 1$$
$${}^{t}A = A^{-1}$$

The Lie algebra is formed by the antisymmetrical matrices of zero trace; they are matrices associated with angular momentum of dimension n.

A9.3.4 SU(2)

For the special unitary group the equations of constraint are

$$\det A = 1 \qquad \text{(zero trace)}$$
$$A^{+} = A^{-1} \qquad \text{(antihermitian)}$$

The Lie algebra corresponds to an algebra of 2×2 matrices of zero trace that are antihermitian:

$$X = iB$$

where B is a hermitian matrix. B can be developed on a basis of the algebra made up of the identity

$$I = \begin{bmatrix} 1 & 0 \\ 0 & 1 \end{bmatrix}$$

and the three Pauli matrices

$$\sigma_1 = \begin{bmatrix} 0 & 1 \\ 1 & 0 \end{bmatrix} \qquad \sigma_2 = \begin{bmatrix} 0 & i \\ -i & 0 \end{bmatrix} \qquad \sigma_3 = \begin{bmatrix} 1 & 0 \\ 0 & -1 \end{bmatrix}$$

which satisfy Lie's law:

$$\left[\frac{\sigma_i}{2}, \frac{\sigma_j}{2}\right] = i\epsilon_{ijk}\frac{\sigma_k}{2}$$

APPENDIX 10 GENERATORS OF ROTATION GROUP

In a trirectangular referential $O\vec{e}_1\vec{e}_2\vec{e}_3$, we carry out the successive rotations of the axes $R_{\vec{e}_3}(\alpha)$, $R_{\vec{e}_2'}(\beta)$, and $R_{\vec{e}_3''}(\gamma)$ defined by Figure A10.1, and write

$$R(\alpha\beta\gamma) = R_{\vec{e}_3''}(\gamma)R_{\vec{e}_2'}(\beta)R_{\vec{e}_3}(\alpha)$$

If we use in a fixed system of reference $(Oxyz)$

$$R_{\vec{e}_3}(\alpha) = R_z(\alpha)$$

$$R_{\vec{e}_2'}(\beta) = R_z(\alpha)R_y(\beta)R_z(\alpha)^{-1}$$

$$R_{\vec{e}_3'}(\gamma) = R_z(\alpha)R_z(\gamma)R_z(\alpha)^{-1}$$

$$R_{\vec{e}_3''} = R_z(\alpha)R_y(\beta)R_z(\gamma)[R_z(\alpha)R_y(\beta)]^{-1}$$

therefore

$$R(\alpha\beta\gamma) = R_z(\alpha)R_y(\beta)R_z(\gamma)$$

The angles α, β, and γ are the three Euler angles, variables associated with three independent rotations. To be precise, to rotations around the axes O_x, O_y, and O_z are associated the matrices in the fixed referential:

$$R_x(\omega) = \begin{bmatrix} 1 & 0 & 0 \\ 0 & \cos\omega & -\sin\omega \\ 0 & \sin\omega & \cos\omega \end{bmatrix}$$

$$R_y(\omega) = \begin{bmatrix} \cos\omega & 0 & \sin\omega \\ 0 & 1 & 0 \\ -\sin\omega & 0 & \cos\omega \end{bmatrix}$$

$$R_z(\omega) = \begin{bmatrix} \cos\omega & -\sin\omega & 0 \\ \sin\omega & \cos\omega & 0 \\ 0 & 0 & 1 \end{bmatrix}$$

Let's consider an element of the group of rotations for which the matrix representation is $R_{\vec{X}_i}(\omega')$, the rotation of angle ω' around the axis Ox_i if

$$\omega' = \omega + \epsilon \qquad \text{with} \quad \epsilon \ll \omega$$

$$R_{\vec{X}_i}(\omega + \epsilon) = R_{\vec{X}_i}(\omega) + \epsilon\left(\frac{\partial R_{\vec{X}_i}(\omega')}{\partial\omega'}\right)_{\omega'=\omega} + \cdots$$

In the case of an infinitesimal rotation, we once again see the result of Appendix 9. The generators of the rotation group are thus defined by the

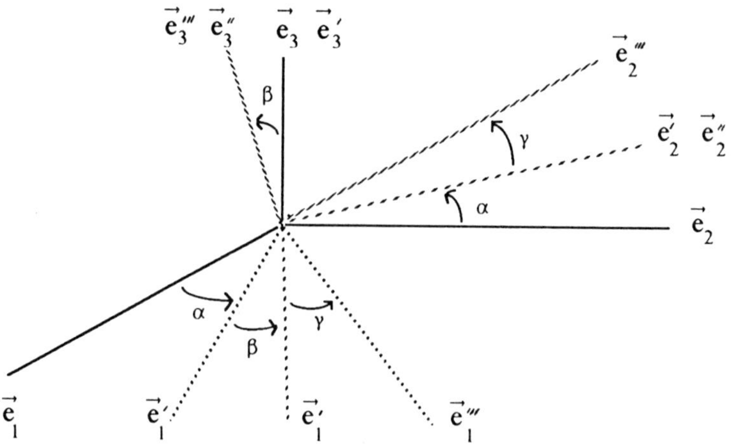

Figure A10.1 Rotations.

relation $\partial R_{\vec{X}_i}(\omega)/\partial\omega$:

$$L_x = \begin{bmatrix} 0 & 0 & 0 \\ 0 & 0 & -1 \\ 0 & 1 & 0 \end{bmatrix}$$

$$L_y = \begin{bmatrix} 0 & 0 & 1 \\ 0 & 0 & 0 \\ -1 & 0 & 0 \end{bmatrix}$$

$$L_z = \begin{bmatrix} 0 & -1 & 0 \\ 1 & 0 & 0 \\ 0 & 0 & 0 \end{bmatrix}$$

Lie's algebra,

$$[L_i, L_j] = \epsilon_{ijk} L^k$$

can easily be confirmed.

APPENDIX 11 SU(2): UNFAITHFUL REPRESENTATION OF SO(3)

A11.1 Homomorphism SU(2)–SO(3)

We showed earlier that for the SO(3) group, the generators L_i comply with the relation of the Lie algebra

$$[L_i, L_j] = \epsilon_{ijk} L^k$$

and the group SU(2) the generators σ_i are such that

$$\left[\frac{\sigma_i}{2}, \frac{\sigma_j}{2}\right] = i\epsilon_{ij}{}^{k} \frac{\sigma_k}{2}$$

The algebras of SO(3) and SU(2) have the same structure constants. Compare the two algebras. A 2×2 matrix M can always be associated with a vector of R^3 denoted $\vec{V}(X,Y,Z)$ such that

$$M = X^i\sigma_i = \begin{vmatrix} Z & X + iY \\ X - iY & -Z \end{vmatrix}$$

The properties of this matrix are

$$\begin{aligned} M &= M^+ \qquad \text{(hermiticity)} \\ \operatorname{tr} M &= 0 \\ \det M &= -V^2 \end{aligned}$$

Inversely, the components of $\vec{V}$ starting from M are given by

$$X_i = \tfrac{1}{2} \operatorname{tr}(\sigma_i M)$$

Now take a matrix P that belongs to the algebra of SU(2) and build the matrix

$$M' = PMP^+ \tag{A11.1}$$

M' is the matrix transformed from M by the algebra of SU(2). We check that M' complies with the same requirements as M:

$$\begin{aligned} M'^+ &= (PMP^+)^+ = PM^+P^+ = M' \\ \text{trace } M' &= \text{trace}(PMP^+) = \text{trace}(P^+PM) = \operatorname{tr} M = 0 \\ \det M' &= \det(PMP^+) = \det P \det M \det P^+ = \det M = -V^2 \end{aligned}$$

since

$$P^+ = P^{-1}$$

We now have a vector $\vec{V}'$ of components X_i' such that

$$M' = X'^i\sigma_i$$

which leads to

$$\det M' = -V'^2$$

and also that

$$V'^2 = V^2$$

The norm of vector V is preserved; it is the characteristic of a rotation. There is a correspondence between SO(3) and SU(2). The relation (A11.1)

$$M' = PMP^+$$

can also be put in the form

$$M' = (-P)M(-P)^+$$

The matrices P and $(-P)$ comply with the same relation and induce the same transformation:

$$M \xrightarrow[\text{a rotation}]{\text{SO(3)}} M'$$

$$M \xrightarrow[(P,-P)]{\text{SU(2)}} M'$$

There is no bijection; the representation of the rotations of SO(3) by the matrix of SU(2) is not regular since two matrices correspond to one rotation. If we write Z_n as the group of nth roots of the unity

$$Z_1 = [1]$$
$$Z_2 = [1,-1]$$

and so on, we have SO(3) $\sim$ SU(2)/Z_2 (equivalent classes and factor group).

A11.2 Identity Matrix of SO(3)

The matrix M can be written precisely

$$M = e^{i[(\vec{\lambda}\cdot\vec{\sigma})/2]}$$

similarly, for a rotation matrix,

$$R = e^{i\vec{\lambda}\cdot\vec{L}}$$

Let's chose a rotation around Oz and angle $\|\vec{\lambda}\| = 2\pi$, or

$$R = e^{i\cdot 2\pi L_z}$$
$$M = e^{i\pi\sigma_z}$$

with

$$\sigma_z = \begin{bmatrix} 1 & 0 \\ 0 & -1 \end{bmatrix} \quad \text{and} \quad L_z = \begin{bmatrix} 0 & 1 & 0 \\ -1 & 0 & 0 \\ 0 & 0 & 0 \end{bmatrix}$$

Now

$$M = \sum_{n=0}^{\infty} \frac{1}{n!} (i\pi\sigma_z)^n = \begin{vmatrix} e^{i\pi} & 0 \\ 0 & e^{-i\pi} \end{vmatrix} = -1$$

where 1 is the identity matrix, and

$$R = \sum_{n=0}^{\infty} \frac{1}{n!} (2\pi i L_z)^n$$

As L_z is not a diagonal matrix, there is no simple relation for the exponentials. We are also going to put the matrix associated with L_z into a diagonal form by means of a unitary transformation S such that

$$L_z = PL_z^{(\text{diagonal})}P^+ = P \begin{bmatrix} -1 & 0 & 0 \\ 0 & 0 & 0 \\ 0 & 0 & 1 \end{bmatrix} P^+$$

and

$$R = Pe^{i2\pi L_z^{(\text{diagonal})}} P^+ = P \begin{bmatrix} e^{-2i\pi} & 0 & 0 \\ 0 & e^0 & 0 \\ 0 & 0 & e^{2i\pi} \end{bmatrix} P^+ = 1$$

To sum up, the identity matrix corresponds to a rotation of angle 2π in SO(3), while in SU(2) we find the matrix opposed to the identity.

A11.3 Geometrical Representation

The parameters of the group SO(3) define a sphere in the space of three dimensions. We are looking for the space defined by the parameters of the group SU(2):

$$\det A = 1$$

that is,

$$ad - bc = 1 \tag{A11.2}$$

$$A^+ = A^{-1} \tag{A11.3}$$

where $A \in$ SU(2). We write

$$A = \begin{bmatrix} a & b \\ c & d \end{bmatrix}$$

The relation of constraint (A11.3) requires that

$$d^* = a$$

$$c^* = -b$$

Otherwise,

$$A = \begin{bmatrix} a & b \\ -b^* & a^* \end{bmatrix}$$

and with (A11.2),

$$|a|^2 + |b|^2 = 1 \tag{A11.4}$$

Now a and b are complex numbers. We can write

$$a = x + iy$$

$$b = z + it$$

and (A11.4) gives

$$x^2 + y^2 + z^2 + t^2 = 1$$

an equation of a sphere of four dimensions.

The correspondence SO(3)–SU(2) can be represented as the projection of the sphere of dimension 4 on the sphere of dimension 3 or (cf. Figure A11.1) of a sphere of three dimensions on the plane of the equator. The northern hemisphere, for example, corresponds to the group SO(3), whereas the two hemispheres must be associated with SU(2).

APPENDIX 12 LORENTZ AND POINCARÉ GROUPS

A12.1 Lorentz Group

We limit ourselves to the proper group $L_+^\uparrow$ (cf. Section 2.11.13); for every quadrivector

$$X^\mu = (X^0, X^i)$$

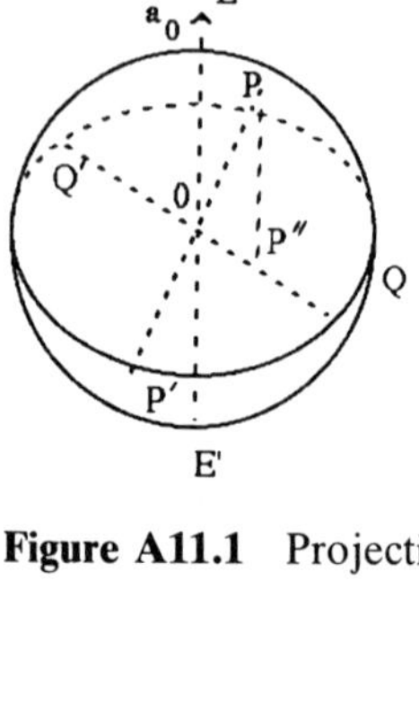

Figure A11.1 Projection.

we associate a matrix M such that

$$M = \sigma_\mu X^\mu \qquad (\mu = 0,1,2,3)$$

where the σ_μ are, respectively, the 2×2 identity matrix and the Pauli matrices $\sigma_i (i = 1,2,3)$. The metric is given by:

$$g_{\mu\nu} = \begin{bmatrix} 1 & 0 & 0 & 0 \\ 0 & -1 & 0 & 0 \\ 0 & 0 & -1 & 0 \\ 0 & 0 & 0 & -1 \end{bmatrix}$$

so

$$M = \begin{bmatrix} X^0 + X^3 & X^1 - iX^2 \\ X^1 + iX^2 & X_0 - X^3 \end{bmatrix}$$

It follows that

$$X^\mu = \tfrac{1}{2}\,\mathrm{tr}(\tilde{\sigma}^\mu M)$$

with

$$\tilde{\sigma}^\mu = \sigma_\mu = (\sigma_0, \sigma_i)$$

$$\tilde{\sigma}_\mu = \sigma^\mu = (\sigma_0, -\sigma_i)$$

We can easily check that

$$\det(M) = (X^0)^2 - (X^1)^2 - (X^2)^2 - (X^3)^2 = X^\mu X_\mu$$

The M matrices are transformed by an element A of SL(2,**C**), a special linear group with complex elements, into matrix M' such that

$$M' = AMA^+$$

We confirm that

$$\det(M') = \det M$$

that is, that

$$X^\mu X_\mu = X'^{\mu'} X'_{\mu'}$$

We come back to the invariance of Minkowski's norm; the matrices M are associated with a Lorentz transformation L.

For an element of the Lorentz matrix we have

$$L^\mu{}_\nu = \tfrac{1}{2}\,\mathrm{tr}(\tilde{\sigma}^\mu A \sigma_\nu A^+) \tag{1}$$

The relation (1) is checked again if $A \to -A$. Two matrices, A and $-A$, of complex elements correspond to a matrix L. A homomorphism exists between the Lorentz group $L_+^\Uparrow$ and group SL(2,**C**).

We saw earlier that SU(2) is simply connected and that SO(3) is doubly connected (cf. Appendix 13). We could also show that $L_+^\uparrow$ is twice connected and SL(2,**C**) is simply connected. We say that SL(2,**C**) is the universal covering group of $L_+^\uparrow$.

A12.2 Generators of the Lorentz Group

The Lorentz group has six parameters, three for the rotations and three for the boosts. Indeed, for a Lorentz infinitesimal transformation L,

$$L^\mu{}_\nu = \delta^\mu{}_\nu + \epsilon^\mu{}_\nu$$

and

$$g = Lg\tilde{L} \qquad \text{(Section 2.3)}$$

that is,

$$L^\mu{}_\alpha g_{\mu\nu} L^\nu{}_\beta = g_{\alpha\beta}$$

or

$$(\delta^\mu{}_\alpha + \epsilon^\mu{}_\alpha) g_{\mu\nu} (\delta_\beta{}^\nu + \epsilon_\beta{}^\nu) = g_{\alpha\beta}$$

which means that

$$\epsilon^\mu{}_\alpha = -\epsilon_\alpha{}^\mu$$

(antisymmetry). The number of parameters defining a 4×4 matrix is 16; in the case of antisymmetry only six terms are independent:

$$R_1 = R(\varphi_1) = \begin{bmatrix} 1 & 0 & 0 & 0 \\ 0 & 1 & 0 & 0 \\ 0 & 0 & \cos\varphi_1 & -\sin\varphi_1 \\ 0 & 0 & \sin\varphi_1 & \cos\varphi_1 \end{bmatrix}$$

$$R_2 = R(\varphi_2) = \begin{bmatrix} 1 & 0 & 0 & 0 \\ 0 & \cos\varphi_2 & 0 & -\sin\varphi_2 \\ 0 & 0 & 1 & 0 \\ 0 & \sin\varphi_2 & 0 & \cos\varphi_2 \end{bmatrix}$$

$$R_3 = R(\varphi_3) = \begin{bmatrix} 1 & 0 & 0 & 0 \\ 0 & \cos\varphi_3 & -\sin\varphi_3 & 0 \\ 0 & \sin\varphi_3 & \cos\varphi_3 & 0 \\ 0 & 0 & 0 & 1 \end{bmatrix}$$

$$B_1 = B(v_1) = \begin{bmatrix} \Gamma_1 & -\beta_1\Gamma_1 & 0 & 0 \\ -\beta_1\Gamma_1 & \Gamma_1 & 0 & 0 \\ 0 & 0 & 1 & 0 \\ 0 & 0 & 0 & 1 \end{bmatrix}$$

$$B_2 = B(v_2) = \begin{bmatrix} \Gamma_2 & 0 & -\beta_2\Gamma_2 & 0 \\ 0 & 1 & 0 & 0 \\ -\beta_2\Gamma_2 & 0 & \Gamma_2 & 0 \\ 0 & 0 & 0 & 1 \end{bmatrix}$$

$$B_3 = B(v_3) = \begin{bmatrix} \Gamma_3 & 0 & 0 & -\beta_3\Gamma_3 \\ 0 & 1 & 0 & 0 \\ 0 & 0 & 1 & 0 \\ -\beta_3\Gamma_3 & 0 & 0 & \Gamma_3 \end{bmatrix}$$

and when φ_i is infinitesimal ($i = 1,2,3$),

$$R_1 \simeq \begin{bmatrix} 1 & 0 & 0 & 0 \\ 0 & 1 & 0 & 0 \\ 0 & 0 & 1 & -\varphi_1 \\ 0 & 0 & \varphi_1 & 1 \end{bmatrix}$$

$$R_2 = \begin{bmatrix} 1 & 0 & 0 & 0 \\ 0 & 1 & 0 & \varphi_2 \\ 0 & 0 & 1 & 0 \\ 0 & \varphi_2 & 0 & 1 \end{bmatrix}$$

$$R_3 = \begin{bmatrix} 1 & 0 & 0 & 0 \\ 0 & 1 & \varphi_3 & 0 \\ 0 & \varphi_3 & 1 & 0 \\ 0 & 0 & 0 & 1 \end{bmatrix}$$

According to Section 2.2,

$$\Gamma_l = \cosh \alpha_l$$

$$\beta_l = \tanh \alpha_l$$

and for example, when α_1 is infinitesimal,

$$B_1 = \begin{bmatrix} \cosh \alpha_1 & -\sinh \alpha_1 & 0 & 0 \\ -\sinh \alpha_1 & \cosh \alpha_1 & 0 & 0 \\ 0 & 0 & 1 & 0 \\ 0 & 0 & 0 & 1 \end{bmatrix} \simeq \begin{bmatrix} 1 & -\alpha_1 & 0 & 0 \\ -\alpha_1 & 1 & 0 & 0 \\ 0 & 0 & 1 & 0 \\ 0 & 0 & 0 & 1 \end{bmatrix}$$

As for the group of rotations, the infinitesimal generators are introduced:

$$J_l = i\left(\frac{\partial R_l}{\partial \varphi_l}\right)_{\varphi_l=0}$$

$$K_l = i\left(\frac{\partial B_l}{\partial \alpha_l}\right)_{\alpha_l=0}$$

or

$$J_1 = \begin{bmatrix} 0 & 0 & 0 & 0 \\ 0 & 0 & 0 & 0 \\ 0 & 0 & 0 & -i \\ 0 & 0 & i & 0 \end{bmatrix}$$

$$J_2 = \begin{bmatrix} 0 & 0 & 0 & 0 \\ 0 & 0 & 0 & i \\ 0 & 0 & 0 & 0 \\ 0 & -i & 0 & 0 \end{bmatrix}$$

$$J_3 = \begin{bmatrix} 0 & 0 & 0 & 0 \\ 0 & 0 & -i & 0 \\ 0 & i & 0 & 0 \\ 0 & 0 & 0 & 0 \end{bmatrix}$$

$$K_1 = \begin{bmatrix} 0 & -i & 0 & 0 \\ -i & 0 & 0 & 0 \\ 0 & 0 & 0 & 0 \\ 0 & 0 & 0 & 0 \end{bmatrix}$$

$$K_2 = \begin{bmatrix} 0 & 0 & -i & 0 \\ 0 & 0 & 0 & 0 \\ -i & 0 & 0 & 0 \\ 0 & 0 & 0 & 0 \end{bmatrix}$$

$$K_3 = \begin{bmatrix} 0 & 0 & 0 & -i \\ 0 & 0 & 0 & 0 \\ 0 & 0 & 0 & 0 \\ -i & 0 & 0 & 0 \end{bmatrix}$$

In general, a Lorentz matrix L is expressed by

$$L = e^{-i(\vec{\alpha}\cdot\vec{K} + \vec{\varphi}\cdot\vec{J})} \tag{2}$$

where

$$\vec{\alpha} = \alpha \frac{\vec{v}}{\|\vec{v}\|} \qquad \text{and} \qquad \vec{\varphi} = \varphi\vec{n}$$

A12.3 Lie Algebra of the Lorentz Group

The values of the following commutators can easily be confirmed:

$$[J_i, J_j] = i\epsilon_{ij}{}^{k} J_k$$

$$[K_i, K_j] = -i\epsilon_{ij}{}^{k} J_k$$

$$[J_i, K_j] = i\epsilon_{ij}{}^{k} K_k$$

By introducing the dual tensor M_{ij} such that

$$M_{ij} = \epsilon_{ij}{}^{k} J_k$$
$$M_{0i} = \epsilon_{0i}{}^{k} K_k$$

(using the antisymmetry of the J_i and K_i), the former algebra is shortened to

$$[M_{\lambda\rho}, M_{\mu\nu}] = -i(g_{\lambda\mu} M_{\rho\nu} + g_{\rho\nu} M_{\lambda\mu} - g_{\lambda\nu} M_{\rho\mu} - g_{\rho\mu} M_{\lambda\nu})$$

The Lie algebra of the Lorentz group and formula (2) of Section A12.2 become

$$L = e^{-(1/2)\epsilon^{\mu\nu} M_{\mu\nu}}$$

There are two Casimir operators (operators that commute with all the generators of the group):

$$\tfrac{1}{2} M_{\mu\nu} M^{\mu\nu} = J^2 - K^2$$
$$\tfrac{1}{4} \epsilon^{\mu\nu\sigma\tau} M_{\mu\nu} M_{\sigma\tau} = -\vec{J}\cdot\vec{K}$$

Similar relations were obtained in Chapter 3:

$$F_{\mu\nu} F^{\mu\nu} = -2\left(\frac{E^2}{c^2} - B^2\right)$$
$$F^{*}_{\mu\nu} F^{\mu\nu} = 4\,\frac{\vec{E}\cdot\overleftrightarrow{B}}{c}$$

The number of Casimir operators gives the rank of the algebra; the Lie algebra of the Lorentz group is therefore of rank 2.

A12.4 Representation of the Lorentz Group

If the following two operators are introduced:

$$\mathbf{M}_i = \tfrac{1}{2}(J_i + iK_i)$$
$$\mathbf{N}_i = \tfrac{1}{2}(J_i - iK_i)$$

we have the relations

$$[\mathbf{M}_i, \mathbf{M}_j] = i\epsilon_{ij}{}^{k} \mathbf{M}_k \tag{3a}$$
$$[\mathbf{N}_i, \mathbf{N}_j] = i\epsilon_{ij}{}^{k} \mathbf{N}_k \tag{3b}$$
$$[\mathbf{M}_i, \mathbf{N}_j] = 0 \tag{4}$$

and

$$[\mathbf{M}^2, \mathbf{M}_i] = 0 \tag{5a}$$
$$[\mathbf{N}^2, \mathbf{N}_i] = 0 \tag{5b}$$

The relations (3a) and (3b) show that the operators $\mathbf{M}_i$ and $\mathbf{N}_i$ behave as do the angular momentums [algebra of SO(3)]. The relations (5a) and (5b) indicate that the quantities $\mathbf{M}^2$ and $\mathbf{N}^2$ are Casimir operators of algebra defined by (3a) and (3b). Relation (4) confirms that the two algebras are independent.

The irreducible representations have been divided into two; the algebra of the Lorentz group has been divided into two independent subalgebras su(2) $\otimes$ su(2)'. The two subalgebras are different:

$$\text{su}(2) \neq \text{su}(2)'$$

This stems from the fact that the parity operator does not alter the J_i, yet alters the sign of the K_i. Note that

$$|\psi_{jm_j,j'm'_{j'}}\rangle = |jm_j\rangle \otimes |j'm'_{j'}\rangle = |jm_j,j'm'_{j'}\rangle$$

the eigenvector common to the operators $\mathbf{M}^2$, $\mathbf{N}^2$, $\mathbf{N}_3$, and $\mathbf{M}_3$; the indexes of the eigenvector are counted as four, by the degrees of freedom in Minkowski space. Since the subalgebras have been divided in two, we take j and m_j for su(2) and j', $m'_{j'}$ for su(2)'; m_j is joined to $\mathbf{N}_3$ and $m'_{j'}$ to $\mathbf{M}_3$:

$$\mathbf{N}^2|jm_j,j'm'_{j'}\rangle = j(j+1)|jm_j,j'm'_{j'}\rangle$$

$$\mathbf{M}_2|jm_j,j'm'_{j'}\rangle = j'(j'+1)|jm_j,j'm'_{j'}\rangle$$

$$\mathbf{N}_3|jm_j,j'm'_{j'}\rangle = m_j|jm_j,j'm'_{j'}\rangle$$

$$\mathbf{M}_3|jm_j,j'm'_{j'}\rangle = m'_{j'}|jm_j,j'm'_{j'}\rangle$$

By restriction to the subgroup of rotations, an irreducible representation can be developed in the form of a tensor product of two irreducible representations of SO(3):

$$D^{j+j'}(R) = D^{j}(R) \otimes D^{j'}(R) = D^{j+j'}(R) \otimes \cdots \otimes D^{|j-j'|}(R)$$

If

$$j + j' = n$$

n is a whole number we have a tensor representation of single value; if

$$j + j' = \frac{n}{2}$$

the representation is a spinor representation of double value, written (j,j').

Note. In the case of elementary particles of matter, the spinor representation $(\frac{1}{2},0)$ describes the left-handed spinors and the representation $(0,\frac{1}{2})$ describes the right-handed spinors. These two spinor components are called

Weyl spinors. The sum

$$(\tfrac{1}{2},0) \oplus (0,\tfrac{1}{2})$$

represents a *Dirac spinor*.

A12.5 Poincaré Group

A quadrivector is said to be transformed by a nonhomogeneous element of a Poincaré or Lorentz group if its components are such that

$$X'^{\mu} = L^{\mu}{}_{\nu} X^{\nu} + a^{\mu}$$

The first term of the right-hand member is associated with a Lorentz transformation, whereas the second term is a translation of four dimensions.

This group is denoted $\mathbf{P} = (a,L)$ with $(a,L) = (a,1)\,(0,L)$. Indeed, we show that if

$$X' = L'X + a'$$

and

$$X'' = L''X' + a''$$

then

$$X'' = L''L'X + L''a' + a''$$

that is,

$$(a'',L'')(a',L') = (a'' + L''a',L''L')$$

(the product of two Poincaré transformations is a Poincaré transformation).

Note. The order of the operations is important; a Lorentz transformation does not commute with a translation.

Poincaré's group is doubly connected and has 10 parameters (six for the Lorentz group and four for the translation). Four sets $P_{+}^{\uparrow}$, $P_{-}^{\uparrow}$, $P_{+}^{\downarrow}$, and $P_{-}^{\downarrow}$ are associated with the four Lorentz transformations.

A universal covering group is (a,A), where A is a matrix belonging to SL(2,C). A Lie algebra associated with the Poincaré group contains generators J_i and K_i of the Lorentz group plus generators P_μ (impulsions) of the translation group:

$$[P_\mu,P_\nu] = 0 \qquad [J_i,P_0] = 0 \qquad [J_i,P_j] = i\epsilon_{ij}{}^{k}P^{k}$$

$$[K_i,P_0] = -iP_i \qquad [K_i,P_i] = -iP_0 \qquad [K_i,K_j] = -i\epsilon_{ij}{}^{k}J_k$$

$$[J_i,J_j] = i\epsilon_{ij}{}^{k}J_k \qquad [J_i,K_j] = i\epsilon_{ij}{}^{k}K_k$$

By introducing the quantities $M_{\mu\nu}$ and the former relations, form three identities:

$$[P_\mu, P_\nu] = 0$$

$$[M_{\mu\nu}, P_\rho] = -i(g_{\mu\rho}P_\nu - g_{\nu\rho}P_\mu)$$

$$[M_{\rho\lambda}, M_{\mu\nu}] = -i(g_{\rho\mu}M_{\lambda\nu} + g_{\lambda\nu}M_{\rho\mu} - g_{\rho\nu}M_{\lambda\mu} - g_{\lambda\mu}M_{\rho\nu})$$

The *Pauli–Lubanski vector W* is such that

$$W_\mu = \tfrac{1}{2}\epsilon_{\mu\nu\rho\sigma}M^{\nu\rho}P^\sigma$$

it is orthogonal to the quadri-impulsion P:

$$W_\mu P^\mu = 0$$

and complies with the following relation:

$$[P_\mu, W_\nu] = 0$$

$$[M_{\mu\nu}, W_\sigma] = -i(g_{\nu\sigma}W_\mu - g_{\mu\sigma}W_\nu)$$

$$[W_\mu, W_\nu] = i\epsilon_{\mu\nu\rho\sigma}W^\rho P^\sigma$$

The Casimir operators of the group are

$$W^2 = W_\mu W^\mu$$

$$P^2 = P_\mu P^\mu$$

We can easily show that

$$W_0 = P_i J^i$$

$$W_i = P_0 J_i - \epsilon_{ijk}P^j K^k$$

and a general transformation of Poincaré $U(a,L)$ is written

$$U(a,L) = e^{-ia_\mu P^\mu - (i/2)\epsilon^{\mu\nu}M_{\mu\nu}}$$

Note. In the proper referential of the particle of mass m_0,

$$P^\mu P_\mu = m_0^2 c^2$$

and thus

$$W_0 = 0$$

$$W_i = P_0 J_i = m_0 c J_i$$

The state of the particle is represented by the eigenfunction

$$|\psi\rangle = |m_0, j, m_j\rangle$$

such that

$$W^2|\psi\rangle = +m_0^2c^2j(j+1)|\psi\rangle$$
$$W_3|\psi\rangle = m_0cm_j|\psi\rangle$$

APPENDIX 13 HOMOTOPY GROUP

A13.1 Introduction

The aim of this section is to provide a geometric representation of the homomorphism su(2) → so(3) and therefore of transformation (5.10) of Section 5.2 (Cayley–Klein parameters). Let's consider two sets (or manifolds) that we write E_1 and E_2; f and g are two continuous mappings of E_1 in E_2:

$$f\colon E_1 \longrightarrow E_2$$
$$g\colon E_1 \longrightarrow E_2$$

By definition, f and g are said to be homotopic if f can be continually changed into g, and vice versa. This means that a family of intermediary mappings $H(x,t)$ exists with $x \in E_1$ and $t \in I = [0,1]$ and such that

$$H(x,0) = f(x)$$
$$H(x,1) = g(x)$$

or

$$E_1XI \xrightarrow{H} E_2$$

which we can visualize with the help of Figure A13.1. To simplify understanding, we can envisage that t is the parameter and that the functions g

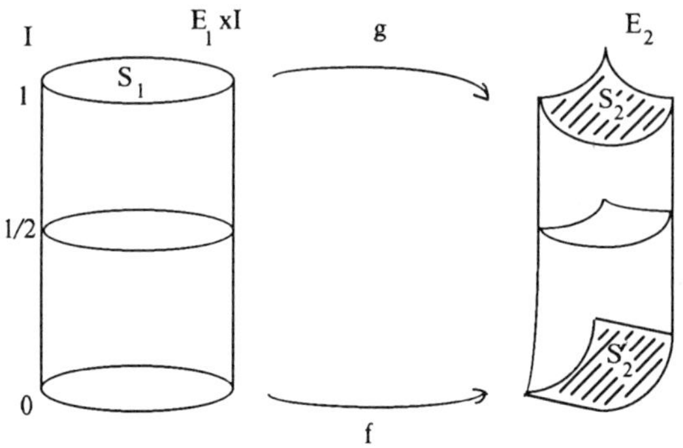

Figure A13.1 Homotopy.

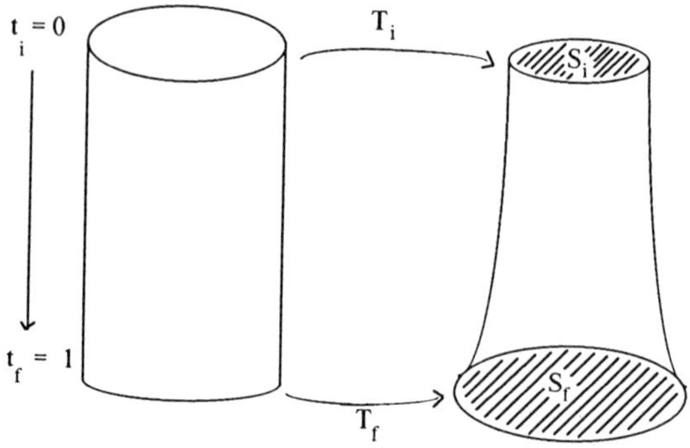

Figure A13.2 Special case.

and f correspond to the different temperatures applied to the surface S (Figure A13.2).

$$H(S,0) = T_{t_i}(S) = S_i$$

$$H(S,1) = T_{t_f}(S) = S_f$$

For the moment we show that the homotopy written $f \sim g$ is an equivalence relation:

1. *Reflexive.* $f \sim f$ (no change in the surface S).
2. *Symmetric.* $f \sim g$ implies that $g \sim f$ (a reversible transformation). In fact, $f \sim g$ implies that a family of functions F exists such that

$$F(x,0) = f(x)$$

$$F(x,1) = g(x)$$

We now build a family of functions G such that

$$G(x,0) = F(x,1) = g(x)$$

$$G(x,1) = F(x,0) = f(x)$$

It defines $g \sim f$.

3. *Transitive.* It is supposed that $f \sim g$ and $g \sim h$; thus $f \sim g$ brings us to

$$F(x,0) = f(x)$$

$$F(x,1) = g(x)$$

and $g \sim h$ leads to

$$G(x,0) = g(x)$$

$$G(x,1) = h(x)$$

A family H is built such that

$$H(x,t) = F(x,2t) \qquad \text{with } t \in [0,\tfrac{1}{2}]$$

that is,

$$H(x,0) = F(x,0) = f(x)$$
$$H(x,\tfrac{1}{2}) = F(x,1) = g(x)$$

and

$$H(x,t) = G(x, 2t - 1) \qquad \text{with } t \in [\tfrac{1}{2},1]$$

which is

$$H(x,\tfrac{1}{2}) = G(x,0) = g(x)$$
$$H(x,1) = G(x,1) = h(x)$$

from which it can be deduced that

$$H(x,0) = f(x)$$
$$H(x,1) = h(x)$$

which means that $f \sim h$.

We have just defined the equivalence classes on the mappings of $E_1 \rightarrow E_2$. These classes form a group with the multiplication law, namely, the homotopy group.

A13.2 Homotopy Group $\pi_1(U_1)$

Let's consider the set E_1 of the points M situated on the circle S_1 of radius unity. The polar coordinates of M are

$$\{(\cos\theta, \sin\theta) | \theta \in [0,2\pi]\}$$

The chosen set E_2 is that of the 1×1, unitary matrix a, that is, such that

$$a^+ = a^{-1}$$

Chosen for the representation of a is (Figure A13.3)

$$a = \{e^{i\alpha} | \alpha \in R\}$$

The linear mapping f is such that

$$\theta \in [0,2\pi] \longrightarrow U_1$$

that is,

$$\theta \longrightarrow f(\theta) = e^{i\alpha(\theta)}$$

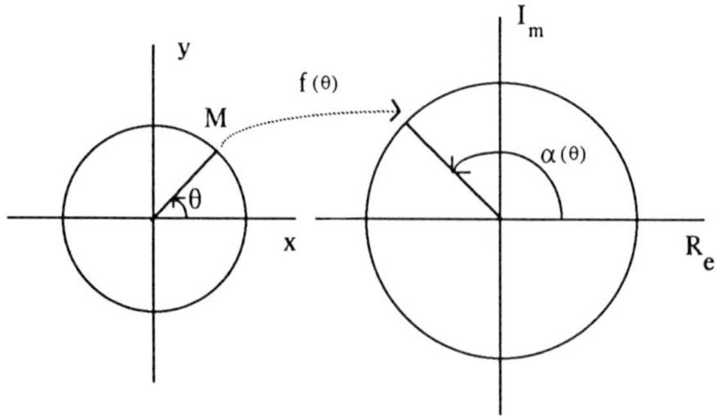

Figure A13.3 Group $\pi_1(U_1)$.

We have

$$f(0) = e^{i\alpha(0)}$$
$$f(2\pi) = e^{i\alpha(2\pi)}$$

but also $f(0) = f(2\pi)$, which leads to

$$\alpha(2\pi) = \alpha(0) + 2n_f\pi$$

$n_f \in Z$ any whole number. The mappings f are thus broken down into homotopy classes labeled by n:

$$\pi_1(U_1) = Z$$

A13.3 Example

We choose the two mappings f and g in such a way that (Figure A13.4)

$$f(\theta) = e^{i\sin\theta}$$

that is,

$$\alpha_f(\theta) = \sin\theta \qquad \theta \in [0,2\pi]$$
$$g(\theta) = 1 \qquad \text{i.e., } \alpha_g(\theta) = 0$$
$$\alpha_f(2\pi) = \alpha_f(0) \qquad \text{if } n_f = 0$$
$$\alpha_g(2\pi) = \alpha_{g(0)} \qquad \text{if } n_g = 0$$

We are for the moment trying to find a family of continuous functions $H(\theta,t)$ that continuously transform f into g. Take

$$H(\theta,t) = e^{it\sin\theta}$$

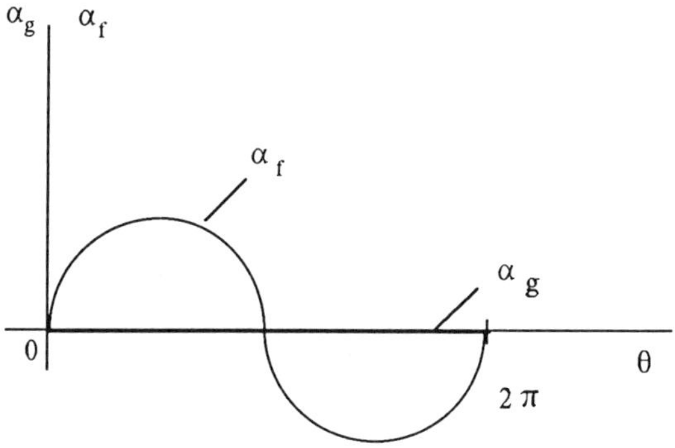

Figure A13.4 Mappings α_f and α_g.

We verify that

$$H(\theta,0) = 1 = g(\theta)$$
$$H(\theta,1) = e^{i\sin\theta} = f(\theta)$$

The functions f and g belong to the same homotopy class: They are homotopic.

Note. If we take (Figure A13.5)

$$f(\theta) = e^{i\cos\theta} \qquad \alpha_f = \cos\theta \qquad \theta \in [0,2\pi]$$
$$g(\theta) = 1 \qquad \alpha_g = 0$$

It is evident that we cannot transform f into g in a continuous way on the interval $[0,2\pi]$; f and g can be made homotopic by taking $\theta \in$

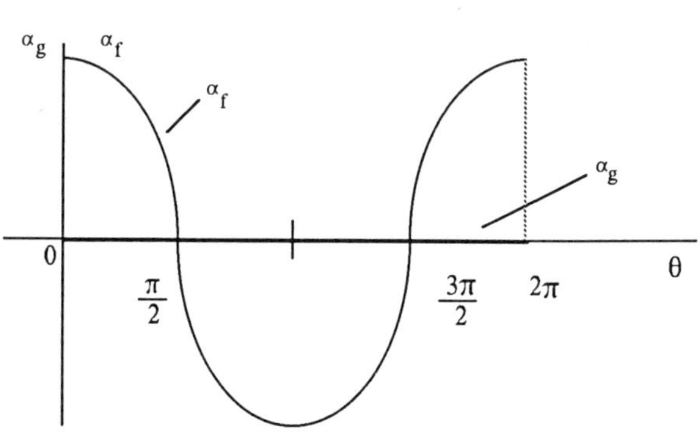

Figure A13.5 Example.

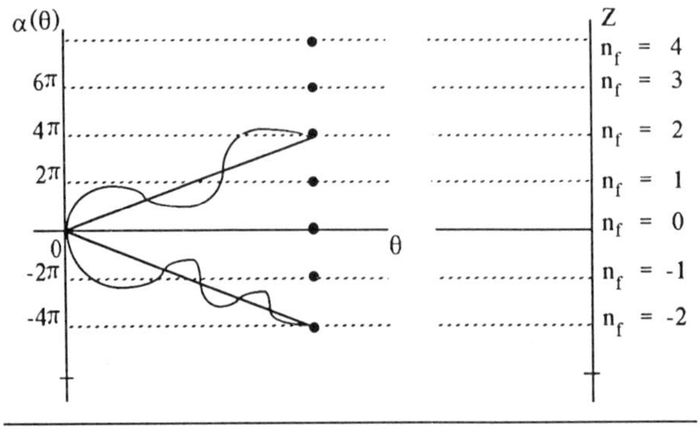

Figure A13.6 Classes of homotopy.

$[\pi/2, 5\pi/2]$. In $\pi_1(U_1)$, in order that f and g belong to the same homotopy class, the curves $\alpha_g(\theta)$ and $\alpha_f(\theta)$ must have the same extremities on an interval of definition of θ.

In general, the relation

$$\alpha(2\pi) = \alpha(0) + 2n_f\pi$$

brings us to [if $\alpha(0) = 0$]

$$\alpha(2\pi) = 2n_f\pi \qquad \text{(cf. Figure A13.6)}$$

The curves $\alpha(\theta)$ have the same extremities but for the same value of n_f; n_f is called the winding number.

A13.4 Homotopy Group of Order *n*

$\pi_n(E)$ is written as the nth homotopy group, defined by the set of equivalence classes of the mappings of the unit sphere S^n into the set E. We can show that

$$\pi_1(\mathrm{SU}(2)) = \pi_1(S^3)$$

All the mappings are homotopic and we can also show that SU(2) is simply connected, as well as showing that

$$\pi_1(\mathrm{SO}(3)) = Z_2 = (1, -1)$$

We have two homotopy classes and SO(3) is doubly connected.

APPENDIX 14 CARTAN EQUATION

A14.1 Review of Definitions

If G is a group, a subset H of elements of G will form a subgroup of G if the elements of H have the structure of a group. We say that H is an invariant subgroup of G if

$$\forall h \in H \text{ and } g \in G \text{ then } ghg^{-1} \in H$$

A group G is said to be *simple* if it has no invariant subgroup (apart from itself and the identity). A *semisimple* group has no invariant abelian subgroup.

If A and B are two algebras, B is said to be a *subalgebra* of A if B is a subset of A. If $\forall a \in A$, $b \in B$ and $[a,b] \in B$ where $[a,b] = ab - ba$, B will be a *subinvariant algebra* of A.

An algebra is said to be *simple* if it has no invariant subalgebra and *semisimple* if it has no invariant abelian subalgebra.

A14.2 Cartan Problem

Take A a semisimple Lie algebra of dimension n on $\mathbf{R}$. A basis of the algebra or generators of the associated group is formed by $X_1, \ldots, X_n$ with

$$[X_i, X_j] = c_{ij}{}^k X_k \tag{A14.1}$$

If $a \in A$, we are trying to find the eigenvectors (operators) X such that

$$[a, Y] = \rho Y \tag{A14.2}$$

(Cartan equation), where ρ is a constant. Project on the $\{X_i\}$ basis

$$a = a^i X_i \tag{A14.3}$$

$$Y = y^j X_j \tag{A14.4}$$

from which we get the form of (A14.2):

$$a^i y^j c_{ij}{}^k X_k = \rho y^j X_j \tag{A14.5a}$$

or even

$$a^i y^j c_{ij}{}^k = \rho y^k \tag{A14.5b}$$

Nonzero solutions exist if

$$\det(a^i c_{ij}{}^k - \rho \delta_j^k) = 0 \tag{A14.6}$$

a polynominal of order n in ρ; there are n solutions in $\mathbf{C}$.

A14.3 Cartan Theorem

A is a semisimple algebra; we chose compliance with condition (A14.6) in such a way that the determinant gives the maximum number of distinct roots, so:

1. The root $\rho = 0$ exists and is r times degenerate and there are r linearly independent elements written $H_1, \ldots, H_r$, such that

$$[a,H_i] = 0 \tag{A14.7}$$

$$[K_i,H_j] = 0 \qquad 1 \leqslant i \leqslant r \tag{A14.8}$$

2. The other roots are nondegenerate; their number is $n - r$.

   If E_α are the eigenvectors associated with the eigenvalue α,

$$[a,E_\alpha] = \alpha E_\alpha \tag{A14.9}$$

the r elements H_i and the $n - r$ elements E_α are then linearly independent and form a basis of A, said to be Cartan.

Note. The degree of degeneracy on r of the eigenvalue $\rho = 0$ is typical of the algebra and allows classification of the semisimple algebra; r is called the *rank* of the algebra and is equal to the number of Casimir operators of the algebra.

We can prove the following relations:

$$[H_i,E_\alpha] = \alpha_i E_\alpha \tag{A14.10}$$

$$[E_\alpha,E_{-\alpha}] = \alpha^i H_i \tag{A14.11}$$

$$[E_\alpha,E_\beta] = \begin{cases} N_{\alpha\beta}E_{\alpha+\beta} & \text{if } \alpha + \beta \text{ is a nonzero root} \\ 0 & \text{if } \alpha + \beta \text{ is not a root} \end{cases} \tag{A14.12}$$

According to (A14.10), E_α is the eigenvector of H_i associated with the eigenvalue α_i. The α_i are called *roots* and can be represented with the help of the vector $(\alpha_i, \ldots, \alpha_r)$ belonging to a space of dimension r.

The quantities H_i form the Cartan subalgebra. According to (A14.7), they have an eigenvalue of zero.

A14.4 Example of SU(2)

The three generators of the group are written J_1, J_2, and J_3. There is, in view of what was seen earlier, only one Casimir operator $C_1 = J^2$. The rank is $r = 1$; the root $\rho = 0$ is not degenerate. In (A14.2) we arbitrarily choose $a = J_3$:

$$[J_3,Y] = \rho Y$$

and according to the well-known results of quantum mechanics,

$$[J_3,J_\pm] = \pm i\hbar J_\pm$$

The J_+ (respectively, J_-) play the role of E_α (respectively, $E_{-\alpha}$), but equally,

$$J_\pm = J_1 \pm iJ_2$$

The diagram of the roots is of dimension 1 (as $r = 1$):

$$\xrightarrow{\qquad\qquad\qquad} \; J_3$$

$$J_- \qquad 0 \qquad J_+$$

The results obtained are also valid for SO(3).

A14.5 Example of SU(3)

There are $N^2 - 1$ generators of group SU(N) or 8 for SU(3), noted J_1, J_2, . . . , J_8. We have two Casimir operators:

$$C_1 = J^2$$

$$C_2 = f_{ijk}J_iJ_jJ_k$$

The f_{ijk} are structure constants. We choose the operators J_1 and J_8 such that

$$[J_1,J_8] = 0$$

We want to find the solutions of

$$[J_3,E_\alpha] = \alpha_3 E_\alpha$$

$$[J_8,E_\alpha] = \alpha_8 E_\alpha$$

which are given by

$$T_\pm = J_1 \pm iJ_2$$

$$U_\pm = J_6 \pm iJ_7$$

$$V_\pm = J_4 \pm iJ_5$$

and such that

$$[J_3,T_\pm] = \pm T_\pm$$

$$[J_3,U_\pm] = \mp\tfrac{1}{2}U_\pm$$

$$[J_3,V_\pm] = \pm\tfrac{1}{2}V_\pm$$

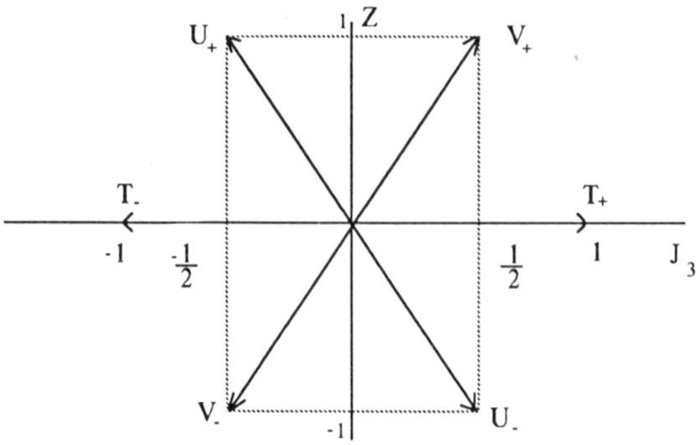

Figure A14.1 Diagram of roots.

We define $Z = (2/\sqrt{3})J_8$, while

$$[Z,T_\pm] = 0$$

$$[Z,U_\pm] = \pm U_\pm$$

$$[Z,V_\pm] = \pm V_\pm$$

The diagram of roots is given in Figure A14.1

A14.6 Diagram of Weights

Take a hermitian representation of a semisimple algebra. The Cartan subalgebra is then represented by the hermitian operators H_i. We should have

$$[H_i,H_j] = 0$$

and thus an eigenvector $|m\rangle$ exists such that

$$H_i|m\rangle = m_i|m\rangle$$

The vector of components m_i is indicated by m. This vector, similar to a root, is called the *weight* of the eigenvector $|m\rangle$. Moreover, we know that

$$[H_i,E_\alpha] = \alpha_i E_\alpha$$

and thus by applying it to $|m\rangle$,

$$(H_i E_\alpha - E_\alpha H_i)|m\rangle = \alpha_i E_\alpha|m\rangle$$

from which it can be deduced that

$$H_i E_\alpha|m\rangle = (m_i + \alpha_i)E_\alpha|m\rangle$$

which we translate in the form of a theorem: If $|m\rangle$ is an eigenvector of the H_i of weight m, then $E_\alpha|m\rangle$ is an eigenvector of the same operators H_i but of weight $m + \alpha$.

Example of SU(2). We are working on the representations D_j:

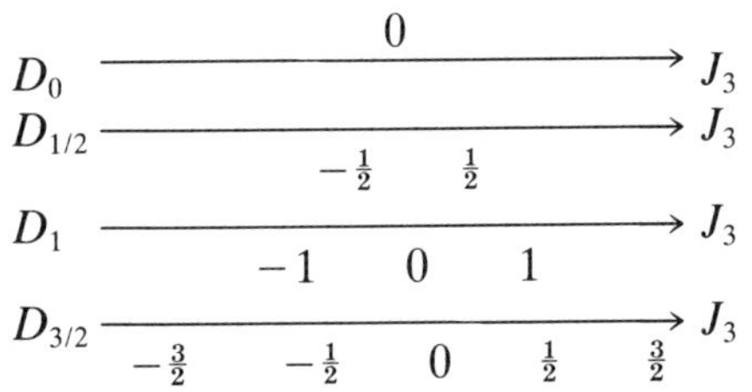

Example of the Product of Representations $D_{1/2} \otimes D_1$

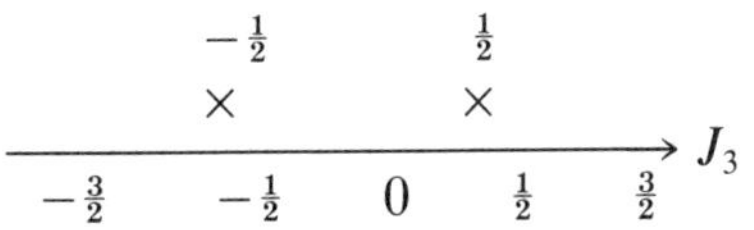

Example of SU(3) We have

$$J_3 = \frac{1}{2}\begin{bmatrix} 1 & 0 & 0 \\ 0 & -1 & 0 \\ 0 & 0 & 0 \end{bmatrix} \qquad Z = \frac{1}{\sqrt{3}}\begin{bmatrix} 1 & 0 & 0 \\ 0 & -1 & 0 \\ 0 & 0 & 0 \end{bmatrix}$$

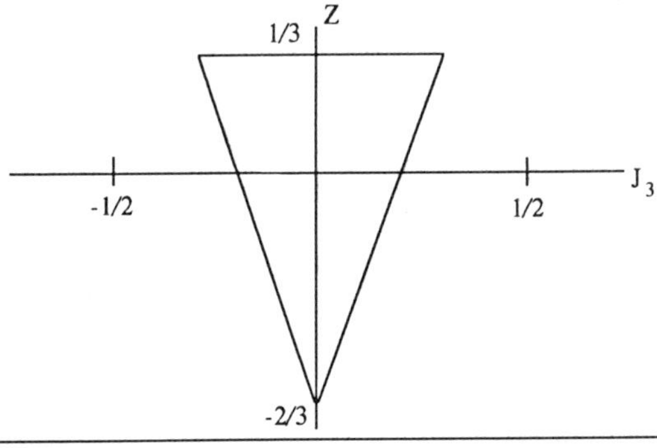

Figure A14.2 Diagram of the weights of SU(3).

The fundamental representation of dimension 3 has as basis vectors,

$$\{|J_3,Z\rangle\} = |\tfrac{1}{2},\tfrac{1}{3}\rangle,\ |-\tfrac{1}{2},\tfrac{1}{3}\rangle,\ |0,-\tfrac{2}{3}\rangle$$

Figure A14.2 gives the diagram of the weight of SU(3).

In the physics of elementary particles we associate three quarks (u,d,s) with this representation:

u = up
d = down
s = strange
Z = hypercharged

The antiparticles correspond to the dual diagram obtained by reversing the quantum numbers. We get the other particles by the products of these representations.

5
Vector Algebra

Calculation using the concepts of geometry in relation to the rules of an algebra has for many years been the aim of the research of many mathematicians, Leibnitz and Carnot being two of them. W. R. Hamilton (1805–1865) was the first to make systematic constructions and he created the quaternions. He generalized the complex numbers to that space and was obliged to abandon commutativity for the product. H. Grassmann (1809–1877) defined the interior and exterior products of the vectors. The unification of the two former points of view was performed by W. K. Clifford (1845–1879) in the same algebra.

5.1 INNER PRODUCT

If a^2 is a quadratic form evaluated from a vector a from E_n, a vector space of dimension n:

$$(a + b)^2 = a^2 + ab + ba + b^2 \tag{5.1}$$

from which we deduce that

$$ab + ba = (a + b)^2 - a^2 - b^2 \tag{5.2}$$

which is a scalar and is written

$$a \cdot b = \tfrac{1}{2}(ab + ba) \tag{5.3}$$

a definition of the inner product of the vectors a and b.

It can be seen that

$$a \cdot a = a^2 \tag{5.4}$$

and that

$$a \cdot b = 0$$

which implies [cf. (5.3)] that

$$ab + ba = 0$$

a and b are anticommutative numbers or Grassmann numbers, in which case a and b are said to be conjugate in relation to the quadratic form or orthogonal. We have, according to (5.3),

$$a \cdot b = b \cdot a$$

The inner product is commutative.

5.2 EXTERIOR PRODUCT

The quantity ab can still be written as the sum of a symmetric term and of an antisymmetric term,

$$ab = \tfrac{1}{2}(ab + ba) + \tfrac{1}{2}(ab - ba) \tag{5.5}$$

The antisymmetrical part $\tfrac{1}{2}(ab - ba)$, which is written $a \wedge b$, is called the *exterior product* of a and b:

$$a \wedge b = \tfrac{1}{2}(ab - ba) = ab - a \cdot b \tag{5.6}$$

According to (5.3),

$$a \wedge b = ab - a \cdot b$$

or

$$ab = a \wedge b + a \cdot b \tag{5.7}$$

By definition,

$$b \wedge a = \tfrac{1}{2}(ba - ab) = -a \wedge b \tag{5.8}$$

The condition

$$a \wedge b = 0 \tag{5.9}$$

implies that [cf. (5.7)]

$$a \cdot b = ab \tag{5.10}$$

Otherwise,

$$a \cdot b = \tfrac{1}{2}(ab + ba)$$

which means that we should have

$$ab = ba$$

a relation that is checked only if the vectors a and b are collinear.

5.3 SPACE ALGEBRA

The algebra of space is the algebra of the three-dimensional space of euclidean geometry. Consider the Pauli matrices

$$\sigma_1 = \begin{bmatrix} 0 & 1 \\ 1 & 0 \end{bmatrix} \qquad \sigma_2 = \begin{bmatrix} 0 & -i \\ i & 0 \end{bmatrix} \qquad \sigma_3 = \begin{bmatrix} 1 & 0 \\ 0 & -1 \end{bmatrix}$$

It was seen earlier that these matrices confirm the relations

$$\sigma_j\sigma_k + \sigma_k\sigma_j = 2\delta_{jk} \tag{5.11}$$

$k, j = 1, 2, 3$ and $\delta_{jk} = 0$ if $j \neq k$, $\delta_{jj} = 1$; but also,

$$\sigma_i\sigma_j = i\sigma_k \tag{5.12}$$

$$\sigma_i^2 = I_2 \tag{5.13}$$

if I_2 is the 2×2 unit matrix and i is an index that runs from 1 to 3. It is obvious that

$$\sigma_i\sigma_j\sigma_k = i\sigma_k\sigma_k = iI_2 \tag{5.14}$$

We often use the notation

$$I_2 \equiv 1$$

$$iI_2 \equiv i$$

We build a vector space $\mathbf{C}_3$ from the complex matrices with σ_1, σ_2, σ_3, I_2, $i\sigma_1$, $i\sigma_2$, $i\sigma_3$, and iI_2. Now according to the relations (5.11) to (5.14), this set of matrices is also written σ_1, σ_2, σ_3, I_2, $\sigma_1\sigma_2$, $\sigma_2\sigma_3$, $\sigma_3\sigma_1$, $\sigma_1\sigma_2\sigma_3$.

Due to their construction, the matrices I_2, σ_1, σ_2, σ_3, $\sigma_1\sigma_2$, $\sigma_2\sigma_3$, $\sigma_3\sigma_1$, and $\sigma_1\sigma_2\sigma_3$ form a basis of this space. In can easily be proved that

$$a_0I_2 + a_1\sigma_1 + a_2\sigma_2 + a_3\sigma_3 + a_4\sigma_1\sigma_2 + a_5\sigma_2\sigma_3 + a_6\sigma_3\sigma_1 + a_7\sigma_1\sigma_2\sigma_3 = 0$$

leads to

$$a_0 = a_1 = a_2 = a_3 = a_4 = a_5 = a_6 = a_7 = 0$$

The matrices σ_1, σ_2, and σ_3 are henceforth considered as the vectors of space, of square 1, of ordinary geometry and we identify the vector $\vec{r}$ denoted r:

$$r = x_1\sigma_1 + x_2\sigma_2 + x_3\sigma_3 \tag{5.15}$$

with the vector of components x_1, x_2, and x_3 on an orthonormal basis of the geometric space (cf. Chapter 4).

5.3.1 Fundamental Operations

Any vector of $\mathbf{C}_3$ can be considered as the sum of:

A scalar: $A_s = \lambda_0 I_2$ (5.16)
A vector: $A_v = \lambda_1\sigma_1 + \lambda_2\sigma_2 + \lambda_3\sigma_3$ (5.17)
A bivector: $A_b = \lambda_4\sigma_1\sigma_2 + \lambda_5\sigma_2\sigma_3 + \lambda_6\sigma_3\sigma_1$ (5.18)
A pseudoscalar: $A_p = \lambda_7\sigma_1\sigma_2\sigma_3 = \lambda_7 iI_2$ (5.19)

(a scalar that changes sign when the direction of the axes is changed in exchange, for example, σ_1 in σ_2).

Therefore, in $\mathbf{C}_3$ every element is written

$$A = A_s + A_v + A_b + A_p \tag{5.20}$$

We can define the degree of A_s, A_v, A_b, and A_p by the number of Pauli matrices occurring in their formulas [cf. (5.16) to (5.19)]:

0 for A_s

1 for A_v

2 for A_b

3 for A_p

Introduce three fundamental operations.

Change of Direction

Beginning with A, this operation gives the element $\overline{A}$ by changing the direction of all the vectors that occur in A:

$$\overline{A} = A_s - A_v + A_b - A_p \tag{5.21}$$

If $\overline{A} = A$, it is said that A is even; if $\overline{A} = -A$, it is said that A is odd.

Inversion of the Order

The inversion of the order in the products transforms A into $\tilde{A}$ such that

$$\tilde{A} = A_s + A_v - A_b - A_p \tag{5.22}$$

Indeed,

$$\tilde{\sigma}_i\sigma_j = -\sigma_j\sigma_i = i\sigma_k$$

and

$$\widetilde{\sigma_i\sigma_j} = \sigma_j\sigma_i = -\sigma_i\sigma_j$$

$$\widetilde{\sigma_1\sigma_2\sigma_3} = \sigma_3\sigma_2\sigma_1 = -i\sigma_1\sigma_2 = -iI_2 = -\sigma_1\sigma_2\sigma_3$$

Dual

We can define the dual of A by its multiplication with iI_2. It is obvious that

$$iI_2A = AiI_2$$

The dual of a scalar $\lambda_0 I_2$ is a pseudoscalar, since

$$iI_2A_s = iI_2\lambda_0 I_2 = \lambda_0 iI_2 = -\lambda_0\sigma_1\sigma_2\sigma_3 = A_p$$

The dual of a vector is a bivector and reciprocally as

$$iI_2A_v = iI_2(\lambda_1\sigma_1 + \lambda_2\sigma_2 + \lambda_3\sigma_3) = (\lambda_1 i\sigma_1 + \lambda_2 i\sigma_2 + \lambda_3 i\sigma_3)$$

$$= \lambda_1\sigma_2\sigma_3 + \lambda_2\sigma_3\sigma_1 + \lambda_3\sigma_1\sigma_2 = A_b$$

$$iI_2A_b = iI_2(\lambda_4\sigma_1\sigma_2 + \lambda_5\sigma_2\sigma_3 + \lambda_6\sigma_3\sigma_1)$$

$$= i[\lambda_4(i\sigma_3) + \lambda_5(i\sigma_1) + \lambda_6(i\sigma_2)]$$

$$= -\lambda_4\sigma_3 - \lambda_5\sigma_1 - \lambda_6\sigma_2 = A_v$$

The dual of a pseudoscalar is a scalar

$$iI_2A_p = iI_2\lambda_7\sigma_1\sigma_2\sigma_3 = iI_2\lambda_7(iI_2) = -\lambda_7 I_2 = A_s$$

By going back to the rules of definition of the dual in Chapter 4, the degree of the dual is the dimension of the vector space $\mathbf{C}_3$, that is, 3, reduced by the degree of the element A_s, A_v, A_b, or A_p. For example, the dimension of the dual of A_s is $3 - 0 = 3$; it is therefore a pseudoscalar A_p.

5.3.2 Quaternions

By definition, A is said to be a *quaternion*, noted A_q, if the condition

$$A = \overline{A} \tag{5.23}$$

is proved, A being an element of $\mathbf{C}_3$; otherwise,

$$A_q = A_s + A_b \tag{5.24}$$

or even

$$A_q = \lambda_0 I_2 + \lambda_4 \sigma_1 \sigma_2 + \lambda_5 \sigma_2 \sigma_3 + \lambda_6 \sigma_3 \sigma_1 \tag{5.25}$$

which we write in the general form

$$A_q = \lambda_0 I_2 + i\lambda_k \sigma_k \tag{5.26}$$

since

$$\sigma_i \sigma_j = i\sigma_k$$

or even

$$A_q = A_s + i\vec{a}\cdot\vec{\sigma} \tag{5.27}$$

where A_s is a scalar ($A_s = \lambda_0 I_2$), $\vec{a}$ is a vector of real components, and $\vec{\sigma}$ is a vector of norm equal to 3.

The set of the quaternions has a vector space structure of dimension 4, a subspace of $\mathbf{C}_3$, as the elements I_2, $i\sigma_1$, $i\sigma_2$, and $i\sigma_3$ form a free system, the equality

$$\lambda_0 I_2 + a_1 i\sigma_1 + a_2 i\sigma_2 + a_3 i\sigma_3 = 0$$

requiring on the one hand,

$$\lambda_0 = 0$$

and on the other hand,

$$a_1 = a_2 = a_3 = 0$$

This vector subspace of $\mathbf{C}_3$ is indicated by Q_4. Moreover, it can be normalized by writing, for example,

$$\text{norm of } A_q = \|A_q\| = (\lambda_0^2 + a^2)^{1/2} \tag{5.28}$$

Note that following (5.26),

$$\tilde{A}_q = \lambda_0 I_2 + \vec{a}\cdot\widetilde{i\vec{\sigma}} = \lambda_0 I_2 - \vec{a}\cdot i\vec{\sigma} \tag{5.29}$$

since

$$i\sigma_k = \sigma_i\sigma_j$$

and thus

$$A_q\tilde{A}_q = \lambda_0^2 + a^2 \tag{5.30}$$

The tilde plays the role of complex conjugate.

If we indicate by A_q and $A_{q'}$ two quaternions, their dot product is defined by

$$\langle A_q|A_{q'}\rangle = \tfrac{1}{2}(A_q\tilde{A}_{q'} + A_{q'}\tilde{A}_q) \tag{5.31}$$

If A_q is a nonzero quaternion, an inverse A_q^{-1} exists such that

$$A_q^{-1} = \frac{A_s - i\vec{a}\cdot\vec{\sigma}}{\|A_q\|^2}$$

as then

$$A_q^{-1}A_q = \frac{(A_s - i\vec{a}\cdot\vec{\sigma})(A_s + i\vec{a}\cdot\vec{\sigma})}{\|A_q\|^2} = \frac{\lambda_s^2 + a^2}{\|A_q\|^2} = 1$$

The field of quaternions is not commutative:

$$A_qA_{q'} \neq A_{q'}A_q \tag{5.32}$$

The equality, thus the commutativity, would require that

$$A_qA_{q'} = A_{q'}A_q$$

that is, the condition

$$A_q \wedge A_{q'} = 0 \qquad \text{[cf. (5.6)]}$$

which would lead to the collinearity of A_q and $A_{q'}$.

The quantity

$$\begin{aligned} e^{i\vec{a}\cdot\vec{\sigma}} &= I_2 + i\vec{a}\cdot\vec{\sigma} - \tfrac{1}{2}(\vec{a}\cdot\vec{\sigma})^2 + \cdots \\ &= I_2(1 - \tfrac{1}{2}a^2 + \cdots) + i\vec{\sigma}\cdot[\vec{a} + \tfrac{1}{3}(\vec{a})^3 + \cdots] \\ &= I_2 \cos\vec{a} + i\vec{\sigma}\cdot\sin\vec{a} \end{aligned} \tag{5.33}$$

is such that

$$\|e^{i\vec{a}\cdot\vec{\sigma}}\| = cos^2\vec{a} + \sin^2\vec{a} = 1 \tag{5.34}$$

and therefore defines the unit quaternion $(A_q)_u$ by

$$(A_q)_u = e^{i\vec{a}\cdot\vec{\sigma}} \tag{5.35}$$

5.4 REAL ALGEBRA OF SPACE-TIME

We build this algebra of space-time with the help of four vectors γ_μ or γ_0 and γ_i, $i = 1, 2, 3$. The vector γ_0 is a temporal vector of square 1 and the γ_k three vectors of space of square -1. These vectors are also orthogonal; that is, that they satisfy the relation (cf. Section 5.1)

$$\gamma_i\gamma_j + \gamma_j\gamma_i = 0 \qquad i \neq j, \quad i, j = 1, 2, 3 \tag{5.36a}$$

which we write as

$$[\gamma_i, \gamma_j]_+ = -2\delta_{ij} \tag{5.36b}$$

We show that such vectors can exist representing them by the matrices

$$\gamma_0 = \begin{bmatrix} I_2 & \emptyset \\ \emptyset & -I_2 \end{bmatrix} \tag{5.37}$$

$$\gamma_i = \begin{bmatrix} \emptyset & -\sigma_i \\ \sigma_i & \emptyset \end{bmatrix} \tag{5.38}$$

where I_2 is the 2×2 unit matrix and σ_i is the Pauli matrix.

To prove that

$$\gamma_0^2 = \begin{bmatrix} I_2 & \emptyset \\ \emptyset & -I_2 \end{bmatrix} \begin{bmatrix} I_2 & \emptyset \\ \emptyset & -I_2 \end{bmatrix} = \begin{bmatrix} I_2 & \emptyset \\ \emptyset & I_2 \end{bmatrix} = I_4 \tag{5.39}$$

$$\gamma_k^2 = \begin{bmatrix} \emptyset & -\sigma_k \\ \sigma_k & \emptyset \end{bmatrix} \begin{bmatrix} \emptyset & -\sigma_k \\ \sigma_k & \emptyset \end{bmatrix} = \begin{bmatrix} -\sigma_k^2 & \emptyset \\ \emptyset & -\sigma_k^2 \end{bmatrix} = -I_4 \tag{5.40}$$

the relations (5.39) and (5.40) are written by introducing the metric tensor $g_{\mu\nu}$:

$$[\gamma_\mu, \gamma_\nu]_+ = 2g_{\mu\nu}$$

with, by definition of the anticommutator,

$$[\gamma_\mu, \gamma_\nu]_+ = \gamma_\mu\gamma_\nu + \gamma_\nu\gamma_\mu$$

It is said that γ_0 is a temporal vector and the γ_k vectors of space by analogy between the norm in Minkovski space:

$$X^2 = c^2t^2 - \vec{r}^{\,2}$$

and the relations (5.39) and (5.40). The following results can easily be confirmed:

$$\gamma_0\gamma_k = \begin{bmatrix} I_2 & \emptyset \\ \emptyset & -I_2 \end{bmatrix} \begin{bmatrix} \emptyset & -\sigma_k \\ \sigma_k & \emptyset \end{bmatrix} = -\begin{bmatrix} \emptyset & \sigma_k \\ \sigma_k & \emptyset \end{bmatrix} = -\Sigma_k \tag{5.41}$$

$$\begin{aligned} \gamma_i\gamma_j &= \begin{bmatrix} \emptyset & -\sigma_i \\ \sigma_i & \emptyset \end{bmatrix} \begin{bmatrix} \emptyset & -\sigma_j \\ \sigma_j & \emptyset \end{bmatrix} \\ &= \begin{bmatrix} -\sigma_i\sigma_j & \emptyset \\ \emptyset & -\sigma_i\sigma_j \end{bmatrix} = -i\begin{bmatrix} \sigma_k & \emptyset \\ \emptyset & \sigma_k \end{bmatrix} \end{aligned} \tag{5.42}$$

with

$$\Sigma_k = \begin{bmatrix} \emptyset & \sigma_k \\ \sigma_k & \emptyset \end{bmatrix} \tag{5.43}$$

There is a Clifford algebra. The product of two vectors is the product of two matrices. The associated space $\mathbf{C}_4$ is of dimension 16 on the field of the real numbers since I_4, γ_0, γ_1, γ_2, γ_3, $\gamma_1\gamma_0$, $\gamma_2\gamma_0$, $\gamma_3\gamma_0$, $\gamma_1\gamma_2$, $\gamma_2\gamma_3$, $\gamma_3\gamma_1$, $\gamma_0\gamma_1\gamma_2$, $\gamma_0\gamma_2\gamma_3$, $\gamma_0\gamma_3\gamma_1$, $\gamma_1\gamma_2\gamma_3$, and $\gamma_0\gamma_1\gamma_2\gamma_3$ form a basis of the vector space $\mathbf{C}_4$. It can easily be shown that

$$(\gamma_0\gamma_1\gamma_2\gamma_3)^2 = -I_4 \tag{5.44}$$

To be exact,

$$I_4 = \begin{bmatrix} 1 & 0 & 0 & 0 \\ 0 & 1 & 0 & 0 \\ 0 & 0 & 1 & 0 \\ 0 & 0 & 0 & 1 \end{bmatrix} \tag{5.45}$$

$$\gamma_0 = \begin{bmatrix} 1 & 0 & 0 & 0 \\ 0 & 1 & 0 & 0 \\ 0 & 0 & -1 & 0 \\ 0 & 0 & 0 & -1 \end{bmatrix} \tag{5.46}$$

$$\gamma_1 = \begin{bmatrix} 0 & 0 & 0 & -1 \\ 0 & 0 & -1 & 0 \\ 0 & 1 & 0 & 0 \\ 1 & 0 & 0 & 0 \end{bmatrix} \tag{5.47}$$

$$\gamma_2 = \begin{bmatrix} 0 & 0 & 0 & i \\ 0 & 0 & -i & 0 \\ 0 & -i & 0 & 0 \\ i & 0 & 0 & 0 \end{bmatrix} \tag{5.48}$$

$$\gamma_3 = \begin{bmatrix} 0 & 0 & -1 & 0 \\ 0 & 0 & 0 & 1 \\ 1 & 0 & 0 & 0 \\ 0 & -1 & 0 & 0 \end{bmatrix} \tag{5.49}$$

$$\gamma_1\gamma_0 = \begin{bmatrix} 0 & 0 & 0 & 1 \\ 0 & 0 & 1 & 0 \\ 0 & 1 & 0 & 0 \\ 1 & 0 & 0 & 0 \end{bmatrix} = \begin{bmatrix} \emptyset & \sigma_1 \\ \sigma_1 & \emptyset \end{bmatrix} = \Sigma_1 \tag{5.50}$$

$$\gamma_2\gamma_0 = \begin{bmatrix} 0 & 0 & 0 & -i \\ 0 & 0 & i & 0 \\ 0 & -i & 0 & 0 \\ i & 0 & 0 & 0 \end{bmatrix} = \begin{bmatrix} \emptyset & \sigma_2 \\ \sigma_2 & \emptyset \end{bmatrix} = \Sigma_2 \tag{5.51}$$

$$\gamma_3\gamma_0 = \begin{bmatrix} 0 & 0 & 1 & 0 \\ 0 & 0 & 0 & -1 \\ 1 & 0 & 0 & 0 \\ 0 & -1 & 0 & 0 \end{bmatrix} = \begin{bmatrix} \emptyset & \sigma_3 \\ \sigma_3 & \emptyset \end{bmatrix} = \Sigma_3 \tag{5.52}$$

$$\gamma_2\gamma_1 = \begin{bmatrix} i & 0 & 0 & 0 \\ 0 & -i & 0 & 0 \\ 0 & 0 & i & 0 \\ 0 & 0 & 0 & -i \end{bmatrix} \tag{5.53}$$

$$\gamma_3\gamma_1 = \begin{bmatrix} 0 & -1 & 0 & 0 \\ 1 & 0 & 0 & 0 \\ 0 & 0 & 0 & -1 \\ 0 & 0 & 1 & 0 \end{bmatrix} \tag{5.54}$$

$$\gamma_3\gamma_2 = \begin{bmatrix} 0 & i & 0 & 0 \\ i & 0 & 0 & 0 \\ 0 & 0 & 0 & i \\ 0 & 0 & i & 0 \end{bmatrix} \tag{5.55}$$

$$\gamma_0\gamma_1\gamma_2\gamma_3 = \begin{bmatrix} 0 & 0 & i & 0 \\ 0 & 0 & 0 & i \\ i & 0 & 0 & 0 \\ 0 & i & 0 & 0 \end{bmatrix} \tag{5.56}$$

and

$$(\gamma_0\gamma_1\gamma_2\gamma_3)^2 = -I_4 \tag{5.57}$$

Thus if A_d is an element of $\mathbf{C}_4$ called Dirac's *d-number* or simply a *d*-number, we can write

$$A_d = A_s + A_v + A_b + A_t + A_p \tag{5.58}$$

that is, the sum of

A scalar: $A_s = \lambda_0 I_4$
A vector: $A_v = \lambda_1\gamma_1 + \lambda_2\gamma_2 + \lambda_3\gamma_3$
A bivector: $A_b = \lambda_4\gamma_1\gamma_0 + \lambda_5\gamma_2\gamma_0 + \lambda_6\gamma_3\gamma_0 + \lambda_7\gamma_1\gamma_2 + \lambda_8\gamma_2\gamma_3 + \lambda_9\gamma_3\gamma_1$
A trivector: $A_t = \lambda_{10}\gamma_0\gamma_1\gamma_2 + \lambda_{11}\gamma_0\gamma_1\gamma_3 + \lambda_{12}\gamma_0\gamma_2\gamma_3 + \lambda_{13}\gamma_1\gamma_2\gamma_3$
A pseudoscalar: $A_p = \lambda_{14}\gamma_0\gamma_1\gamma_2\gamma_3$

Finally, the quadratic form associated with this algebra is obtained by writing

$$X = x^0\gamma_0 + x^1\gamma_1 + x^2\gamma_2 + x^3\gamma_2 = x^\mu\gamma_\nu \qquad \text{with } \mu = 0, 1, 2, 3$$

and by squaring

$$X^2 = (x^0)^2 - (x^1)^2 - (x^2)^2 - (x^3)^2$$

(the Minkowski norm). By writing $x^0 = x_0 = ct$, where t is the time and c the speed of propagation of light, we obtain

$$X^2 = c^2t^2 - \vec{r}^2 = s^2 \tag{5.59}$$

It is the form of restricted relativity. The vectors are introduced for convenience

$$\gamma^0 = \gamma_0 \qquad \gamma^i = -\gamma_i \qquad i = 1, 2, 3$$

(conform to $x^0 = x_0$ and $x^i = -x_i$). We then call the gradient the vector $\gamma^\mu\partial_\mu$ and

$$(\gamma^\mu\partial_\mu)^2 = \partial_0^2 - \partial_1^2 - \partial_2^2 - \partial_3^2 = \square \tag{5.60}$$

the d'Alembertian.

5.5 LORENTZ TRANSFORMATION

5.5.1 *d*-Numbers, Odd and Even

From the d-number A_d we can define an even d-number $(A_d)_p$ such that

$$(A_d)_p = A_s + A_b + A_p = \lambda_0\gamma_0 + \lambda_1\gamma_1\gamma_0 + \lambda_2\gamma_2\gamma_0 + \lambda_3\gamma_3\gamma_0 + \lambda_4\gamma_1\gamma_2 + \lambda_5\gamma_2\gamma_3 + \lambda_6\gamma_3\gamma_1 + \lambda_7\gamma_0\gamma_1\gamma_2\gamma_3 \tag{5.61}$$

We have

$$\begin{aligned}(A_d)_p(\tilde{A}_d)_p &= (A_s + A_b + A_p)(A_s - A_b + A_p)\\ &= (A_s + A_p)^2 - A_b^2\end{aligned} \tag{5.62}$$

and an odd d-number $(A_d)_i$ such that

$$(A_d)_i = A_v + A_t = \lambda_1\gamma_1 + \lambda_2\gamma_2 + \lambda_3\gamma_3 + \lambda_4\gamma_0\gamma_1\gamma_2 + \lambda_5\gamma_0\gamma_1\gamma_3 + \lambda_6\gamma_0\gamma_2\gamma_3 + \lambda_7\gamma_0\gamma_1\gamma_3 \tag{5.63}$$

5.5.2 Writing an Even d-Number

On the right-hand side of (5.62), the quantity A_b^2 appears. We evaluate it by taking

$$A_b = \lambda_1\gamma_1\gamma_0 + \lambda_2\gamma_2\gamma_0 + \lambda_3\gamma_3\gamma_0 + \lambda_4\gamma_1\gamma_2 + \lambda_5\gamma_2\gamma_3 + \lambda_6\gamma_3\gamma_1 \tag{5.64}$$

from which

$$A_b^2 = (\lambda_1^2 + \lambda_2^2 + \lambda_3^2 - \lambda_4^2 - \lambda_5^2 - \lambda_6^2)I_4 - 2(\lambda_1\lambda_5 + \lambda_2\lambda_6 + \lambda_3\lambda_4)\gamma_0\gamma_1\gamma_2\gamma_3$$

Now

$$\gamma_0\gamma_1\gamma_2\gamma_3 = i\begin{bmatrix} 0 & 0 & 1 & 0 \\ 0 & 0 & 0 & 1 \\ 1 & 0 & 0 & 0 \\ 0 & 1 & 0 & 0 \end{bmatrix} = i\begin{bmatrix} \emptyset & I_2 \\ I_2 & \emptyset \end{bmatrix}$$

It is therefore the sum of a scalar and a pseudoscalar:

$$A_b^2 = A_s + A_p \tag{5.65}$$

We reach the same conclusion if (5.64) is written in the form

$$A_b = \lambda_1\Sigma_1 + \lambda_2\Sigma_2 + \lambda_3\Sigma_3 + \lambda_4\gamma_1\gamma_2 + \lambda_5\gamma_2\gamma_3 + \lambda_6\gamma_3\gamma_1$$

and

$$A_b^2 = (\lambda_1\Sigma_1 + \lambda_2\Sigma_2 + \lambda_3\Sigma_3)^2 - (\lambda_5\Sigma_1 + \lambda_6\Sigma_2 + \lambda_4\Sigma_3)^2 - i[(\lambda_1\Sigma_1 + \lambda_2\Sigma_2 + \lambda_3\Sigma_3)(\lambda_5\Sigma_1 + \lambda_6\Sigma_2 + \lambda_4\Sigma_3) + (\lambda_5\Sigma_1 + \lambda_6\Sigma_2 + \lambda_4\Sigma_3)(\lambda_1\Sigma_1 + \lambda_2\Sigma_2 + \lambda_3\Sigma_3)]$$

from which it can be deducted that [cf. (5.52), (5.65), (5.57)]

$$(A_d)_p(\tilde{A}_d)_p = A_s' + A_p' = (a + ia')I_4 \tag{5.66}$$

which since it is a complex, if

$$(A_d)_p(\tilde{A}_d)_p \neq 0$$

and by noting that

$$a = r\cos\phi$$
$$a' = r\sin\phi \qquad r > 0 \text{ and } \phi \text{ modulo } 2\pi.$$

is written in the form

$$(A_d)_p(\tilde{A}_d)_p = (re^{i\phi})I_4 \tag{5.67}$$

5.5.3 Operator *R*

R is defined by

$$R = (A_d)_p(re^{i\phi})^{-1/2} \tag{5.68a}$$

The properties of this element are:

Change in direction. $\overline{R} = R$; in fact,

$$(A_d)_p = A_s + A_b + A_p$$

$$(\overline{A}_d)_p = A_s + A_b + A_p = (A_d)_p$$

and

$$\overline{R} = (\overline{A_d})_p(re^{i\phi})^{-1/2} = R \tag{5.69}$$

Inversion of the order. We have seen that

$$(\widetilde{A_d})_p = A_s - A_b + A_p$$

With (5.62) and (5.67),

$$(A_d)_p(\widetilde{A_d})_p = (A_s + A_p)^2 - A_b^2 = re^{i\phi}I_4$$

Thus

$$R\tilde{R} = (re^{i\phi})^{-1}re^{i\phi}I_4 = I_4 \tag{5.70}$$

It can be concluded that all Dirac even numbers are put into canonical form,

$$(A_d)_p = (re^{i\phi})^{1/2}R \tag{5.68b}$$

if

$$(A_d)_p(\widetilde{A_d})_p \neq 0 \qquad \text{with } R = \overline{R} \quad \text{and} \quad \tilde{R}R = 1$$

5.5.4 Isometry

Let X, a vector of time-space, and X', its transform, be related by

$$X' = RX\tilde{R} \tag{5.71}$$

As the vector of time-space is such that

$$\overline{\vec{r}} = -\vec{r} \tag{5.72}$$

The temporal component remaining identical, an obvious relation for a change in direction and also

$$\tilde{X} = X \tag{5.73}$$

(the change in order of the terms does not affect X), we see that

$$\overline{\vec{r}'} = R\overline{\vec{r}}\tilde{R} = -R\vec{r}\tilde{R} = -\vec{r}' \tag{5.74}$$

We used the fact that $\bar{\tilde{R}} = \tilde{\bar{R}}$ and

$$\widetilde{X'} = R\tilde{X}\tilde{R} = RX\tilde{R} = X' \tag{5.75}$$

from which we can deduce the nature of X'. Indeed, a d-number is in general written

$$A_d = A_s + A_v + A_b + A_t + A_p$$

with

$$\overline{A_d} = A_s - A_v + A_b - A_t + A_p$$

$$\widetilde{A_d} = A_s + A_v - A_b - A_t + A_p$$

The condition (5.74) means that

$$A_b = A_p = 0$$

and (5.75) means that

$$A_t = 0$$

from which

$$X' = \lambda_0\gamma_0 + \lambda_1\gamma_1 + \lambda_2\gamma_2 + \lambda_3\gamma_3$$

It is thus a vector. Evaluate the quantity X'^2:

$$X'^2 = RX\tilde{R}RX\tilde{R} = RX^2\tilde{R} = X^2$$

(X^2 is a scalar). The transformation defined by (5.71) is therefore an isometry; we have $X'^2 = X^2$. It is also bijective, as can easily be shown that any X' is the image of a single X defined by

$$X = \tilde{R}X'R$$

The transformation preserves the direction of the basis, since we can show that

$$R\gamma_0\gamma_1\gamma_2\gamma_3\tilde{R} = \gamma_0\gamma_1\gamma_2\gamma_3$$

R is a rotation of time-space or a Lorentz rotation.

If we take

$$R = \cosh\frac{\theta}{2} I_4 + \sinh\frac{\theta}{2}\Sigma_3$$

a vector of space-time X then transforms itself into a vector of space-time X' by

$$X' = \left(\cosh\frac{\theta}{2} I_4 + \sinh\frac{\theta}{2}\Sigma_3\right) X \left(\cosh\frac{\theta}{2} I_4 - \sinh\frac{\theta}{2}\Sigma_3\right)$$

with

$$X = ct\gamma_0 + x\gamma_1 + y\gamma_2 + z\gamma_3$$

(which means that the γ_μ are the basis vectors); as $\Sigma_k = \gamma_k\gamma_0$,

$$\gamma_0\Sigma_3 + \Sigma_3\gamma_0 = 0 \qquad \gamma_3\Sigma_3 + \Sigma_3\gamma_3 = 0$$

γ_0 and γ_3 anticommute with Σ_3 but

$$\gamma_1\Sigma_3 + \Sigma_3\gamma_1 = 0 \qquad \gamma_2\Sigma_3 + \Sigma_3\gamma_2 = 0$$

γ_1 and γ_2 commute with Σ_3. It can be seen that

$$X' = \gamma_1 x + \gamma_2 y + (\cosh\theta\, I_4 + \sinh\theta\, \Sigma_3)(\gamma_0 ct + \gamma_3 z)$$

By identification of the coefficients of the γ_μ, we obtain

$$x' = x$$
$$y' = y$$
$$z' = z\cosh\theta - ct\sinh\theta$$
$$ct' = ct\cosh\theta - z\sinh\theta$$

These formulas are no other than the Lorentz special transformation. The argument θ being constant, if

$$z = ct\tanh\theta$$
$$x = y = 0$$

then

$$x' = y' = z' = 0$$

which indicates a rectilinear and uniform translation of the z axis. Similarly, $(c\tanh\theta)$ is nothing other than the speed of the translation:

$$\tanh\theta = \frac{v}{c}$$

We find a notation that has already been used. Note that

$$\cosh\theta = (1 - \tanh^2\theta)^{-1/2} = \left(1 - \frac{v^2}{c^2}\right)^{-1/2}$$

5.6 APPLICATIONS

5.6.1 Quaternions

The aim of vector algebra is to generalize the correspondence

space R^2 (plane geometry) $\longrightarrow$ complex numbers

and

space R^3 (spatial geometry) $\longrightarrow$ quaternions

Consider a field K and a vector space of dimension 4, denoted V_k, built from K. We write $\mathbb{1}$, i, j, and k as the basis vectors of this space. Every element A_q from V_k is then written

$$A_q = a\mathbb{1} + bi + cj + dk \tag{1}$$

a, b, c, and d belong to K. The vectors, i, j, and k can be considered as vectors of space (in relativity) and $\mathbb{1}$ as a vector of time:

$$\mathbb{1}^2 = 1 \qquad i^2 = -1 \qquad j^2 = -1 \qquad k^2 = -1 \tag{2}$$

The basis vectors satisfy the rules of multiplication

$$\begin{aligned} ij &= -ji = -k \\ jk &= -kj = -i \\ ki &= -ik = -j \end{aligned} \tag{3}$$

The laws are identical to those with which the Pauli matrices comply. Bearing in mind (3) the relation of definition (1) is also written

$$A_q = (a\mathbb{1} + bi) + j(c\mathbb{1} + di) \tag{4}$$

The relation (4) defines j as a second complex number such as $j^2 = -1$:

$$A_q = Z + jZ' \tag{5}$$

Form the products

$$\begin{aligned} A_q A_{q'} = {} & (aa' - bb' - cc' - dd')\mathbb{1} \\ & + (ab' + ba' + cd' - dc')i \\ & + (ac' + ca' + da' - ad')j \\ & + (ad' + da' + ba' - ab')k \end{aligned}$$

and if $\overline{A}_q$ is the vector obtained by a change in direction,

$$\overline{A}_q = a\overline{1} + b\overline{i} + c\overline{j} + d\overline{k}$$
$$= a1 - bi - cj - dk$$

The product

$$A_q\overline{A}_q = (a^2 + b^2 + c^2 + d^2)1 = N^2(q)1$$

is called the square of the norm $N(q)$ of A_q. Thus regardless of A_q, an inverse $(A_q)^{-1}$ is such that

$$(A_q)^{-1} = \frac{\overline{A}_q}{N^2(q)}$$

In fact,

$$(A_q)(A_q)^{-1} = \frac{A_q\overline{A}_q}{N^2(q)} = 1$$

Note. The quaternions A_q with coefficients (a,b,c,d) belonging to R form a noncommutative field. A commutative field (complex numbers) is associated with plane geometry, while a noncommutative field corresponds to three-dimensional geometry.

Note the isomorphism between the basis vectors i, j, k and the Pauli matrices

$$e_l = i\sigma_l \qquad l = 1, 2, 3 \quad \text{and} \quad e_i = i, \quad e_2 = j, \quad e_3 = k \tag{6}$$

If we generalize the relation (4) to a space of dimension 8 we define the octonion

$$A_0 = (a1 + bi) + j(c1 + di) + l[(a'1 + b'i) + j(c'1 + d'i)]$$

The products ji, ki, kj, and kji define four new basis vectors, denoted k, m, n, p, respectively, and

$$A_0 = a1 + bi + cj + dk + a'l + b'm + c'n + d'p$$

We obviously have

$$l^2 = -1$$

and the relations generalizing (3) of the type

$$li + il = 0$$

The equivalent of (6) involves the matrix λ_i of Gell-Mann [cf. Lie algebra of su(3)]. The algebra of octonions is noncommutative and nonassociative.

6

Application of Clifford Algebra to Bosons: Klein–Gordon Equation

6.1 FIELD PARTICLES

Up until the experiment of Carlo–Rubbia at CERN in Geneva in 1984, the physicist recognized four fundamental interactions responsible for the structure of matter. They are all characterized by field quanta of spin 1:

Electromagnetic interaction (a photon denoted γ of zero mass)
Weak interaction (intermediary bosons W^+, Z^0, W^- of mass $\simeq$ 100 GeV)
Strong interaction (eight gluons without mass)
Gravitational interaction (graviton)

In the "big-bang theory" these interactions were unified with the creation of the universe.

6.2 ZERO-SPIN PARTICLES

In this chapter we intend only to examine the scalar fields. Particles at zero spin (bosons) are called *mesons*. Their features are given in the following table:

Particle	Symbol	Mass (MeV)	Mean lifetime (s)
Pion	Π^+	139.58	2×10^{-8}
	Π^0	134.97	1×10^{-16}
Kaon	K^+	493.98	1×10^{-8}
	K^0	497.9	10^{-8} to 10^{-10}
Mesons	D^+	1869.4	9.2×10^{-13}
	B^+	5270.8	

The mass of an electron is of 0.5109 MeV.

These particles (pions or kaons) have a positive, negative, or zero charge. A pion has a mass at rest equal to $\simeq 270$ times that of the electron.

The interactions of the mesons with other particles or with the fields are big. From a more general point of view we are going to imagine in what follows that:

The spin of the particles is zero.
The charge of the particles is positive, negative, or zero.
The particles are free or in an electromagnetic field.

6.3 KLEIN–GORDON EQUATION

For a free relativistic particle it has been shown that

$$E^2 = p^2c^2 + m_0^2c^4 \tag{6.1}$$

and the quantity H will be called the hamiltonian:

$$H = \sqrt{p^2c^2 + m_0^2c^4} \tag{6.2}$$

The relation (6.1) can obviously be written with the help of the energy–momentum quadrivector

$$(P) = \left(\frac{H}{c}, \vec{p}\right) \tag{6.3}$$

since

$$P_\mu P^\mu - m_0^2c^2 = 0 \tag{6.4}$$

restores (6.1). Equation (6.4) is the Gordon equation under the covariant form. When the particle is exposed to an electromagnetic field,

$$H \longrightarrow H + qV \tag{6.5}$$

$$\vec{p} \longrightarrow \vec{p} - q\vec{A} \tag{6.6}$$

(cf. A. Charlier, A. Bérard, and M. F. Charlier, *Analytical Mechanics*, Editions Ellipses, Paris, 1989). This means that

$$\vec{p} = m\vec{v} - q\vec{A} = \frac{m_0\vec{v}}{\sqrt{1 - v^2/c^2}} - q\vec{A} \tag{6.7}$$

and

$$H = \sqrt{(\vec{p} - q\vec{A})^2c^2 + m_0^2c^4} + qV \tag{6.8}$$

which replace (6.2) or (6.4). In another form we write that

$$(P_\mu - qA_\mu)(P^\mu - qA^\mu) - m_0^2c^2 = 0 \tag{6.9}$$

with the quadrivector

$$(A) = \left(\frac{V}{c}, \vec{A}\right)$$

Equation (6.9) is often written

$$(P_\mu - qA_\mu)^2 - m_0^2c^2 = 0 \tag{6.10}$$

6.4 EQUATION OF PROPAGATION

Apply Ehrenfest's theorem and the correspondence of the operators (cf. a quantum mechanics text)

$$\hat{E} \longrightarrow i\hbar\frac{\partial}{\partial t} \tag{6.11}$$

$$\hat{p} \longrightarrow -i\hbar\vec{\nabla} \tag{6.12}$$

If we start from (6.1) we see that for a free particle,

$$-\hbar^2\frac{\partial^2}{\partial t^2}|\psi(\vec{r},t)\rangle = -c^2\hbar^2\nabla^2|\psi(\vec{r},t)\rangle + m_0^2c^4|\psi(\vec{r},t)\rangle$$

or

$$\partial_\mu\partial^\mu|\psi\rangle + k^2|\psi\rangle = |0\rangle \tag{6.13}$$

by writing

$$k^2 = \frac{m_0^2c^2}{\hbar^2} \tag{6.14}$$

Otherwise, by using the d'Alembertian

$$\Box = \partial_\mu \partial^\mu \tag{6.15}$$

$$[\Box + k^2]|\psi(\vec{r},t)\rangle = |0\rangle \tag{6.16}$$

Note. The function $\Psi(\vec{r},t)$ represents a scalar field.

We have the Klein–Gordon equation, uniquely valid for a free particle of zero spin. It was established in 1929. The equivalent for a particle of zero spin but in an electromagnetic field is [cf. (6.8)]

$$\left[\left(\partial_\mu + i\frac{q}{\hbar}A_\mu\right)^2 + k^2\right]|\psi\rangle = |0\rangle \tag{6.17}$$

[cf. formula (1.77) of Chapter 1, in which the quantity $i(q/\hbar)A_\mu$ plays the role of the Christoffel symbols for the covariant derivative $D_\mu = \partial_\mu + i(q/\hbar)A_\mu$]. In fact,

$$(E - qV)^2 - c^2(\vec{p} - q\vec{A})^2 - m_0^2c^4 = 0$$

and by developing,

$$E^2 - p^2c^2 + q^2V^2 - q^2c^2A^2 - q(EV + VE) + qc^2(\vec{p}\cdot\vec{A} + \vec{A}\cdot\vec{p}) - m_0^2c^4 = 0$$

or with the usual correspondences of operators and by dividing by $-\hbar^2c^2$:

$$\left[\frac{1}{c^2}\frac{\partial^2}{\partial t^2} - \nabla^2 - \frac{q^2}{\hbar^2}\left(\frac{V^2}{c^2} - A^2\right) + \frac{iq}{\hbar c^2}\left(\frac{\partial}{\partial t}V + V\frac{\partial}{\partial t}\right) + \frac{iq}{\hbar}(\vec{\nabla}\cdot\vec{A} + \vec{A}\cdot\vec{\nabla}) + k^2\right]|\psi\rangle = |0\rangle$$

that is,

$$\left[\partial_\mu\partial^\mu - \frac{q^2}{\hbar^2}A_\mu A^\mu + \frac{iq}{\hbar}(\partial_\mu A^\mu + A_\mu\partial^\mu) + k^2\right]|\psi\rangle = |0\rangle$$

a relation that is identical to (6.17).

6.5 CURRENT OF PROBABILITY

Start from the Klein–Gordon equation of a free particle [cf. (6.13)]:

$$\frac{1}{c^2}\frac{\partial^2}{\partial t^2}\psi - \nabla^2\psi + k^2\psi = 0 \tag{6.18}$$

Multiply (6.18) by ψ^* on the left:

$$\frac{1}{c^2}\psi^*\frac{\partial^2}{\partial t^2}\psi - \psi^*\nabla^2\psi + k^2\psi^*\psi = 0 \tag{6.19}$$

Take the conjugate of (6.19):

$$\frac{1}{c^2}\psi\frac{\partial^2}{\partial t^2}\psi^* - \psi\nabla^2\psi^* + k^2\psi\psi^* = 0 \tag{6.20}$$

Subtract (6.20) from (6.19):

$$\frac{1}{c^2}\left(\psi^*\frac{\partial^2}{\partial t^2}\psi - \psi\frac{\partial^2}{\partial t^2}\psi^*\right) - (\psi^*\nabla^2\psi + \psi\nabla^2\psi^*) = 0 \tag{6.21}$$

Write

$$\vec{j} = \frac{\hbar q}{2m_0 i}(\psi^*\vec{\nabla}\psi - \psi\vec{\nabla}\psi^*) \tag{6.22}$$

the density of current of probability, and

$$\rho = \frac{i\hbar q}{2m_0c^2}\left(\psi^*\frac{\partial}{\partial t}\psi - \psi\frac{\partial}{\partial t}\psi^*\right) \tag{6.23}$$

the density of charge; we then see that

$$\frac{\partial\rho}{\partial t} = \frac{i\hbar q}{2m_0c^2}\left(\psi^*\frac{\partial^2}{\partial t^2}\psi - \psi\frac{\partial^2}{\partial t^2}\psi^*\right) \tag{6.24}$$

and

$$\vec{\nabla}\cdot\vec{j} = \frac{\hbar q}{2m_0 i}(\psi^*\nabla^2\psi - \psi\nabla^2\psi^*) \tag{6.25}$$

and by substitution in (6.21),

$$\vec{\nabla}\cdot\vec{j} = \frac{\partial\rho}{\partial t} = 0 \tag{6.26}$$

we find the law of conservation of charge in electricity, which confirms that $\vec{j}$ can be considered as the density of current (of probability) and ρ as the usual density of charge.

In nonrelativistic mechanics,

$$\vec{j} = \frac{\hbar q}{2m_0 i} (\psi^* \vec{\nabla} \psi - \psi \vec{\nabla} \psi^*) \tag{6.27}$$

identical to (6.22); however,

$$\rho = q\psi^* \psi \tag{6.28}$$

We find the nonrelativistic results if we imagine stationary states, that is, such that

$$\psi(\vec{r},t) = A \exp\left[\frac{i}{\hbar} (\vec{p} \cdot \vec{r} - Et)\right] \tag{6.29}$$

since then

$$\rho = \frac{Eq}{m_0 c^2} |\psi|^2 \tag{6.30}$$

and

$$\vec{j} = q \frac{\vec{p}}{m_0} |\psi|^2 \tag{6.31}$$

In the case of a free particle of zero spin but in an electromagnetic field,

$$\vec{j} = \frac{\hbar q}{2m_0 i} (\psi^* \vec{\nabla} \psi - \psi \vec{\nabla} \psi^*) - \frac{q^2}{m_0 c} \vec{A} \psi^* \psi \tag{6.32}$$

$$\rho = \frac{i\hbar q}{2m_0 c^2} \left(\psi^* \frac{\partial}{\partial t} \psi - \psi \frac{\partial}{\partial t} \psi^*\right) - \frac{q^2 V}{m_0 c^3} \psi^* \psi \tag{6.33}$$

In covariant notation,

$$j_\mu = \frac{\hbar q}{2m_0 i} (\psi^* D_\mu \psi - \psi D_\mu \psi^*)$$

replacing the previous relations.

6.6 SCHRÖDINGER FORM OF KLEIN–GORDON EQUATION

Try to put the Klein–Gordon equation (6.18) in the form

$$\hat{H}_{KG} |\psi\rangle = i\hbar \frac{\partial}{\partial t} |\psi\rangle \tag{6.34}$$

We are therefore trying to find the form of the hamiltonian $\hat{H}_{KG}$. Consider the simple case of the free particle, that is, of the equation

$$\frac{1}{c^2}\frac{\partial^2}{\partial t^2}\psi - \nabla^2 \psi + k^2\psi = 0 \tag{6.35}$$

and write

$$\psi_0 = \frac{i}{k}\frac{\partial\psi}{\partial X_0} = \frac{i}{kc}\frac{\partial\psi}{\partial t} \tag{6.36}$$

It can be seen that

$$\frac{\partial\psi_0}{\partial X_0} = \frac{1}{c}\frac{\partial\psi_0}{\partial t} = \frac{i}{kc^2}\frac{\partial^2\psi}{\partial t^2} \tag{6.37}$$

Equation (6.35) is thus replaced by (6.36) and (6.37). Introduce (6.37) into (6.35):

$$-ik\frac{\partial\psi_0}{\partial X_0} - \nabla^2\psi + k^2\psi = 0 \tag{6.38}$$

We write for the moment

$$\varphi = \frac{1}{2}\left(\psi + \frac{i\hbar}{m_0 c}\frac{\partial\psi}{\partial X_0}\right) \tag{6.39}$$

$$\chi = \frac{1}{2}\left(\psi - \frac{i\hbar}{m_0 c}\frac{\partial\psi}{\partial X_0}\right) \tag{6.40}$$

or

$$\psi = \varphi + \chi \tag{6.41}$$

Equations (6.39) and (6.40) show that

$$\varphi - \chi = 2\frac{i\hbar}{m_0 c}\frac{\partial\psi}{\partial X_0}$$

and according to (6.36),

$$\frac{\partial\psi}{\partial X_0} = -ik\psi_0$$

Otherwise,

$$\psi_0 = \varphi - \chi \tag{6.42}$$

Rewrite (6.36) and (6.38) with the help of (6.41) and (6.42):

$$\frac{\partial}{\partial t}(\varphi + \chi) = -ikc(\varphi - \chi) \tag{6.43}$$

$$\frac{\partial}{\partial t}(\varphi - \chi) = \frac{ic}{k}\nabla^2(\varphi + \chi) - ikc(\varphi + \chi) \tag{6.44}$$

By adding (6.44) to (6.43) and by multiplying by $i\hbar/2$, we see that [cf. (6.14)]

$$i\hbar\frac{\partial\varphi}{\partial t} = -\frac{\hbar^2}{2m_0}\nabla^2(\varphi + \chi) + m_0c^2\varphi \tag{6.45}$$

By subtracting (6.44) from (6.43) and by multiplying by $i\hbar/2$, we reach

$$i\hbar\frac{\partial\chi}{\partial t} = \frac{\hbar^2}{2m_0}\nabla^2(\varphi + \chi) - m_0c^2\chi \tag{6.46}$$

We can assemble equations (6.45) and (6.46), which are equivalent to the Klein–Gordon equation, in a single equation in matrix form:

$$i\hbar\frac{\partial}{\partial t}\begin{bmatrix}\varphi\\ \chi\end{bmatrix} = -\frac{\hbar^2}{2m_0}\nabla^2\begin{bmatrix}1 & 1\\ -1 & -1\end{bmatrix}\begin{bmatrix}\varphi\\ \chi\end{bmatrix} + m_0c^2\begin{bmatrix}1 & 0\\ 0 & -1\end{bmatrix}\begin{bmatrix}\varphi\\ \chi\end{bmatrix} \tag{6.47}$$

We see that when the Pauli matrices are introduced,

$$\sigma_1 = \begin{bmatrix}0 & 1\\ 1 & 0\end{bmatrix} \qquad \sigma_2 = \begin{bmatrix}0 & -i\\ i & 0\end{bmatrix} \qquad \sigma_3 = \begin{bmatrix}1 & 0\\ 0 & -1\end{bmatrix} \qquad I = \begin{bmatrix}1 & 0\\ 0 & 1\end{bmatrix}$$

vectors of basis of su(2), then

$$\sigma_3 + i\sigma_2 = \begin{bmatrix}1 & 1\\ -1 & -1\end{bmatrix}$$

Write

$$\Psi = \begin{bmatrix}\varphi\\ \chi\end{bmatrix} \tag{6.48}$$

one of Weyl's spinors; thus

$$i\hbar\frac{\partial}{\partial t}\Psi = -\frac{\hbar^2}{2m_0}(\sigma_3 + i\sigma_2)\nabla^2\Psi + \sigma_3 m_0c^2\Psi \tag{6.49}$$

This means that the Klein–Gordon equation has been transformed into an equation of Schrödinger type:

$$\hat{H}_{\text{KG}} = -\frac{\hbar^2}{2m_0}(\sigma_3 + i\sigma_2)\nabla^2 + \sigma_3 m_0 c^2$$

$$= \frac{1}{2m_0}(\sigma_3 + i\sigma_2)\vec{p}^{\,2} + \sigma_3 m_0 c^2 \tag{6.50}$$

This allows us to generalize the result to the case of the electromagnetic field

$$\hat{H}_{\text{KG}} = \frac{1}{2m_0}(\sigma_3 + i\sigma_2)(\vec{p} - q\vec{A})^2 + \sigma_3 m_0 c^3 + qVI \tag{6.51}$$

or even

$$\hat{H}_{\text{KG}} = \frac{1}{2m_0}(\sigma_3 + i\sigma_2)\left(\frac{\hbar}{i}\vec{\nabla} - q\vec{A}\right)^2 + \sigma_3 m_0 c^2 + qVI \tag{6.52}$$

As

$$\Psi = \begin{bmatrix} \varphi \\ \chi \end{bmatrix}$$

we see clearly that the relativistic writing has created new degrees of freedom.

6.7 DENSITIES OF CHARGE AND CURRENT

It was seen earlier that

$$\rho = \frac{i\hbar q}{2m_0c^3}\left(\psi^* \frac{\partial}{\partial t}\psi - \psi \frac{\partial}{dt}\psi^*\right) \tag{6.23}$$

since

$$\frac{\partial}{\partial t}\psi = -ikc\psi_0 \tag{6.36}$$

The electronic density is written [cf. (6.14) for the value of k]

$$\rho = \frac{i\hbar q}{2m_0c^2}(-ikc)(\psi^*\psi_0 + \psi\psi_0^*) = \frac{q}{2}(\psi^*\psi_0 + \psi\psi_0^*) \tag{6.53}$$

with

$$\psi = \varphi + \chi \tag{6.41}$$

$$\psi_0 = \varphi - \chi \tag{6.42}$$

from which

$$\rho = q(\varphi\varphi^* - \chi\chi^*) = q(\varphi^*,\chi^*)\begin{bmatrix}1 & 0\\ 0 & -1\end{bmatrix}\begin{bmatrix}\varphi\\ \chi\end{bmatrix} \tag{6.54}$$

We will write

$$\Psi^+ = (\varphi^*,\chi^*) \tag{6.55}$$

the hermitian conjugate of Ψ if

$$\Psi = \begin{bmatrix}\varphi\\ \chi\end{bmatrix} \tag{6.48}$$

from which

$$\rho = q\Psi^+\sigma_3\Psi \tag{6.56}$$

in matrix notation. In this way the density of current $\vec{j}$ can be evaluated:

$$\vec{j} = \frac{\hbar q}{2im_0}\left[\Psi^+\sigma_3(\sigma_3 + i\sigma_2)\vec{\nabla}\Psi - (\vec{\nabla}\Psi^+)\sigma_3(\sigma_3 + i\sigma_2)\Psi\right] \tag{6.57}$$

We know that [cf. (6.56)]

$$q = \int \rho\, d^3r = q\int \Psi^+\sigma_3\Psi\, d^3r \tag{6.58}$$

This integral of normalization thus depends on the charge ($q = \pm e$):

$$\int \Psi^+\sigma_3\Psi\, d^3r = \pm 1 = \int (\varphi^*\varphi - \chi^*\chi)\, d^3r \tag{6.59}$$

6.8 STATIONARY STATES OF A FREE PARTICLE

If we take the waveform of Louis de Broglie for the spinor

$$\Psi = \frac{1}{\sqrt{V}}\begin{bmatrix}\varphi_0\\ \chi_0\end{bmatrix}\exp\left[\frac{i}{\hbar}(\vec{p}\cdot\vec{r} - Et)\right] \tag{6.60}$$

and substitute in

$$i\hbar\frac{\partial\varphi}{\partial t} = -\frac{\hbar^2}{2m_0}\nabla^2(\varphi + \chi) + m_0c^2\varphi \tag{6.45}$$

$$i\hbar\frac{\partial\chi}{\partial t} = \frac{\hbar^2}{2m_0}\nabla^2(\varphi + \chi) - m_0c^2\chi \tag{6.46}$$

we obtain

$$E\varphi_0 = m_0c^2\varphi_0 + \frac{p}{2m_0}(\varphi_0 + \chi_0) \tag{6.61}$$

$$E\chi_0 = -m_0c^2\chi_0 - \frac{p^2}{2m_0}(\varphi_0 + \chi_0) \tag{6.62}$$

or even in matrix form

$$\begin{bmatrix} E - m_0c^2 - \dfrac{p^2}{2m_0} & -\dfrac{p^2}{2m_0} \\ \dfrac{p^2}{2m_0} & E + m_0c^2 + \dfrac{p^2}{2m_0} \end{bmatrix} \begin{bmatrix} \varphi_0 \\ \chi_0 \end{bmatrix} = 0$$

a system that accepts the nonzero solutions if the determinant associated with the matrix is equal to zero, which gives

$$E = \pm E_p \tag{6.63a}$$

with

$$E_p = c\sqrt{p^2 + m_0^2c^2} \tag{6.63b}$$

an equation of Einstein.

If $E = E_p$, then

$$\Psi_0(+E_p) = \begin{bmatrix} \varphi_0(+E_p) \\ \chi_0(+E_p) \end{bmatrix}$$

By adding (6.61) and (6.62), it can be seen that

$$-(E_p - m_0c^2)\varphi_0 = (E_p + m_0c^2)\chi_0$$

We take

$$\varphi_0 = \alpha(E_p + m_0c^2)$$

$$\chi_0 = -\alpha(E_p - m_0c^2)$$

and with (6.60), condition (6.59) gives

$$\alpha = \frac{1}{2\sqrt{m_0c^2E_p}}$$

or

$$\varphi_0(+E_p) = \frac{E_p + m_0c^2}{2\sqrt{m_0c^2E_p}} \tag{6.64}$$

$$\chi_0(+E_p) = \frac{m_0c^2 - E_p}{2\sqrt{m_0c^2E_p}} \tag{6.65}$$

and

$$\varphi_0(+E_p)\varphi_0(+E_p) - \chi_0(+E_p)\chi_0(+E_p) = 1 \tag{6.66}$$

We are discussing positive solution; the state is associated with a charge $+e$.

If $E = -E_p$, then

$$\Psi_0(-E_p) = \begin{bmatrix} \varphi_0(-E_p) \\ \chi_0(-E_p) \end{bmatrix}$$

with, using the same method as before,

$$\varphi_0(-E_p) = \frac{m_0c^2 - E_p}{2\sqrt{m_0c^2E_p}} \tag{6.67}$$

$$\chi_0(-E_p) = \frac{m_0c^2 + E_p}{2\sqrt{m_0c^2E_p}} \tag{6.68}$$

and

$$\varphi_0(-E_p)\varphi_0(-E_p) - \chi_0(-E_p)\chi_0(-E_p) = -1 \tag{6.69}$$

We are discussing negative solution; the state is associated with a charge $-e$.

6.9 ANTIPARTICLE OR CHARGE CONJUGATION

As a hypothesis, we choose for Ψ a solution of positive charge written in the form of a Broglie wave:

$$\Psi = \begin{bmatrix} \varphi \\ \chi \end{bmatrix} = \frac{1}{\sqrt{V}} \begin{bmatrix} \varphi_0 \\ \chi_0 \end{bmatrix} \exp\left[\frac{i}{\hbar}(\vec{p}\cdot\vec{r} - Et)\right] \tag{6.70}$$

with

$$\varphi_0(+E_p)\varphi_0(+E_p) - \chi_0(+E_p)\chi_0(+E_p) = 1 \tag{6.71}$$

$$\varphi_0(-E_p)\varphi_0(-E_p) - \chi_0(-E_p)\chi_0(-E_p) = -1 \tag{6.72}$$

The function Ψ_c defined by

$$\Psi_c = \sigma_1\Psi^* = \begin{bmatrix} 0 & 1 \\ 1 & 0 \end{bmatrix} \begin{bmatrix} \varphi^* \\ \chi^* \end{bmatrix} = \begin{bmatrix} \chi^* \\ \varphi^* \end{bmatrix} \tag{6.73}$$

is a solution of negative charge. Indeed,

$$\Psi_c = \begin{bmatrix} \chi^* \\ \varphi^* \end{bmatrix} = \frac{1}{\sqrt{V}} \begin{bmatrix} \chi_0 \\ \varphi_0 \end{bmatrix} \exp\left[-\frac{i}{\hbar}(\vec{p}\cdot\vec{r} - Et)\right] \tag{6.74}$$

which means, by comparing (6.72) and (6.73) that we go from Ψ to Ψ_c by

$$\vec{p} \longrightarrow -\vec{p}$$
$$E \longrightarrow -E$$
$$\varphi_0 \longrightarrow \chi_0$$
$$\chi_0 \longrightarrow \varphi_0$$

Equations (6.53) and (6.56) become, by substitution,

$$\chi_0(-E_p)\chi_0(-E_p) - \varphi_0(-E_p)\varphi_0(-E_p) = 1$$
$$\chi_0(+E_p)\chi_0(+E_p) - \varphi_0(+E_p)\varphi_0(+E_p) = -1$$

The last relation clearly shows that Ψ_c is a solution of negative charge for the same value of energy.

If we consider the movement of a particle described by a function Ψ, the particle corresponding to the function conjugated by the charge Ψ_c is called an *antiparticle*.

Example. If a meson Π^+ is the particle, the meson Π^- is the antiparticle, and vice versa. For neutral particles such as the pion (meson Π^0) or the photon,

$$\Psi_c = \sigma_1\Psi^* = \alpha\Psi$$

Now, making the conjugation of the charge equivalent to keeping the same charge,

$$|\alpha| = 1 \qquad \text{or} \qquad \alpha = \pm 1$$

If $\alpha = 1$, $\Psi_c = \Psi$, that is,

$$\begin{bmatrix} \chi^* \\ \varphi^* \end{bmatrix} = \begin{bmatrix} \varphi \\ \chi \end{bmatrix} \qquad \text{or} \qquad \varphi = \chi^*$$

If $\alpha = -1$, $\Psi_c = -\Psi$; that is,

$$\begin{bmatrix} \chi^* \\ \varphi^* \end{bmatrix} = \begin{bmatrix} -\varphi \\ -\chi \end{bmatrix} \qquad \text{or} \qquad \varphi = -\chi^*$$

This means that for $\alpha = 1$ we have a neutral particle of charge parity $+1$, and for $\alpha = -1$, a parity of charge -1.

Note. The meson Π^0 has a value of α equal to 1 (it can be broken down into two photons) and the photon a value of α equal to -1.

7
Fermions: Dirac Equation

7.1 PARTICLES OF MATTER

We consider elementary particles of spin $\frac{1}{2}$, constituents of the matter. Their principal features are given in the following table:

Particles sensitive to strong interaction			Particles nonsensitive to strong interaction		
Quark	Charge	Mass (MeV)	Lepton	Charge	Mass (MeV)
u (up)	$\frac{2}{3}$	$\simeq 336$	Electron e^-	-1	0.5111
d (down)	$-\frac{1}{3}$	$\simeq 336$	Neutrino ν_e	0	$\simeq 0(?)$
c (charm)	$\frac{2}{3}$	$\simeq 1500$	Muon μ^-	-1	105.65
s (strange)	$-\frac{1}{3}$	$\simeq 540$	Neutrino ν_μ	0	$\simeq 0$
t (top)	$\frac{2}{3}$	$\geqslant 40{,}000$	Tau τ^-	-1	1784.2
b (bottom)	$-\frac{1}{3}$	$\simeq 5100$	Neutrino ν_τ	0	$\simeq 0(?)$

7.2 PARTICLES OF SPIN $\frac{1}{2}$

In 1928, Dirac formulated a satisfactory theory for particles of spin $\frac{1}{2}$. Here we have a list of these particles, called baryons, formed from three quarks.

Particle	Symbol	Charge	Mass (MeV)	Mean lifetime (s)
Nucleon	p (proton)	1	938.21	Stable
	n (neutron)	0	939.50	1.01×10^{3}
Hyperon Λ	Λ^0	0	1115	2.57×10^{-10}
Σ	Σ^+	1	1189	0.78×10^{-10}
	Σ^0	0	1191	10^{-22} to 10^{-11}
	Σ^-	−1	1195	1.59×10^{-10}
Hyperon Θ	Θ^0	0	1316	3.9×10^{-10}
	Θ^-	−1	1320	1.4×10^{-10}

7.3 DIRAC RELATIVISTIC EQUATION

Dirac's initial idea was that the differential equation satisfied by the wave function must lead to an equation of continuity associated with a density of probability defined as being positive. He also introduced in place of a unique function ψ a set of functions $\psi_i(\vec{r},t)$, $i = 1, 2, \ldots$, which determine the density of the electric charge ρ:

$$\rho = \sum_i \psi_i^* \psi_i \tag{7.1}$$

If there is no density of current, the equation of continuity

$$\frac{\partial \rho}{\partial t} + \text{div}\, \vec{j} = 0 \tag{7.2}$$

leads to

$$\frac{\partial}{\partial t} \int \rho \, d^3r = q \int \sum_i \left(\frac{\partial \psi_i^*}{\partial t} \psi_i + \psi_i^* \frac{\partial \psi_i}{\partial t} \right) d^3r \tag{7.3}$$

which is the law of preservation of the charge.

To satisfy (7.3), $\partial\psi_i/\partial t$ must be determined by the values of ψ_i, at a given moment. In a very general way this is written in Hilbert space:

$$\frac{\partial \psi_\nu}{\partial t} + \sum_{\mu, k} \alpha_{\nu\mu}^{(k)} \frac{\partial \psi_\mu}{\partial X_k} + \sum_\mu \beta_{\nu\mu} \psi \mu = 0 \tag{7.4}$$

the coefficients $\alpha_{\nu\mu}^{(k)}$ and $\beta_{\nu\mu}$ being such that (7.4) is an equation of continuity and the functions ψ_ν all satisfy the equation of Klein–Gordon.

7.3.1 Linearization

We may use an intuitive formulation. Consider a free particle. The Klein–Gordon equation is the initial formula [cf. (6.4)]

$$(P_\mu P^\mu - m_0^2c^2)|\psi\rangle = |0\rangle \tag{7.5}$$

This relation is expressed by

$$(p_0^2 - p_1^2 - p_2^2 - p_3^2 - m_0^2c^2)|\psi\rangle \tag{7.6}$$

with

$$p_0^2 = \left(\frac{E}{c}\right)^2$$

Try to put it under an equivalent linear form,

$$(p_0 - \gamma_1 p_1 - \gamma_2 p_2 - \gamma_3 p_3 - \chi)|\psi\rangle = |0\rangle \tag{7.7}$$

If

$$\vec{\gamma} = \vec{i}\gamma_1 + \vec{j}\gamma_2 + \vec{k}\gamma_3$$

then

$$p_0|\psi\rangle = \frac{E}{c}|\psi\rangle = \frac{i\hbar}{c}\frac{\partial}{\partial t}|\psi\rangle = (\vec{\gamma}\cdot\vec{p} + \chi)|\Psi\rangle \tag{7.8}$$

or

$$i\hbar\frac{\partial}{\partial t}|\psi\rangle = H_D|\psi\rangle \tag{7.9}$$

with

$$H_D = c\vec{\gamma}\cdot\vec{p} + c\chi \tag{7.10}$$

This is Dirac's hamiltonian, the γ_i and χ being independent of E and the p_i. As we are considering the case where there is no field, all the points of space-time should be equivalent; the operator of the wave equation (7.9) should therefore not depend on x, y, z, or t. The γ_i and χ should also be independent of x, y, z, and t.

These magnitudes subsequently indicate certain variables that have not yet been considered. It is supposed that they describe a certain internal movement of the electron. We will see later that in fact they describe its proper rotation. The γ_i and χ should then commute with E, p, x, y, z, and t.

For the moment we multiply (7.7) by $(p_0 + \vec{\gamma}\cdot\vec{p} + \chi)$:

$$(p_0 - \gamma_1 p_1 - \gamma_2 p_2 - \gamma_3 p_3 - \chi) \times (p_0 + \gamma_1 p_1 + \gamma_2 p_2 + \gamma_3 p_3 + \chi)|\psi\rangle = |0\rangle \tag{7.11}$$

or

$$\begin{aligned}[p_0^2 &- \gamma_1^2 p_1^2 - \gamma_2^2 p_2^2 - \gamma_3^2 p_3^2 - \chi^2 \\ &- (\gamma_1\gamma_2 + \gamma_2\gamma_1)p_1 p_2 - (\gamma_1\gamma_3 + \gamma_3\gamma_1)p_1 p_3 \\ &- (\gamma_2\gamma_3 + \gamma_3\gamma_2)p_2 p_3 - (\gamma_1\chi + \chi\gamma_1)p_1 \\ &- (\gamma_2\chi + \chi\gamma_2)p_2 - (\gamma_3\chi + \chi\gamma_3)p_3]|\psi\rangle = |0\rangle\end{aligned} \tag{7.12}$$

By identifying with (7.6) the conditions to be imposed on γ_i and χ are

$$\gamma_1^2 = \gamma_2^2 = \gamma_3^2 = 1$$

$$\gamma_1\gamma_2 + \gamma_2\gamma_1 = 0$$

$$\gamma_2\gamma_3 + \gamma_3\gamma_2 = 0$$

$$\gamma_1\gamma_3 + \gamma_3\gamma_1 = 0$$

The γ_i cannot be numbers; they are matrices that anticommute and the former formulas can be assembled into

$$\gamma_i\gamma_j + \gamma_j\gamma_i = 2\delta_{ij} \tag{7.13}$$

The other conditions are

$$\chi^2 = m_0^2 c^2 \tag{7.14}$$

$$\begin{aligned}\gamma_1\chi + \chi\gamma_1 &= 0 \\ \gamma_2\chi + \chi\gamma_2 &= 0 \\ \gamma_3\chi + \chi\gamma_3 &= 0\end{aligned} \tag{7.15}$$

Thus χ is equally a matrix that anticommutes with the γ_i matrix; we write

$$\chi = m_0 c \gamma_0 \tag{7.16}$$

In this case (7.15) becomes

$$\gamma_0^2 = 1 \tag{7.17}$$

and the formulas (7.15) are written

$$\gamma_i\gamma_0 + \gamma_0\gamma_i = 0$$

whatever i may be; thus, to sum up,

$$\gamma_\mu\gamma_\nu + \gamma_\nu\gamma_\mu = 2\delta_{\mu,\nu} \qquad \mu,\nu = 0, 1, 2, 3$$

$$\chi = m_0 c \gamma_0$$

Go back to (7.10):

$$H_D = c\vec{\gamma}\cdot\vec{p} + c\chi = c\vec{\gamma}\cdot\vec{p} + m_0 c^2 \gamma_0$$

Dirac's hamiltonian.

7.3.2 Matrices γ_μ

We choose for γ_0 the simple form

$$\gamma_0 = \begin{bmatrix} I & \emptyset \\ \emptyset & -I \end{bmatrix} \tag{7.18}$$

where I is the 2×2 unit matrix and, $\emptyset$ is the 2×2 zero matrix.

$$\gamma_0 = \begin{bmatrix} 1 & 0 & 0 & 0 \\ 0 & 1 & 0 & 0 \\ 0 & 0 & -1 & 0 \\ 0 & 0 & 0 & -1 \end{bmatrix}$$

We prove that

$$\gamma_0^2 = \begin{bmatrix} I & \emptyset \\ \emptyset & -I \end{bmatrix}\begin{bmatrix} I & \emptyset \\ \emptyset & -I \end{bmatrix} = \begin{bmatrix} I & \emptyset \\ \emptyset & -I \end{bmatrix} = 1I$$

Put

$$\gamma_k = \begin{bmatrix} \mathbf{A}_k & \mathbf{B}_k \\ \mathbf{C}_k & \mathbf{D}_k \end{bmatrix} \tag{7.19}$$

where $\mathbf{A}_k$ is a 2×2 matrix, $k = 1, 2, 3$. It can be seen that

$$\gamma_0\gamma_k + \gamma_k\gamma_0 = \begin{bmatrix} 2\mathbf{A}_k & \emptyset \\ \emptyset & -2\mathbf{D}_k \end{bmatrix}$$

and according to (7.15),

$$\gamma_0\gamma_k + \gamma_k\gamma_0 = 0$$

Therefore,

$$\mathbf{A}_k = \mathbf{D}_k = \emptyset$$

or

$$\gamma_k = \begin{bmatrix} \emptyset & \mathbf{B}_k \\ \mathbf{C}_k & \emptyset \end{bmatrix} \tag{7.20}$$

It was also seen that

$$\gamma_i\gamma_k + \gamma_k\gamma_i = 2\delta_{ik}I$$

otherwise

$$\gamma_i\gamma_k + \gamma_k\gamma_i = \begin{bmatrix} \mathbf{B}_i\mathbf{C}_k + \mathbf{B}_k\mathbf{C}_i & \emptyset \\ \emptyset & \mathbf{C}_i\mathbf{B}_k + \mathbf{C}_k\mathbf{B}_i \end{bmatrix}$$

from which it can be concluded that

$$\mathbf{C}_i\mathbf{B}_i = \mathbf{B}_i\mathbf{C}_i = \mathbf{I} \tag{7.21}$$

$$\mathbf{B}_i\mathbf{C}_k + \mathbf{B}_k\mathbf{C}_i = \mathbf{C}_i\mathbf{B}_k + \mathbf{C}_k\mathbf{B}_i = \emptyset \qquad \text{if } i \neq k \tag{7.22}$$

The matrixes **B** and **C** anticommute. This is one of the properties of the Pauli matrix σ_k:

$$\sigma_i\sigma_k + \sigma_k\sigma_i = 0 \qquad \sigma_i\sigma_j = i\sigma_k \qquad \sigma_i^2 = 1$$

We thus take

$$\mathbf{B}_k = a\sigma_k \tag{7.23}$$

$$\mathbf{C}_k = b\sigma_k \tag{7.24}$$

As $\mathbf{B}_i\mathbf{C}_i = \mathbf{I}$, it follows that

$$ab = 1 \tag{7.25}$$

The form of the matrix γ_k is therefore

$$\gamma_k = \begin{bmatrix} \emptyset & b^{-1}\sigma_k \\ b\sigma_k & \emptyset \end{bmatrix} \tag{7.26}$$

There are several possibilities: If $b = i$, $b^{-1} = -i$,

$$\gamma_k = \begin{bmatrix} \emptyset & -i\sigma_k \\ i\sigma_k & \emptyset \end{bmatrix} \tag{7.27}$$

$$\gamma_0 = \begin{bmatrix} I & \emptyset \\ \emptyset & -I \end{bmatrix} = \begin{bmatrix} 1 & 0 & 0 & 0 \\ 0 & 1 & 0 & 0 \\ 0 & 0 & -1 & 0 \\ 0 & 0 & 0 & -1 \end{bmatrix}$$

$$\gamma_1 = \begin{bmatrix} \emptyset & -i\sigma_1 \\ i\sigma_1 & \emptyset \end{bmatrix} = \begin{bmatrix} 0 & 0 & 0 & -i \\ 0 & 0 & -i & 0 \\ 0 & i & 0 & 0 \\ i & 0 & 0 & 0 \end{bmatrix}$$

$$\gamma_2 = \begin{bmatrix} \emptyset & -i\sigma_2 \\ i\sigma_2 & \emptyset \end{bmatrix} = \begin{bmatrix} 0 & 0 & 0 & -1 \\ 0 & 0 & 1 & 0 \\ 0 & 1 & 0 & 0 \\ -1 & 0 & 0 & 0 \end{bmatrix}$$

$$\gamma_3 = \begin{bmatrix} \emptyset & -i\sigma_3 \\ i\sigma_3 & \emptyset \end{bmatrix} = \begin{bmatrix} 0 & 0 & -i & 0 \\ 0 & 0 & 0 & i \\ i & 0 & 0 & 0 \\ 0 & -i & 0 & 0 \end{bmatrix}$$

If $a = b = 1$,

$$\gamma_k = \begin{bmatrix} \emptyset & \sigma_k \\ \sigma_k & \emptyset \end{bmatrix} \tag{7.28}$$

$$\gamma_0 = \begin{bmatrix} I & \emptyset \\ \emptyset & -I \end{bmatrix} = \begin{bmatrix} 1 & 0 & 0 & 0 \\ 0 & 1 & 0 & 0 \\ 0 & 0 & -1 & 0 \\ 0 & 0 & 0 & -1 \end{bmatrix}$$

$$\gamma_1 = \begin{bmatrix} \emptyset & -\sigma_1 \\ \sigma_1 & \emptyset \end{bmatrix} = \begin{bmatrix} 0 & 0 & 0 & 1 \\ 0 & 0 & 1 & 0 \\ 0 & 1 & 0 & 0 \\ 1 & 0 & 0 & 0 \end{bmatrix}$$

$$\gamma_2 = \begin{bmatrix} \emptyset & \sigma_2 \\ \sigma_2 & \emptyset \end{bmatrix} = \begin{bmatrix} 0 & 0 & 0 & -i \\ 0 & 0 & i & 0 \\ 0 & -i & 0 & 0 \\ i & 0 & 0 & 0 \end{bmatrix}$$

$$\gamma_3 = \begin{bmatrix} \emptyset & \sigma_3 \\ i\sigma_3 & \emptyset \end{bmatrix} = \begin{bmatrix} 0 & 0 & 1 & 0 \\ 0 & 0 & 0 & -1 \\ 1 & 0 & 0 & 0 \\ 0 & -1 & 0 & 0 \end{bmatrix}$$

This is called the Pauli representation.

The Kramers representation is

$$\gamma_k = \begin{bmatrix} \sigma_k & \emptyset \\ \emptyset & -\sigma_k \end{bmatrix} \tag{7.29}$$

$$\gamma_0 = \begin{bmatrix} I & \emptyset \\ \emptyset & -I \end{bmatrix} = \begin{bmatrix} 1 & 0 & 0 & 0 \\ 0 & 1 & 0 & 0 \\ 0 & 0 & -1 & 0 \\ 0 & 0 & 0 & -1 \end{bmatrix}$$

$$\gamma_1 = \begin{bmatrix} \sigma_1 & \emptyset \\ \emptyset & -\sigma_1 \end{bmatrix} = \begin{bmatrix} 0 & 1 & 0 & 0 \\ 1 & 0 & 0 & 0 \\ 0 & 0 & 0 & 1 \\ 0 & 0 & 1 & 0 \end{bmatrix} \tag{7.30}$$

$$\gamma_2 = \begin{bmatrix} \sigma_2 & \emptyset \\ \emptyset & -\sigma_2 \end{bmatrix} = \begin{bmatrix} 0 & -i & 0 & 0 \\ i & 0 & 0 & 0 \\ 0 & 0 & 0 & i \\ 0 & 0 & -i & 0 \end{bmatrix}$$

$$\gamma_3 = \begin{bmatrix} \sigma_3 & \emptyset \\ \emptyset & -\sigma_3 \end{bmatrix} = \begin{bmatrix} 1 & 0 & 0 & 0 \\ 0 & -1 & 0 & 0 \\ 0 & 0 & -1 & 0 \\ 0 & 0 & 0 & 1 \end{bmatrix} \tag{7.31}$$

The Majorana representation is

$$\gamma_2 = \begin{bmatrix} \emptyset & -I \\ -I & \emptyset \end{bmatrix} = \begin{bmatrix} 0 & 0 & -1 & 0 \\ 0 & 0 & 0 & -1 \\ -1 & 0 & 0 & 0 \\ 0 & -1 & 0 & 0 \end{bmatrix}$$

$$\gamma_1 = \begin{bmatrix} \sigma_1 & \emptyset \\ \emptyset & -\sigma_1 \end{bmatrix} = \begin{bmatrix} 0 & 1 & 0 & 0 \\ 1 & 0 & 0 & 0 \\ 0 & 0 & 0 & -1 \\ 0 & 0 & -1 & 0 \end{bmatrix}$$

$$\gamma_0 = \begin{bmatrix} \sigma_2 & \emptyset \\ \emptyset & -\sigma_2 \end{bmatrix} = \begin{bmatrix} 0 & -i & 0 & 0 \\ i & 0 & 0 & 0 \\ 0 & 0 & 0 & i \\ 0 & 0 & -i & 0 \end{bmatrix}$$

$$\gamma_3 = \begin{bmatrix} \sigma_3 & \emptyset \\ \emptyset & -\sigma_3 \end{bmatrix} = \begin{bmatrix} 1 & 0 & 0 & 0 \\ 0 & -1 & 0 & 0 \\ 0 & 0 & -1 & 0 \\ 0 & 0 & 0 & 1 \end{bmatrix}$$

7.3.3 Covariant Formulation

According to (7.8) and (7.16),

$$\left(\frac{E}{c} - \vec{\gamma}\cdot\vec{p} - m_0 c\gamma_0\right)|\psi\rangle = |0\rangle \tag{7.32}$$

If we write α_0 as the 4×4 unit matrix

$$\alpha_0 = \begin{bmatrix} I & \emptyset \\ \emptyset & I \end{bmatrix} = \begin{bmatrix} 1 & 0 & 0 & 0 \\ 0 & 1 & 0 & 0 \\ 0 & 0 & 1 & 0 \\ 0 & 0 & 0 & 1 \end{bmatrix}$$

and

$$\vec{\alpha} = \begin{bmatrix} \gamma_1 \\ \gamma_2 \\ \gamma_3 \end{bmatrix} \qquad \alpha_\mu = (\alpha_0, \vec{\alpha}) \tag{7.33}$$

the previous equation becomes

$$\left(\alpha_0 \frac{E}{c} - \vec{\alpha}\cdot\vec{p} - m_0 c \gamma_0\right) |\psi\rangle = |0\rangle$$

or

$$(\alpha_\mu P_\mu - m_0 c \gamma_0)|\psi\rangle = |0\rangle \tag{7.34}$$

a formula that is true for a free particle and can be generalized to a particle in an electromagnetic field in

$$\left[\frac{\alpha_0}{c}(E - qV) - \vec{\alpha}\cdot(\vec{p} - q\vec{A}) - m_0 c \gamma_0\right] |\psi\rangle = |0\rangle$$

or

$$[\alpha_\mu(P_\mu - qA_\mu) - m_0 c \gamma_0]|\psi\rangle = |0\rangle \tag{7.35}$$

a formula that is invariant in relation to Lorentz transformation.

7.4 PROVING ELECTRON SPIN

Compare the classical relativistic hamiltonian of a particle in an electromagnetic field [cf. (6.8)],

$$[(E - qV)^2 - (\vec{p} - q\vec{A})^2 c^2 - m_0^2 c^4]|\psi\rangle = |0\rangle \tag{7.36}$$

to Dirac's formula (7.35):

$$[E - qV - c\vec{\alpha}\cdot(\vec{p} - q\vec{A}) - m_0 c^2 \gamma_0]|\psi\rangle = |0\rangle \tag{7.37}$$

Multiply on the right (7.37) by $[E - qV + c\vec{\alpha}\cdot(\vec{p} - q\vec{A}) + m_0 c^2 \gamma_0]$:

$$\begin{aligned}&[E - qV - c\vec{\alpha}\cdot(\vec{p} - q\vec{A}) - m_0 c^2 \gamma_0] \\ &\quad \times [E - qV + c\vec{\alpha}\cdot(\vec{p} - q\vec{A}) + m_0 c^2 \gamma_0]|\psi\rangle = |0\rangle\end{aligned} \tag{7.38}$$

Note that

$$(E - qV)\gamma_0 = \gamma_0(E - qV)$$

since the γ_μ commute with P^μ, X^μ, A^μ and that

$$\vec{\alpha}\gamma_0\cdot(\vec{p} - q\vec{A}) + \gamma_0\vec{\alpha}\cdot(\vec{p} - q\vec{A}) = 0$$

as

$$\vec{\alpha}\gamma_0 + \gamma_0\vec{\alpha} = \vec{0}$$

and therefore (7.38) can be simplified to

$$\begin{aligned}\{(E - qV)^2 &- c^2[\vec{\alpha}\cdot(\vec{p} - q\vec{A})][\vec{\alpha}\cdot(\vec{p} - q\vec{A})] \\ &+ c\vec{\alpha}\cdot(E - qV)(\vec{p} - q\vec{A}) \\ &- (\vec{p} - q\vec{A})(E - qV) - m_0^2c^4\}|\psi\rangle = |0\rangle\end{aligned} \tag{7.39}$$

Annex Calculations

1. If $\vec{B}$ and $\vec{C}$ commute with $\vec{\alpha}$,

$$\begin{aligned}(\vec{\alpha}\cdot\vec{B})(\vec{\alpha}\cdot\vec{C}) &= \sum_{\text{permutations } 1,2,3} (\alpha_1^2 B_1C_1 + \alpha_1\alpha_2 B_1C_2 + \alpha_2\alpha_1 B_2C_1) \\ &= \vec{B}\cdot\vec{C} + i \sum_{\text{permutations } 1,2,3} \alpha_3(B_1C_2 - B_2C_1) \\ &= \vec{B}\cdot\vec{C} + i\vec{\alpha}\cdot\vec{B}\wedge\vec{C}\end{aligned} \tag{7.40}$$

We used the fact that $\alpha_k^2 = I$ and that $\alpha_1\alpha_2 = i\alpha_3$, from which

$$\begin{aligned}[\vec{\alpha}\cdot(\vec{p} - q\vec{A})][\vec{\alpha}\cdot(\vec{p} - q\vec{A})] &= (\vec{p} - q\vec{A})^2 \\ &+ i\vec{\alpha}\cdot[(\vec{p} - q\vec{A})\wedge(\vec{p} - q\vec{A})]\end{aligned} \tag{7.41}$$

Now

$$\begin{aligned}(\vec{p} - q\vec{A})\wedge(\vec{p} - q\vec{A}) &= -q(\vec{p}\wedge\vec{A} + \vec{A}\wedge\vec{p}) \\ &= i\hbar q(\vec{\nabla}\wedge\vec{A} + \vec{A}\wedge\vec{\nabla})\end{aligned} \tag{7.42}$$

Evaluate precisely the term in parentheses on the right-hand side of (7.42):

$$\begin{aligned}&\vec{\nabla}\wedge(\vec{A}\psi) + \vec{A}\wedge\vec{\nabla}\psi \\ &= \vec{i}\left[\frac{\partial}{\partial y}(A_z\psi) - \frac{\partial}{\partial z}(A_y\psi) + A_y\frac{\partial\psi}{\partial z} - A_z\frac{\partial\psi}{\partial y}\right] \\ &\quad + \vec{j}[\cdots] + \vec{k}[\cdots] \\ &= (\vec{\nabla}\wedge\vec{A})\psi = \vec{B}\psi\end{aligned} \tag{7.43}$$

since $\vec{B} = \vec{\nabla} \wedge \vec{A}$. Rewrite (7.41) as

$$[\vec{\alpha}\cdot(\vec{p} - q\vec{A})][\vec{\alpha}\cdot(\vec{p} - q\vec{A})] = (\vec{p} - q\vec{A})^2 - \hbar q\vec{\alpha}\cdot\vec{B} \tag{7.44}$$

2. We are at present evaluating

$$\begin{aligned}&\vec{\alpha}\cdot(E - qV)(\vec{p} - q\vec{A}) - \vec{\alpha}\cdot(\vec{p} - q\vec{A})(E - qV)\\ &\quad = E\vec{\alpha}\cdot(\vec{p} - q\vec{A}) - \vec{\alpha}\cdot(\vec{p} - q\vec{A})E\\ &\qquad + q\vec{\alpha}\cdot(\vec{p} - q\vec{A})V - qV\vec{\alpha}\cdot(\vec{p} - q\vec{A})\\ &\quad = q\vec{\alpha}\cdot\vec{A}E - qE\vec{\alpha}\cdot\vec{A} + q\vec{\alpha}\cdot\vec{p}V - qV\vec{\alpha}\cdot\vec{p}\end{aligned}$$

as

$$E\vec{\alpha}\cdot\vec{p} = \vec{\alpha}\cdot\vec{p}E$$

since $\partial/\partial t$ and $\partial/\partial x$ commute and

$$V\vec{A} = \vec{A}V$$

from

$$\begin{aligned}&\vec{\alpha}\cdot(E - qV)(\vec{p} - q\vec{A}) - \vec{\alpha}\cdot(\vec{p} - q\vec{A})(E - qV)\\ &\quad = q\vec{\alpha}\cdot(\vec{A}E - E\vec{A} + \vec{p}V - V\vec{p})\end{aligned} \tag{7.45}$$

Now

$$\begin{aligned}\vec{A}E - E\vec{A} + \vec{p}V - V\vec{p} &= i\hbar\left(\vec{A}\frac{\partial}{\partial t} - \frac{\partial}{\partial t}\vec{A}\right) - i\hbar\vec{\nabla}V + i\hbar V\vec{\nabla}\\ &= -i\hbar\frac{\partial\vec{A}}{\partial t} - i\hbar(\vec{\nabla}V) = i\hbar\vec{E}\end{aligned}$$

We can for the moment therefore rewrite (7.39) as

$$\begin{aligned}&\{(E - qV)^2 - c^2[(\vec{p} - q\vec{A})^2] + \hbar qc^2\vec{\alpha}\cdot\vec{B}\\ &\quad + i\hbar cq\vec{\alpha}\cdot\vec{E} - m_0^2c^4\}|\psi\rangle = |0\rangle\end{aligned} \tag{7.46}$$

If we make a comparison with the classical relativistic expression of (7.36), we observe that there are two additional terms:

$$T_1 = \hbar qc^2\vec{\alpha}\cdot\vec{B}$$

$$T_2 = i\hbar qc\vec{\alpha}\cdot\vec{E}$$

These two terms are seen as coming from the electron, which has a magnetic momentum and an electric momentum. The magnetic momentum is provoked by either an internal movement of the electron or by the spin of the electron. The electric momentum is an imaginary quantity, and consequently, it cannot be considered as having any meaning in physics.

The hamiltonian of our equation is real, and the imaginary term was introduced after our artificial operation of multiplication.

Imagine for the moment that the particle movement is nonrelativistic and in the weak electromagnetic field. In classical theory

$$E = qV + \sqrt{(\vec{p} - q\vec{A})^2 + m_0c^4}$$

or

$$E = E' + m_0c^2$$

with

$$(E' - qV) \ll m_0c^2$$

from which, for (7.45),

$$(E' - qV + m_0c^2)^2 - c^2(\vec{p} - q\vec{A})^2 + \hbar qc^2\vec{\alpha}\cdot\vec{B} + i\hbar cq\vec{\alpha}\cdot\vec{E} - m_0^2c^4 = 0 \qquad (7.47)$$

with

$$(E' - qV + m_0c^2)^2 = (E' - qV)^2 + m_0^2c^4 + 2(E' - qV)m_0c^2$$
$$\simeq m_0^2c^4 + 2(E' - qV)m_0c^2$$

The relation (7.47) then becomes, after being divided by $2m_0c^2$,

$$E' = qV + \frac{1}{2m_0}(\vec{p} - q\vec{A})^2 - \frac{\hbar}{2m_0}q\vec{\alpha}\cdot\vec{B} - \frac{i\hbar}{2m_0c}q\vec{\alpha}\cdot\vec{E} \qquad (7.48)$$

The additional term T_1 is thus equivalent to

$$T_1 = -\frac{\hbar}{2m_0}q\vec{\alpha}\cdot\vec{B} = -\vec{\mu}\cdot\vec{B} \qquad (7.49)$$

if

$$\vec{\mu} = \frac{\hbar}{2m_0}q\vec{\alpha} \qquad (7.50)$$

is the magnetic momentum of spin, that is, of the internal movement of the electron. If the magnetic field is directed according to the axis $0z$,

$$T_1 = -\mu_zB \qquad (7.51)$$

with

$$\mu_z = \frac{\hbar}{2m_0}q\alpha_3 \qquad (7.52)$$

The quantity

$$\mu_B = \frac{\hbar}{2m_0} q \tag{7.53}$$

is called the Bohr magneton, so

$$\mu_z = \mu_B \alpha_3$$

Now

$$\alpha_3 = \begin{bmatrix} \emptyset & \sigma_3 \\ \sigma_3 & \emptyset \end{bmatrix}$$

a matrix for which the eigenvalues are the same as those of σ_3, that is, ± 1. The eigenvalues of μ_z are thus $\pm\, (\hbar/2m_0)q$. Recall that

$$H_D = c\vec{\gamma}\cdot\vec{p} + m_0 c^2 \gamma_0 \tag{7.54}$$

Then evaluate the commutator

$$[L_z, H_D] = [L_z, c\vec{\gamma}\cdot\vec{p} + m_0 c^2 \gamma_0] = c\vec{\gamma}\cdot[L_z, \vec{p}]$$
$$= c\gamma_x[L_z, p_x] + c\gamma_y[L_z, p_y]$$

In fact,

$$[L_z, \gamma_0] = 0 \qquad \text{and} \qquad [L_z, p_z] = 0$$

Now

$$[L_i, p_j] = i\hbar p_k$$

Otherwise,

$$[L_z, H_D] = i\hbar c\gamma_x p_y - i\hbar c\gamma_y p_x \tag{7.55}$$

If we define the internal magnetic momentum s_z by

$$s_z = \tfrac{1}{2}\hbar\Sigma_z$$

it can be seen that

$$[s_z, H_D] = \tfrac{1}{2}\hbar[\Sigma_z, c\vec{p}\cdot\vec{p} + m_0 c^2 \gamma_0] \tag{7.56}$$

As

$$\Sigma_z = \begin{bmatrix} \sigma_z & \emptyset \\ \emptyset & \sigma_z \end{bmatrix}$$

and

$$\sigma_z = \begin{bmatrix} 1 & 0 \\ 0 & -1 \end{bmatrix}$$

then

$$\Sigma_z\gamma_0 = \begin{bmatrix} \sigma_z & \emptyset \\ \emptyset & \sigma_z \end{bmatrix} \begin{bmatrix} I & \emptyset \\ \emptyset & -I \end{bmatrix} = \begin{bmatrix} \sigma_z & \emptyset \\ \emptyset & -\sigma_z \end{bmatrix}$$

Now

$$\gamma_0\Sigma_z = \begin{bmatrix} I & \emptyset \\ \emptyset & -I \end{bmatrix} \begin{bmatrix} \sigma_z & \emptyset \\ \emptyset & -\sigma_z \end{bmatrix} = \begin{bmatrix} \sigma_z & \emptyset \\ \emptyset & -\sigma_z \end{bmatrix}$$

Then

$$[s_z, H_D] = \tfrac{1}{2}\hbar c[\sigma_z, \vec{\gamma}\cdot\vec{p}] = \tfrac{1}{2}\hbar c[\sigma_z, \vec{\gamma}]\cdot\vec{p}$$

or

$$\sigma_z\gamma_z = \begin{bmatrix} \sigma_z & \emptyset \\ \emptyset & \sigma_z \end{bmatrix} \begin{bmatrix} \emptyset & -i\sigma_1 \\ i\sigma_1 & \emptyset \end{bmatrix} = \begin{bmatrix} \emptyset & \sigma_y \\ -\sigma_y & \emptyset \end{bmatrix} = i\gamma_y$$

$$\gamma_x\sigma_z = -i\gamma_y$$

since

$$\gamma_y = \begin{bmatrix} \emptyset & -i\sigma_2 \\ i\sigma_2 & \emptyset \end{bmatrix}$$

Thus

$$\sigma_z\gamma_x = -\gamma_x\sigma_z = i\gamma_y$$

$$\sigma_z\gamma_y = -\gamma_y\sigma_z = -i\gamma_x$$

and finally,

$$\begin{aligned}[s_z, H_D] &= \tfrac{1}{2}\hbar c[(\sigma_z\gamma_x - \gamma_x\sigma_z)p_x + (\sigma_z\gamma_y - \gamma_y\sigma_z)p_y] \\ &= i\hbar c[\gamma_y p_x - \gamma_x p_y] \end{aligned} \tag{7.57}$$

By adding (7.55) and (7.57), we see that

$$[L_z + s_z, H_D] = 0 \tag{7.58}$$

which means that the quantity J_z defined by

$$J_z = L_z + s_z = L_z + \tfrac{1}{2}\hbar\Sigma_z$$

is a constant of motion. In general, we write

$$\vec{J} = \vec{L} + \vec{S}$$

$\vec{J}$ being the total angular momentum and $\vec{S}$ the momentum of spin, such that

$$\vec{S} = \tfrac{1}{2}\hbar\vec{\sigma}$$

These calculations prove the existence of spin, spin being an additional degree of freedom that works in an internal space. The spin is a logical consequence of writing the Dirac equation.

7.5 DIRAC EQUATION AND CLIFFORD ALGEBRA

We can prove that the Dirac equation for a free particle in Clifford algebra is written in the form

$$\Box\Psi = m_0c^2\Psi\gamma_0\gamma_2\gamma_1$$

where Ψ is a biquaternion (the equivalent of the wave function) of definition

$$A_{bq} = A_q + iA'_q$$

A_q and A'_q being quaternions (cf. Chapter 5). The γ_μ are the vectors of a fixed basis and were defined in Chapter 5.

In the presence of an electromagnetic field,

$$\Box\Psi = (m_0c^2\Psi\gamma_0 + qA\Psi)\gamma_2\gamma_1$$

$A = A^\mu\gamma_\mu$ is a vector of space-time (potential quadrivector). Recall that in our basis all the vectors are put in the form $V = V^\mu\gamma_\mu$.

7.6 APPLICATIONS

7.6.1 Pauli Equation

We begin with the Schrödinger equation of an electron placed in a region of space where an electromagnetic field $\vec{E}, \vec{B}$ exists:

$$\frac{1}{2m}\{[(-i\hbar\vec{\nabla} - e\vec{A})]^2 - eV\}|\psi(\vec{r},t)\rangle = i\hbar\frac{\partial}{\partial t}|\psi(\vec{r},t)\rangle \tag{1}$$

Proceeding, it can be seen that

$$\frac{1}{2m}[-i\hbar^2\nabla^2 + 2i\hbar e(\vec{A}\cdot\vec{\nabla} + \vec{\nabla}\cdot\vec{A}) + e^2A^2]|\psi(\vec{r},t)\rangle$$

$$= \left(i\hbar\frac{\partial}{\partial t} + eV\right)|\psi(\vec{r},t)\rangle \tag{2}$$

This is an equation that does not involve the spin of the electron. Write $\vec{\pi}$ as the generalized momentum such that

$$\vec{\pi} = \vec{p} - e\vec{A} \tag{3}$$

$$[\pi_i,\pi_j] = ie\hbar B_k \tag{4}$$

If the σ_i are the matrix of Pauli and $\vec{C}$ is a vector,

$$(\vec{\sigma}\cdot\vec{C})^2 = C^2 \tag{5}$$

The matrix σ_i belong to the Clifford algebra:

$$[\sigma_i,\sigma_j]_+ = 2\delta_{ij} \tag{6}$$

They also belong to the Lie algebra SU(2):

$$\left[\frac{\sigma_i}{2},\frac{\sigma_j}{2}\right] = i\epsilon_{ij}{}^{k}\,\frac{\sigma_k}{2} \tag{7}$$

from which it can be deduced that

$$\sigma_i\sigma_j = \delta_{ij} + i\epsilon_{ij}{}^{k}\sigma_k \tag{8}$$

Evaluate the quantity $(\vec{\sigma}\cdot\vec{\pi})^2$:

$$\begin{aligned}(\vec{\sigma}\cdot\vec{\pi})^2 &= (\sigma_i p^i - e\sigma_i A^i)(\sigma_j p^j - e\sigma_j A^j)\\ &= p^2 + e^2A^2 - e\sigma_i\sigma_j(p^iA^j + A^ip^j)\\ &= p^2 + e^2A^2 - e(\delta_{ij} + i\epsilon_{ij}{}^{k}\sigma_k)(p^iA^j + A^ip^j)\\ &= p^2 + e^2A^2 - e(\vec{p}\cdot\vec{A} + \vec{A}\cdot\vec{p}) - ie\epsilon_{ij}{}^{k}(\sigma_k p^iA^j + \sigma_k A^ip^j)\\ &= \pi^2 - ie[\epsilon_{ij}{}^{k}(\sigma_k p^iA^j) + \epsilon_{ij}{}^{k}(\sigma_k A^jp^i)]\\ &= \pi^2 - ie\epsilon_{ij}{}^{k}(\sigma_k)[p^iA^j - A^jp^i]\end{aligned}$$

Now

$$[p^iA^j - A^jp^i]|\psi\rangle = (p^iA^j)|\psi\rangle$$

from which

$$\begin{aligned}(\vec{\sigma}\cdot\vec{\pi})^2 &= \pi^2 - ie\sigma_k\epsilon_{ij}{}^{k}p_iA^j = \pi^2 - ie\sigma_k(\vec{p}\wedge\vec{A})^k\\ &= \pi^2 - \hbar e\sigma_k(\vec{\nabla}\wedge\vec{A})^k = \pi^2 - \hbar e\sigma_k B^k = \pi^2 - \hbar e(\vec{\sigma}\cdot\vec{B})\end{aligned} \tag{9}$$

Rewrite (1) with the help of (3):

$$\left[\frac{1}{2m}(\vec{\pi}^2) - eV\right]|\psi(\vec{r},t)\rangle = i\hbar\frac{\partial}{\partial t}|\psi(\vec{r},t)\rangle \tag{10}$$

In (10), Pauli replaced π^2 by $(\vec{\sigma}\cdot\vec{\pi})^2$:

$$\left[\frac{1}{2m}\pi^2 - \frac{1}{2m}\hbar e(\vec{\sigma}\cdot\vec{B}) - eV\right]|\psi(\vec{r},t)\rangle = i\hbar\frac{\partial}{\partial t}|\psi(\vec{r},t)\rangle \tag{11}$$

(Pauli's equation) containing the spin by the intermediary of $(\vec{\sigma}\cdot\vec{B})$.

7.6.2 Transition from Dirac Equation to Pauli Equation

Recall Schrödinger's form of the Dirac equation:

$$\hat{H}_D|\psi\rangle = i\hbar \frac{\partial}{\partial t} |\psi\rangle \tag{1}$$

with

$$\hat{H}_D = c\hat{\gamma}\cdot\hat{p} + c\chi \tag{2}$$

where

$$\chi = m_0 c \gamma_0 \tag{3}$$

If for an electron in an electromagnetic field, we replace $\vec{p}$ by $\vec{\pi}$ and add the quantity $-eV$, we have

$$\hat{H}_D = c\vec{\gamma}\cdot\vec{\pi} + m_0 c^2 \gamma_0 - eVI \tag{4}$$

For the stationary states,

$$\psi = \frac{1}{\sqrt{V}} \begin{bmatrix} \varphi \\ \chi \end{bmatrix} e^{-(i/\hbar)Et} \tag{5}$$

with

$$E = m_0 c^2 + T \tag{6}$$

where T is the kinetic energy. Remember that

$$\gamma_0 = \begin{bmatrix} I & \emptyset \\ \emptyset & -I \end{bmatrix} \qquad \gamma_k = \begin{bmatrix} \emptyset & -i\sigma_k \\ i\sigma_k & \emptyset \end{bmatrix} \tag{7.28}$$

and therefore (1) is written

$$E \begin{bmatrix} \varphi \\ \chi \end{bmatrix} = c \begin{bmatrix} \emptyset & -i\vec{\sigma}\cdot\vec{\pi} \\ i\vec{\sigma}\cdot\vec{\pi} & \emptyset \end{bmatrix} \begin{bmatrix} \varphi \\ \chi \end{bmatrix} + m_0 c^2 \begin{bmatrix} I & \emptyset \\ \emptyset & -I \end{bmatrix} \begin{bmatrix} \varphi \\ \chi \end{bmatrix} - eV \begin{bmatrix} I & \emptyset \\ \emptyset & I \end{bmatrix} \begin{bmatrix} \varphi \\ \chi \end{bmatrix} \tag{7}$$

or

$$(m_0 c^2 + T)\varphi = -ic\vec{\sigma}\cdot\vec{\pi}\chi + m_0 c^2 \varphi - eV\varphi \tag{8}$$

$$(m_0 c^2 + T)\chi = +ic\vec{\sigma}\cdot\vec{\pi}\varphi - m_0 c^2 \chi - eV\chi \tag{9}$$

From (9) we obtain

$$\chi = \frac{ic\vec{\sigma}\cdot\vec{\pi}}{2m_0 c^2 + T + eV} \varphi \tag{10a}$$

In the nonrelativistic approximation we take

$$\chi = \frac{ic\vec{\sigma}\cdot\vec{\pi}}{2m_0c^2}\,\varphi \tag{10b}$$

which is introduced in (8):

$$T\varphi = \left[\frac{(\vec{\sigma}\cdot\vec{\pi})^2}{2m_0} - eV\right]\varphi$$

a relation that is equivalent to the Pauli equation, the energy of the particle being purely kinetic.

7.6.3 Invariance of Dirac Equation

Recall Dirac's equation of a particle in an electromagnetic field:

$$\hat{H}_D|\psi\rangle = (c\vec{\gamma}\cdot\vec{p} + m_0c^2\gamma_0)|\psi\rangle = E|\psi\rangle$$

from which we get the relation

$$\left[\gamma_0\left(\vec{\gamma}\cdot\vec{p} - \frac{E}{c}\right) + m_0c\right]|\psi\rangle = |0\rangle$$

Write

$$\gamma_0\vec{\gamma} = \vec{\Gamma}$$

and

$$\gamma_0 = \Gamma_0$$

from which

$$(\Gamma^\mu P_\mu - m_0c)|\psi\rangle = |0\rangle \tag{1}$$

which is Dirac's equation, a covariant of the Lorentz equation.

$$\Gamma_0 = \begin{bmatrix} I & \emptyset \\ \emptyset & -I \end{bmatrix} \tag{2}$$

$$\Gamma_k = -i\begin{bmatrix} \emptyset & \sigma_k \\ \sigma_k & \emptyset \end{bmatrix} \tag{3}$$

We find (7.28) with a different notation. In the case of an electromagnetic field, (1) generalizes to

$$[\Gamma^\mu(P_\mu - eA_\mu) - m_0c]|\psi\rangle = |0\rangle \tag{4}$$

Relation (4) is made invariant under the action of one of Poincaré's transformations:

$$X'^\mu = L^\mu{}_\nu X^\nu + a^\mu \tag{5}$$

Then

$$\psi(X) \longrightarrow \psi'(X') \tag{6}$$

$$A^{\mu}(X) \longrightarrow A'^{\mu'}(X') \tag{7}$$

As

$$\partial_{\mu} = \frac{\partial X'^{\mu'}}{\partial X^{\mu}} \partial_{\mu'} = L^{\mu'}{}_{\mu}\partial_{\mu'} \tag{8}$$

we have

$$A'^{\mu'}(X') = L^{\mu'}{}_{\nu}A^{\nu}(X) \tag{9}$$

For the fields we write

$$\psi'_{\alpha}(X') = D_{\alpha\beta}(L)\psi_{\beta}(X)$$

We are trying to find the matrix $D_{\alpha\beta}(L)$ that leaves (4) invariant. By Poincaré's transformation, (4) becomes

$$[\Gamma^{\mu}(L^{\nu'}{}_{\mu}\partial'_{\nu'} - eL^{\nu'}{}_{\mu}A'_{\nu'}) - mc]D^{-1}\psi'(X') = 0$$

by multiplying on the left by D and by identifying with (4):

$$D\Gamma^{\mu}L^{\nu'}{}_{\mu}D^{-1} = \Gamma^{\nu'} \tag{10}$$

a condition of invariance of Dirac's equation.

7.6.4 Lorentz Transformation

According to (10),

$$D^{-1}\Gamma^{\nu}D = \Gamma^{\mu}L^{\nu}{}_{\mu} \tag{11}$$

with

$$\gamma_k = \begin{bmatrix} \emptyset & -i\sigma_k \\ i\sigma_k & \emptyset \end{bmatrix}$$

$$\gamma_0 = \Gamma_0 = \begin{bmatrix} I & \emptyset \\ \emptyset & -I \end{bmatrix} \qquad \gamma_k = \begin{bmatrix} \emptyset & -i\sigma_k \\ i\sigma_k & \emptyset \end{bmatrix}$$

$$\Gamma_k = \Gamma_0\gamma_k = \begin{bmatrix} \emptyset & -i\sigma_k \\ -i\sigma_k & \emptyset \end{bmatrix}$$

It can easily be proved that

$$\Gamma_{\mu}^{+} = \Gamma_0\Gamma_{\mu}\Gamma_0 \tag{12}$$

Taking the addition from both sides, (11) is thus written

$$(D^{-1}\Gamma^{\nu}D)^{+} = L^{\nu}{}_{\mu}\Gamma^{0}\Gamma^{\mu}\Gamma^{0}$$

and multiplying by Γ^0 on the left and on the right, we obtain

$$\Gamma^0(D^{-1}\Gamma^\nu D)^+\Gamma^0 = L^\nu{}_\mu\Gamma^\mu \tag{13}$$

Equation (11) can also be put into the form

$$\Gamma^\nu D = D\Gamma^\mu L^\nu{}_\mu \tag{14}$$

from which we take the addition

$$D^+\Gamma^0\Gamma^\nu\Gamma^0 = L^\nu{}_\mu\Gamma^0\Gamma^\mu\Gamma^0 D^+ \tag{15}$$

By remultiplying (15) by Γ^0 on the left and right,

$$\Gamma^0 D^+\Gamma^0\Gamma^\nu = L^\nu{}_\mu\Gamma^\mu\Gamma^0 D^+\Gamma^0 \tag{16}$$

taking (11) into consideration,

$$\Gamma^0 D^+\Gamma^0\Gamma^\nu = D^{-1}\Gamma^\nu D\Gamma^0 D^+\Gamma^0$$

then multiplying on the left by D, we have

$$(D\Gamma^0 D^+\Gamma^0)\Gamma^\nu = \Gamma^\nu(D\Gamma^0 D^+\Gamma^0) \tag{17}$$

The term $(D\Gamma^0 D^+\Gamma^0)$ commutes with the Γ^ν matrix; it is, therefore, in accordance with Schur's theorem, a constant that is written λI, where I is the identity matrix:

$$D\Gamma^0 D^+\Gamma^0 = \lambda I \tag{18}$$

Since the determinant of D is equal to 1 (D characterizes the Lorentz proper group),

$$\lambda^4 = 1$$

from where there are four solutions:

$$\lambda = \pm 1$$

$$\lambda = \pm i$$

According to (18),

$$D^+\Gamma^0 = (\Gamma^0)^{-1}D^{-1}\lambda = \Gamma^0 D^{-1}\lambda$$

and

$$D^+\Gamma^0\Gamma^0 D = \lambda\Gamma^0 D^{-1}\Gamma^0 D$$

or

$$D^+D = \lambda\Gamma^0 L^0{}_\mu\gamma^\mu = \lambda\Gamma^0(L^0{}_0 + L^0{}_k\gamma^k)$$

and

$$\mathrm{tr}(D^+D) = 4\lambda L^0{}_0$$

The trace of D^+D is still positive; now, for the proper group, the orthochrone,

$$L^0{}_0 \geqslant 1$$

from which $\lambda \geqslant 0$ and thus we take $\lambda = 1$; it follows [cf. (18)] that

$$D^+\Gamma^0 = \Gamma^0 D^{-1}$$

In conclusion,

$$\psi'(X') = D\psi(X) \tag{19}$$

Then

$$\psi'(X')^+ = \psi(X)^+D^+ = \psi(X)^+\Gamma^0\Gamma^0 D^+$$

and multiplying on the right by Γ^0,

$$\begin{aligned} \psi'(X')^+\Gamma^0 = \bar{\psi}(X) &= \bar{\psi}(X)\Gamma^0 D^+\Gamma^0 \\ &= \bar{\psi}(X)D^{-1} \end{aligned} \tag{20}$$

or

$$\boxed{\begin{aligned} \psi'(X') &= D_{\alpha\beta}(L)\psi_\beta(X) \\ \bar{\psi}_\alpha(X') &= \bar{\psi}_\beta(X)D^{-1}_{\beta\alpha}(L) \end{aligned}} \tag{21}$$

Example of use:

1. *For a scalar*, write

$$S(X) = \bar{\psi}(X)\psi(X)$$

By Lorentz transformation this quantity is transformed into

$$S'(X') = \bar{\psi}'(X')\psi'(X')$$

and with (21),

$$S'(X') = \bar{\psi}(X)D^{-1}(L)D(L)\psi(X) = \bar{\psi}(X)\psi(X) = S(X)$$

We have a scalar.

2. *For a vector*, take

$$\begin{aligned} V'^{\mu'}(X') &= \bar{\psi}'(X')\Gamma^{\mu'}\psi'(X') \\ &= \bar{\psi}(X)D^{-1}(L)\Gamma^{\mu'}D(L\psi(X) \\ &= \bar{\psi}(X)L^{\mu'}{}_\mu\Gamma^\mu\psi(X) = L^{\mu'}{}_\mu V^\mu(X) \end{aligned}$$

We find the definition of the current vector.

3. *For a tensor*, in the same way,

$$T'^{\mu'\nu'}(X') = L^{\mu'}{}_\mu L^{\nu'}{}_\nu T^{\mu\nu}(X)$$

where

$$T^{\mu\nu}(X) = \bar{\psi}\sigma^{\mu\nu}\psi$$

with

$$\sigma^{\mu\nu} = [\Gamma^{\mu},\Gamma^{\nu}]$$

4. *For a pseudoscalar,*

$$P'(X') = \bar{\psi}'(X')\Gamma^5\psi'(X')$$

with

$$\Gamma^5 = \frac{1}{4!}\,\epsilon_{\mu\nu\rho\sigma}\gamma^{\mu}\gamma^{\nu}\gamma^{\rho}\gamma^{\sigma}$$

and

$$\begin{aligned} P'(X') &= \frac{1}{4!}\,\bar{\psi}(X)\epsilon_{\alpha\mu\nu\beta}D^{-1}\Gamma^{\alpha}DD^{-1}\Gamma^{\mu}DD^{-1}\Gamma^{\nu}DD^{-1}\Gamma^{\beta}D\psi(X) \\ &= \frac{1}{4!}\,\bar{\psi}(X)\epsilon_{\alpha\mu\nu\beta}L^{\alpha}{}_{a}L^{\mu}{}_{b}L^{\nu}{}_{c}L^{\beta}{}_{d}\Gamma^{a}\Gamma^{b}\Gamma^{c}\Gamma^{d}\psi(X) \\ &= \det(L)P(X) \end{aligned}$$

5. *For a pseudovector,*

$$A'_{\mu'}(X') = \det(L)L^{\mu}{}_{\nu}A_{\mu}(X)$$

where

$$A_{\mu}(X) = \bar{\psi}(X)\Gamma^{\mu}\Gamma^{5}\psi(X)$$

7.6.5 Form of *D*

Take a Lorentz infinitesimal transformation:

$$L^{\alpha}{}_{\beta} = \delta^{\alpha}{}_{\beta} + \epsilon^{\alpha}{}_{\beta}$$

and

$$D(L) = 1 - \frac{i}{2}\,D_{\alpha\beta}\epsilon^{\alpha\beta}$$

a relation obtained by Taylor's development. It can also be written by introducing the spinor coefficients:

$$D_{ab}(L) = \delta_{ab} - \frac{i}{2}\,(D_{\alpha\beta})_{ab}\epsilon^{\alpha\beta} \tag{22}$$

Starting from

$$D^{-1}\Gamma^{\alpha}D = L^{\alpha}{}_{\beta}\Gamma^{\beta}$$

and bearing (22) in mind, we find that

$$\left(1 + \frac{i}{2} D_{\alpha\beta}\epsilon^{\alpha\beta}\right) \Gamma^{\mu} \left(1 - \frac{i}{2} D_{\alpha\beta}\epsilon^{\alpha\beta}\right) = (\delta^{\alpha}{}_{\beta} + \epsilon^{\alpha}{}_{\beta})\Gamma^{\beta}$$

In first order,

$$\frac{i}{2} [D_{\alpha\beta},\Gamma^{\mu}]\epsilon^{\alpha\beta} = \epsilon^{\mu\alpha}\Gamma_{\alpha}$$

or

$$\frac{i}{2} [D_{\alpha\beta},\Gamma_{\mu}] = \frac{1}{2} (g_{\alpha\mu}\Gamma_{\nu} - g_{\alpha\nu}\Gamma_{\mu})$$

which gives

$$D_{\mu\nu} = \frac{i}{4} [\Gamma_{\mu},\Gamma_{\nu}]$$

By definition we write

$$\sigma_{\mu\nu} = \frac{i}{2} [\Gamma_{\mu},\Gamma_{\nu}]$$

and

$$D_{\mu\nu} = \frac{\sigma_{\mu\nu}}{2}$$

Go back to $\psi'(X')$:

$$\psi'(X') = \left(1 - \frac{i}{4} \sigma_{\mu\nu}\epsilon^{\mu\nu}\right) \psi(X)$$

Otherwise,

$$\psi'(X') = \exp\left(-\frac{i}{4} \sigma_{\mu\nu}L^{\mu\nu}\right) \psi(X)$$

a relation that defines the transformation of one of Dirac's spinors under the action of an element of a Lorentz group.

7.6.6 Left and Right Spinors

Beginning with

$$[\Gamma^\mu P_\mu - m_0 c]|\psi\rangle = |0\rangle \tag{1}$$

we see that

$$[i\hbar\Gamma^\mu \partial_\mu - m_0 c]|\psi\rangle = |0\rangle \tag{2}$$

an equation from which we take the stationary solution in the form of plane waves:

$$\psi(X) = u(P)e^{-P\cdot X} \tag{3}$$

Equation (2), with (3), gives the condition

$$[\Gamma^\mu P_\mu - m_0 c]|u(P)\rangle = |0\rangle$$

which is expressed by

$$(\Gamma^0 P^0 - \Gamma^k P^k - m_0 c)u(P) = 0 \tag{4}$$

Since

$$\Gamma^k = \Gamma^0\Gamma^5\Sigma^K \tag{5}$$

with

$$\Sigma^K = i\epsilon^k{}_{lm}\Gamma^l\Gamma^m \tag{6}$$

equation (4) gives, after multiplication by Γ^0 on the left,

$$(P^0 - \Gamma^5\Sigma^K P^k - \Gamma^0 m_0 c)u(P) = 0 \tag{7}$$

We write

$$u_L(P) = \tfrac{1}{2}(1 - \Gamma^5)u(P) \tag{8}$$

$$u_R(P) = \tfrac{1}{2}(1 + \Gamma^5)u(P) \tag{9}$$

that is,

$$u(P) = u_L(P) + u_R(P) \tag{10}$$

By multiplying (7) by Γ^5 and, with the help of the commutation laws, this relation becomes

$$(P^0 - \Gamma^5\Sigma\cdot P + \Gamma^0 m_0 c)\Gamma^5 u(P) = 0 \tag{11}$$

Adding (7) and (11), we get

$$P^0 - \Gamma^5\vec{\Sigma}\cdot\vec{P})u_R(P) - m_0 c\Gamma^0 u_L(P) = 0 \tag{12}$$

Subtracting (11) from (7), we have

$$(P^0 - \Gamma^5\vec{\Sigma}\cdot\vec{P})u(P)_L - m_0 c\Gamma^0 u_R(P) = 0 \tag{13}$$

According to (8) and (9),

$$\Gamma^5 u_L(P) = -u_L(P) \tag{14}$$

$$\Gamma^5 u_R(P) = u_R(P) \tag{15}$$

The quantities $u_L(P)$ and $u_R(P)$ are eigenvectors of Γ^5 for the eigenvalues -1 and $+1$, respectively.

Equations (14) and (15) transform (12) and (13) into

$$\begin{aligned}(P^0 - \vec{\Sigma}\cdot\vec{P})u_R(P) - m\Gamma^0 u_L(P) &= 0\\ (P^0 + \vec{\Sigma}\cdot\vec{P})u_L(P) - m\Gamma^0 u_R(P) &= 0\end{aligned} \tag{16}$$

a representation in which Γ^5, a diagonal matrix, is used:

$$\Gamma^5 = \begin{bmatrix} 1 & \emptyset \\ \emptyset & -1 \end{bmatrix}$$

In this representation we have

$$\Gamma^0 = \begin{bmatrix} \emptyset & 1 \\ 1 & \emptyset \end{bmatrix}$$

If $u(P)$ is written in the form of one of Dirac's spinors (four components),

$$u(P) = \begin{bmatrix} u_1(P) \\ u_2(P) \\ u_3(P) \\ u_4(P) \end{bmatrix}$$

then in accordance with (8) and (9),

$$u_L(P) = \begin{bmatrix} 0 \\ 0 \\ u_3(P) \\ u_4(p) \end{bmatrix} \qquad u_R(P) = \begin{bmatrix} u_1(P) \\ u_2(P) \\ 0 \\ 0 \end{bmatrix}$$

and we write

$$u(P) = \begin{bmatrix} u_R(P) \\ u_L(P) \end{bmatrix}$$

a relation to be compared with (10).

7.6.7 Neutrino

According to the table in Section 7.1, for a neutrino, we take the mass m_0 equal to zero. The equations (16) in Section 7.6.6 can be simplified to

$$[P^0 - \vec{\sigma}\cdot\vec{P}]u_R(P) = 0 \tag{1}$$

$$[P^0 + \vec{\sigma}\cdot\vec{P}]u_L(P) = 0 \tag{2}$$

which is *Weyl's equation*. They are also separate equations: $u_L(P)$ represents a left-hand neutrino and $u_R(P)$ a right-hand neutrino:

$$u_L(P) = \begin{bmatrix} u_3(P) \\ u_4(P) \end{bmatrix} \tag{3}$$

$$u_R(P) = \begin{bmatrix} u_1(P) \\ u_2(P) \end{bmatrix} \tag{4}$$

They are two-component spinors. We used the fact that

$$\Sigma^K = \begin{bmatrix} \sigma_k & \emptyset \\ \emptyset & \sigma_k \end{bmatrix} \tag{5}$$

Remember that

$$E^2 = m_0^2c^4 + p^2c^2$$

or

$$E = \|\vec{p}\|c = P^0c \tag{6}$$

and equations (1) and (2) are also written

$$u_R(P) = \frac{\vec{\sigma}\cdot\vec{p}}{\|\vec{p}\|}\, u_R(P)$$

and

$$u_L(P) = -\frac{\vec{\sigma}\cdot\vec{p}}{\|\vec{p}\|}\, u_L(P)$$

The quantity denoted λ is called *helicity*:

$$\lambda = \frac{\vec{\sigma}\cdot\vec{p}}{\|\vec{p}\|}$$

and

$$\lambda u_R(P) = u_R(P)$$

$$\lambda u_L(P) = -u_L(P)$$

$u_R(P)$ and $u_L(P)$, associated with the eigenvalues $+1$ and -1, respectively, are the eigenvectors of the helicity.

If the neutrino is massive, in accordance with (16) and (17) in Section 7.6.4, the right-hand neutrino can be transformed into a left-hand neutrino, and vice versa (oscillations of neutrino).

7.68 Gauge Invariance

Take the Dirac equation [formula (1) in Section 7.6.3]

$$(i\hbar\Gamma_\mu\partial^\mu - m_0c)\Psi(X) = 0 \tag{1}$$

By taking the conjugate of this relation, the operator ∂^μ being applied on the left,

$$\Psi^+(X)[-i\hbar(\Gamma_\mu)^+\partial^\mu - m_0c] = 0 \tag{2}$$

with [formula (12) in Section 7.6.4]

$$\Gamma_\mu^+ = \Gamma_0\Gamma_\mu\Gamma_0$$

Using the property $\Gamma_0^2 = 1$, formula (1) is transformed into

$$\Psi^+(X)\Gamma_0[-i\hbar\Gamma_0(\Gamma_\mu)^+\partial^\mu - m_0c\Gamma_0] = 0$$

or

$$\overline{\Psi}(X)(i\hbar\Gamma_\mu\partial^\mu + m_0c) = 0 \tag{3}$$

by writing

$$\overline{\Psi}(X) = \Psi^+(X)\Gamma_0$$

in accordance with the relation (20) in Section 7.6.4.

Equations (1) and (3) are invariant under global first gauge transformation, that is, by the change

$$\Psi(X) \longrightarrow e^{iq\alpha/\hbar}\Psi(X) = \Psi'(X)$$

where q plays the role of a charge and α is a constant. The same equations are not invariant by local transformation or by that of the second kind:

$$\Psi(X) \longrightarrow e^{iq\alpha(X)/\hbar}\Psi(X) \tag{4}$$

since for (1), for example,

$$i\hbar\Gamma_\mu(\partial^\mu e^{iq\alpha(X)/\hbar})\Psi(X) + i\hbar\Gamma_\mu e^{iq\alpha(X)/\hbar}\partial^\mu\Psi(X)$$
$$- m_0ce^{iq\alpha(X)/\hbar}\Psi(X) = 0 \tag{5}$$

The first term on the right-hand side prevents invariance of (1) under the local transformation. In fact, the latter takes away the intrinsic sense from the derivative: It implies a comparison between distinct points. We came across this problem when defining the covariant derivative; see (6.6) in Section 6.3 for generalization of the momentum.

Put

$$\partial_\mu \longrightarrow D_\mu = \partial_\mu + \frac{iq}{\hbar} A_\mu \tag{6}$$

from which, for (5),

$$i\hbar\Gamma_\mu(\partial^\mu e^{iq\alpha(X)/\hbar})\Psi(X) + i\hbar\Gamma_\mu \left(i\frac{q}{\hbar} A^\mu\right) \Psi(X)e^{iq\alpha(X)/\hbar}$$
$$+ i\hbar\Gamma^\mu e^{iq\alpha(X)/\hbar}\partial^\mu\Psi(X) - m_0ce^{iq\alpha(X)/\hbar}\Psi(X) = 0$$

The equation becomes

$$(i\hbar\Gamma_\mu D^\mu - m_0c)\Psi(X) = 0 \tag{7}$$

and by local transformation, since

$$(D^\mu\Psi(X))' = e^{iq\alpha(X)/\hbar}D^\mu\Psi(X) \tag{8}$$

due to fact that the term $(D^\mu\Psi(X))$ is a vector, (8) gives

$$(D^\mu\Psi(X))' = e^{iq\alpha(X)/\hbar}\left(\partial^\mu + \frac{iq}{\hbar} A^\mu\right) \Psi(X)$$

as well as

$$(D^\mu\Psi(X))' = \left(\partial^\mu + \frac{iq}{\hbar} A'^\mu\right) e^{iq\alpha(X)/\hbar}\Psi(X)$$

or, by identification,

$$\partial^\mu(e^{iq\alpha(X)/\hbar}) = \frac{iq}{\hbar}(A^\mu - A'^\mu)e^{iq\alpha(X)/\hbar}$$

and

$$A'^\mu = A^\mu - \partial^\mu\alpha(X) \tag{9}$$

Note. This is the relation of gauge transformation usually used in electromagnetism. It is needed to assure the invariance of Dirac's equation.

In the same way the adjoint equation (3) is written

$$i\hbar\partial^\mu\overline{\Psi}(X)\Gamma_\mu + q\Gamma_\mu A^\mu\overline{\Psi}(X) + m_0c\overline{\Psi}(X) = 0$$

$A^\mu(X)$ can be considered as the quadripotential of electromagnetism; we then have

$$\partial_\nu F^{\mu\nu} = j^\mu$$

$$F^{\mu\nu} = \partial^\mu A^\nu - \partial^\nu A^\mu$$

and the Lagrange density of the system can be written

$$\mathbf{L} = \tfrac{1}{4}F^{\mu\nu}F_{\mu\nu} + \overline{\Psi}(i\hbar\Gamma_\mu D^\mu - m_0c)\Psi$$
$$= \tfrac{1}{4}F^{\mu\nu}F_{\mu\nu} + \overline{\Psi}(i\hbar\Gamma_\mu\partial^\mu - m_0c)\Psi - j^\mu A_\mu$$

where we write

$$j^\mu = q\overline{\Psi}\Gamma^\mu\Psi$$

This expression of **L** gives, with the help of the Lagrange equation for fields, the two Dirac equations (1) and (3).

We see in writing the Lagrange density:

The electromagnetic term (photon) $\tfrac{1}{4}F^{\mu\nu}F_{\mu\nu}$
The fermion term (particle of charged matter) $i\hbar\overline{\Psi}\Gamma_\mu\partial^\mu\Psi - m_0c\overline{\Psi}\Psi$
The interaction term $j^\mu A_\mu$

7.6.9 Generalization

Take the Lie group SU(N), local, and $\Psi(X)$ a spinor function, representing, for example, a particle of matter [quarks in the case of SU(3), a color group]. The action of Su(N) on $\Psi(X)$ transforms this function into $\Psi'(X)$ such that

$$\Psi'(X) = G(X)\Psi(X)$$

where $G(X)$ belongs to SU(N).

The observation made in the preceding section in relation to the covariant derivative is generalized to

$$D_\mu\Psi(X)\,dX^\mu = \Psi(X + dX) - \Psi_P(X)$$

In this last relation the quantity $\Psi_P(X)$ is the parallel transport of $\Psi(X)$ in $X + dX$ (cf. Figure 19). We have

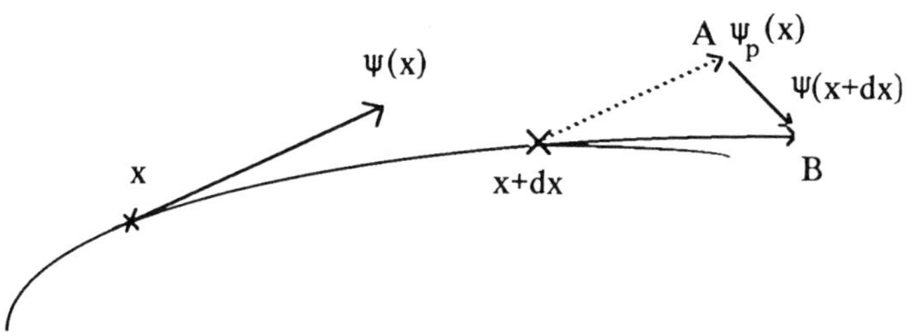

Figure 19 Parallel transport.

$$\Psi(X + dX) - \Psi_P(X) = d\Psi = (D_\mu \Psi(X))\, dX^\mu$$

a definition equivalent to that of formula (1.75) in Chapter 1. Now

$$\Psi(X + dX) = \Psi(X) + dX^\mu \partial_\mu \Psi(X)$$

and $\Psi_P(X)$ is the transform of $\Psi(X)$ by the group SU(N):

$$\Psi_P(X) = G(X)\Psi(X)$$

with

$$G(X) = e^{QA_\mu(X)dX^\mu}$$

where $A_\mu(X)$ belongs to a Lie algebra of su(N) [in the preceding section the quadrivector A belonged to a Lie algebra of $u(1)$]. We can see that

$$\begin{aligned} D_\mu \Psi(X)\, dX^\mu &= (1 + \partial_\mu - e^{QA_\mu})\Psi(X)\, dX^\mu \\ &\simeq (\partial_\mu - QA_\mu)\Psi(X)\, dX^\mu \end{aligned}$$

or

$$D_\mu = \partial_\mu - QA_\mu$$

(in the case of electromagnetism, $Q = iq/\hbar$).

Note. D_μ is a matrix, as is A_μ. We should write

$$D_\mu = \partial_\mu I - QA_\mu$$

or even

$$A_\mu = \sum_{a=1}^{N^2-1} A_\mu^a t_a$$

where the t_a form the basis of a Lie algebra. The transformation requires that

$$(D_\mu \Psi(X))' = G(X) D_\mu \Psi(X)$$

which implies that

$$D'_\mu = G(X) D_\mu G^{-1}(X)$$

and

$$(\partial_\mu - QA'_\mu) G(X)\Psi(X) = G(X)(\partial_\mu - QA_\mu)\Psi(X)$$

Otherwise,

$$A'_\mu = G(X) A_\mu G^{-1}(X) - \frac{1}{Q}(\partial_\mu G(X)) G^{-1}(X)$$

a generalization of the results of the preceding section to a transformation of any gauge. The calculation expressed by the term $[D_\mu, D_\nu]$ gives

$$[D_\mu, D_\nu] = QF_{\mu\nu} = Q(\partial_\mu A_\nu - \partial_\nu A_\mu + [A_\mu, A_\nu]$$

If we make the transformation on the tensor of curvature $F_{\mu\nu}$,

$$QF'_{\mu\nu} = [D'_\mu, D'_\nu] = G(X)D_\mu D_\nu G^{-1}(X) = G(X)F_{\mu\nu}G^{-1}(X)$$

the right-hand side enables us to find the transformation law for a tensor of the second order.

7.6.10 Charge Conjugation

The quantities Ψ, q, $s = \frac{1}{2}$, and m (respectively, Ψ_c, $-q$, $s = \frac{1}{2}$, and m) are associated with a particle (respectively, antiparticle). According to formula (2) in Section 7.6.6,

$$(i\hbar\Gamma^\mu D_\mu - m_0 c)\Psi = 0 \tag{1}$$

Now

$$D_\mu = \partial_\mu + \frac{iq}{\hbar} A_\mu \tag{2}$$

Equation (1) for the antiparticle is written

$$[(i\hbar\Gamma^\mu\partial_\mu + q\Gamma^\mu A_\mu) - m_0 c]\Psi_c = 0 \tag{3}$$

In relation to (1),

$$[(i\hbar\Gamma^\mu\partial_\mu + q\Gamma^\mu A_\mu) - m_0 c]\Psi = 0 \tag{4}$$

and by taking the adjoint,

$$\Psi^+[-i\hbar(\Gamma^\mu)^+\partial_\mu - q\Gamma^\mu A_\mu - m_0 c] = 0$$

Then remembering that

$$\Gamma^0\Gamma^\mu\Gamma^0 = (\Gamma^\mu)^+$$

and that

$$\overline{\Psi} = \Psi^+\Gamma^0$$

(5) gives

$$i\hbar\overline{\Psi}\Gamma^\mu\partial_\mu + q\overline{\Psi}\Gamma^\mu A_\mu + m_0 c\overline{\Psi} = 0 \tag{6}$$

By taking the transpose of the former equation,

$$(i\hbar\tilde{\Gamma}^\mu\partial_\mu + q\tilde{\Gamma}^\mu A_\mu + m_0 c)\tilde{\overline{\Psi}} = 0 \tag{7}$$

Multiplying on the left by a specific matrix M,

$$(i\hbar M\tilde{\Gamma}^{\mu}\partial_{\mu}MM^{-1} + qM\tilde{\Gamma}^{\mu}A_{\mu}MM^{-1} + m_0cM)\tilde{\overline{\Psi}} = 0 \tag{8}$$

which is compared with (3). The two expressions are identical if we write

$$M\tilde{\overline{\Psi}} = \Psi_c \tag{9}$$

and

$$M\tilde{\Gamma}^{\mu}M^{-1} = -\Gamma^{\mu} \tag{10}$$

and similarly for the complex conjugate:

$$-\tilde{\Psi}M^{-1} = \overline{\Psi}_c \tag{11}$$

and

$$M^{-1}\Gamma^{\mu}M = -\tilde{\Gamma}^{\mu} \tag{12}$$

This means that for the matrix M there are properties of antisymmetry; this matrix must be unitary and comply with (12). Take the transformation of the global gauge

$$\Psi'(X) = U^{-1}(\alpha)\Psi(X)U(\alpha) \tag{13}$$

with

$$U = e^{i\hat{q}\alpha/\hbar} \tag{14}$$

$\hat{q}$ being the operator of charge. The result of the operation made by (13) depends on Section 7.6.8.

$$\Psi'(X) = e^{iq\alpha/\hbar}\Psi(X) \tag{15}$$

We also have

$$\overline{\Psi}'(X) = e^{-iq\alpha/\hbar}\overline{\Psi}(X) \tag{16}$$

In accordance with (13), if α is infinitesimal,

$$\left(1 - \frac{i\hat{q}\alpha}{\hbar}\right)\Psi\left(1 + \frac{i\hat{q}\alpha}{\hbar}\right) = \left(1 + \frac{iq\alpha}{\hbar}\right)\Psi$$

or

$$[\Psi,\hat{q}] = q\Psi \tag{17}$$

We also prove that

$$[\overline{\Psi},\hat{q}] = -q\overline{\Psi} \tag{18}$$

the conjugation of charge, by using a unitary matrix M:

$$M^{-1}\hat{q}M = -\hat{q} \tag{19}$$

Majorana's field is, by definition, a Dirac field that coincides with its conjugate of charge, except for a phase factor:

$$\Psi(X) = \eta\Psi_c(X) \tag{20}$$

with

$$\Psi_c(X) = M^{-1}\Psi(X)M \tag{21}$$

and

$$\eta\eta^* = 1 \tag{22}$$

where M is a unitary matrix. The Majorana field can be used to study particles of neutral matter.

7.6.11 Justification of Introduction of Quantum Theory of Fields

We are looking for a current density $(j^\mu)_c$ such that [cf. formulas (9) and (11)]

$$(j^\mu)_c = \overline{\Psi}_c\Gamma^\mu\Psi_c = -\tilde{\Psi}M^{-1}\Gamma^\mu M\tilde{\overline{\Psi}}$$

and with (12),

$$(j^\mu)_c = \tilde{\Psi}\tilde{\Gamma}^\mu\tilde{\overline{\Psi}} = \overline{\Psi}\Gamma^\mu\Psi = (j^\mu)$$

There is, however, a slight problem: $-(j^\mu)$ should have the result of the conjugation of charge. To resolve this problem, the field $\Psi(X)$ must be considered as an operator complying with the rules of anticommutation. The physical magnitudes are then introduced as *normal products*:

$$(j^\mu)_c = :\overline{\Psi}_c\Gamma^\mu\Psi_c: = :\tilde{\Psi}\tilde{\Gamma}^\mu\tilde{\overline{\Psi}}: = -:\overline{\Psi}\Gamma^\mu\Psi: = -(j^\mu)$$

We refer the reader to the literature for this notion.

Bibliography

Arfken, G., *Mathematical Methods for Physicists*, Academic Press, New York, 1970

Bacry, H., *Leçons sur la théorie des groupes et les symétries des particules élémentaires*, Gordon and Breach, Paris, 1967

Bouchiat, C., *Introduction aux symétries des particules*, Ecole Polytechnique, Paris, 1982

Brandt, N. B., Chudinov, S. M., and Ponomarev Y. G., *Semimetals: Graphite and Its Compounds*, North Holland, Amsterdam, 1988

Brillouin, L., *Les tenseurs*, Masson, Paris, 1946

Casanova, G., *Algèbre vectorielle*, P.U.F., Paris, 1976

Charlier, M. F., and Charlier, A., *Chemistry and Physics of Carbon*, Vol 20, Marcel Dekker, New York, 1987

Charlier, A., Bérard, A., and Charlier, M. F., *Mécanique analytique*, Ellipses, Paris, 1989

Choquet-Bruhat, Y., *Géométria différentielle et systèmes extérieurs*, Dunod, Paris, 1968

Cohen Tannoudji, C., Dupont-Roc, J., and Grynberg, G., *Photons et atomes*, Savoirs Actuels, Paris, 1987

Constantinescu, F., and Magyari, E., *Problems in Quantum Mechanics*, Pergamon Press, Oxford, 1976

Davydov, A. S., *Quantum Mechanics*, Pergamon Press, Oxford, 1976

Dirac, P. A. M., *Les principes de la mécanique quantique*, P.U.F., Paris, 1931

Dirac, P. A. M., *The principles of Quantum Mechanics*, Oxford at the Clarendon Press, Oxford, 1976

Denève, P., *Mécanique*, Ellipses, Paris, 1986

Dittrich, W., and Reuter, M., *Selected topics in Gauge Theory*, Springer-Verlag, Berlin, 1986

Doubrovine, B., Novikov, S., and Fomenko, A., *Géométrie contemporaine*, Editions de Moscou (Mir), Moscow, 1979

Elbaz, E., *Relativité générale et gravitation*, Ellipses, Paris, 1986

Felsager, B., *Geometry, Particles and Fields*, Odensee University Press, Gylling, 1981

Feynman, R. P., *Quantum Electrodynamics*, Benjamin, New York, 1962

Fleury, N., *Champs et particules*, Masson, Paris, 1990

Flügge, S., *Practical Quantum Mechanics*, Springer-Verlag, New York, 1974

Georgi, H., *Lie Algebra in Particle Physics*, Frontiers in Physics, The Benjamin/ Cummings Publishing Company, London, 1982

Gourdin, M., *Formalisme lagrangien et lois de symétrie*, Gordon and Breach, Paris, 1967

Grossetête, C., *Relativité restreinte*, Ellipses, Paris, 1985

Huang, K., *Quarks, Leptons and Gauge Field*, World Scientific, Singapour, 1982

Itzykson, C., and Zuber, J. B., *Quantum Field Theory*, McGraw-Hill, New York, 1980

Landau, L., and Lifchitz, E., *Théorie des champs*, Editions de Moscou (Mir), Moscow, 1969

Leité-Lopez, J., *Lectures in Symmetry*, Gordon and Breach, London, 1969

Merzbacher, E., *Quantum Mechanics*, Wiley, New York, 1970

Moeller, C., *The Theory of Relativity*, Oxford University Press, Oxford, 1952

Nye, J. F., *Physical Properties of Crystals*, Oxford Science Publications, Oxford, 1987

Omnes, A., *Introduction à l'étude des particules élémentaires*, Ediscience, Paris, 1970

Roman, P., *Advanced Quantum Theory*, Addison-Wesley, Reading, 1965

Ramon, P., *Field Theory, a Modern Primer*, Frontier in Physics, The Benjamin/Cummings Publishing Company, London, 1981

Schultz, B., *Geometrical Methods of Mathematical Physics*, Cambridge University Press, Cambridge, 1987

Ter Haar, D., *Selected Problems in Quantum Mechanics*, Infosearch Limited, London, 1964

Trautman, A., *Differential Geometry for Physics*, Stony Brook Lectures, Napoli, 1984

Index